FRAGRANCES AND FLAVORS

FRAGRANCES AND FLAVORS

Recent Developments

Edited by S. Torrey

NOYES DATA CORPORATION
Park Ridge, New Jersey, U.S.A.
1980

Library of Congress Catalog Card Number: 80-12954
ISBN: 0-8155-0798-4
Printed in the United States

Published in the United States of America by
Noyes Data Corporation
Noyes Building, Park Ridge, New Jersey 07656

Library of Congress Cataloging in Publication Data

Torrey, S
Fragrances and flavors.

(Chemical technology review ; no. 156)
Includes indexes.
1. Perfumes--Patents. 2. Perfumes, Synthetic--
Patents. I. Title. II. Series.
TP983.T62 668'.54'0272 80-12954
ISBN 0-8155-0798-4

FOREWORD

The detailed, descriptive information in this book is found in U.S. patents issued since January 1978 that deal with the technology of fragrances and flavors.

This book is a data-based publication, providing information retrieved and made available from the U.S. patent literature. It thus serves a double purpose in that it supplies detailed technical information and can be used as a guide to the patent literature in this field. By indicating all the information that is significant, and eliminating legal jargon and juristic phraseology, this book presents an advanced commercially oriented review of the technology of fragrances and flavors. This title contains all recent developments since our previous title *Fragrance Technology,* published in 1975. New in this volume are the recent synthetic techniques and manufacturing processes for flavors, not included heretofore.

The U.S. patent literature is the largest and most comprehensive collection of technical information in the world. There is more practical, commercial, timely process information assembled here than is available from any other source. The technical information obtained from a patent is extremely reliable and comprehensive; sufficient information must be included to avoid rejection for "insufficient disclosure." These patents include practically all of those issued on the subject in the United States during the period under review; there has been no bias in the selection of patents for inclusion.

The patent literature covers a substantial amount of information not available in the journal literature. The patent literature is a prime source of basic commercially useful information. This information is overlooked by those who rely primarily on the periodical journal literature. It is realized that there is a lag between a patent application on a new process development and the granting of a patent, but it is felt that this may roughly parallel or even anticipate the lag in putting that development into commercial practice.

Many of these patents are being utilized commercially. Whether used or not, they offer opportunities for technological transfer. Also, a major purpose of this book is to describe the number of technical possibilities available, which may open up profitable areas of research and development. The information contained in this book will allow you to establish a sound background before launching into research in this field.

Advanced composition and production methods developed by Noyes Data are employed to bring these durably bound books to you in a minimum of time. Special techniques are used to close the gap between "manuscript" and "completed book." Industrial technology is progressing so rapidly that time-honored, conventional typesetting, binding and shipping methods are no longer suitable. We have bypassed the delays in the conventional book publishing cycle and provide the user with an effective and convenient means of reviewing up-to-date information in depth.

The table of contents is organized in such a way as to serve as a subject index. Other indexes by company, inventor and patent number help in providing easy access to the information contained in this book.

Due to data-base publishing techniques several of the structural formulas reproduced in this book were not available in their original form and may be less clear than desired.

15 Reasons Why the U.S. Patent Office Literature Is Important to You –

1. The U.S. patent literature is the largest and most comprehensive collection of technical information in the world. There is more practical commercial process information assembled here than is available from any other source.
2. The technical information obtained from the patent literature is extremely comprehensive; sufficient information must be included to avoid rejection for "insufficient disclosure."
3. The patent literature is a prime source of basic commercially utilizable information. This information is overlooked by those who rely primarily on the periodical journal literature.
4. An important feature of the patent literature is that it can serve to avoid duplication of research and development.
5. Patents, unlike periodical literature, are bound by definition to contain new information, data and ideas.
6. It can serve as a source of new ideas in a different but related field, and may be outside the patent protection offered the original invention.
7. Since claims are narrowly defined, much valuable information is included that may be outside the legal protection afforded by the claims.
8. Patents discuss the difficulties associated with previous research, development or production techniques, and offer a specific method of overcoming problems. This gives clues to current process information that has not been published in periodicals or books.
9. Can aid in process design by providing a selection of alternate techniques. A powerful research and engineering tool.
10. Obtain licenses–many U.S. chemical patents have not been developed commercially.
11. Patents provide an excellent starting point for the next investigator.
12. Frequently, innovations derived from research are first disclosed in the patent literature, prior to coverage in the periodical literature.
13. Patents offer a most valuable method of keeping abreast of latest technologies, serving an individual's own "current awareness" program.
14. Copies of U.S. patents are easily obtained from the U.S. Patent Office at 50¢ a copy.
15. It is a creative source of ideas for those with imagination.

CONTENTS AND SUBJECT INDEX

INTRODUCTION

Man's search for substances which can produce new flavors and fragrances, substitute for expensive and/or scarce ones, or augment and enhance existing desirable ones continues apace.

Perhaps the most interesting results of the research of the last few years in the fragrance and flavor fields are the many compounds described in the first chapter of this book which are of use in both areas. They may be used to engender or augment flavors in foodstuffs, chewing gums, and medicinal products like mouthwash and toothpaste. The same compounds or closely related ones serve also to produce desirable aromas for perfumes, perfumed compositions such as soaps and detergents, and cosmetics. Most of these multipurpose formulations are useful also for treating tobacco, where they are used to improve both taste and aromas of such products as cigarettes and chewing tobacco.

The thrust of fragrance research in the last few years has been towards the synthesis of compounds that can be substituted for existing substances which, in their natural forms, are expensive to procure, are becoming scarce, or vary so much in intensity that producing perfumes in which they are used becomes more a matter of trial and error than of exact formulation. The patented processes described in chapters two through six describe the results of this research.

Chapter seven describes various patented products made with synthesized fragrances. Chapters eight, nine and ten include the patents derived from research in the flavor field. It is notable that the emphasis in the flavor field, to judge by the number of patents granted, has been toward producing meaty-type flavors, many of which are specified as being particularly useful as flavorants for synthesized vegetable protein.

MULTIPURPOSE FLAVORS AND FRAGRANCES

ALICYCLIC COMPOUNDS

Cyclohexadiene Derivatives

Cyclohexadiene derivatives useful for flavoring foodstuffs and tobacco and for perfuming colognes, toiletries, etc., have been described by *K.K. Light, B.M. Spencer, J.F. Vinals, J. Kiwala, M.H. Vock and E.J. Shuster; U.S. Patents 4,076,854; February 28, 1978; 4,130,508; December 19, 1978; 4,154,696; May 15, 1979; and 4,154,763; May 15, 1979; all assigned to International Flavors & Fragrances Inc.*

The cyclohexadiene derivatives have the formula

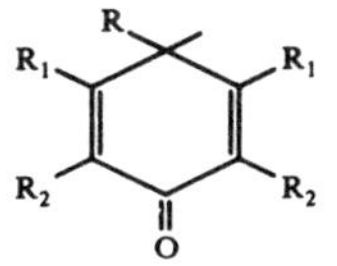

R
R_1 R_1
R_2 R_2
O

wherein R is one of C_{1-5} alkyl, R_1 is methyl when R_2 is hydrogen, and R_1 is hydrogen when R_2 is methyl, and are useful in foodstuffs, chewing gums, toothpastes, medicinal product flavors, perfumes, colognes, perfumed articles, tobaccos and tobacco flavors.

These cyclohexadiene derivatives may be produced according to one of the following processes:

(1) Reacting an alkyl-trimethylcyclohexa-1,3-diene with oxygen in the presence of light, an ultraviolet sensitizer and a basic catalyst and fractionally distilling the resulting product. This process is illustrated by the reaction shown on the following page.

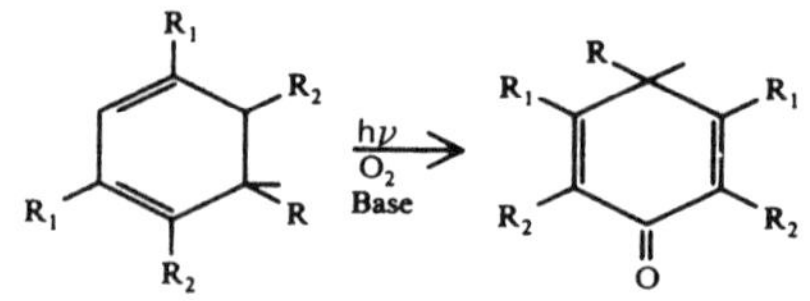

(2) Reacting a 4-alkyl-4-hydroxy-3,5,5-trimethyl-2-cyclohexen-1-one with dilute aqueous mineral acid and distilling the resulting reaction product. This process can be illustrated by the following reaction:

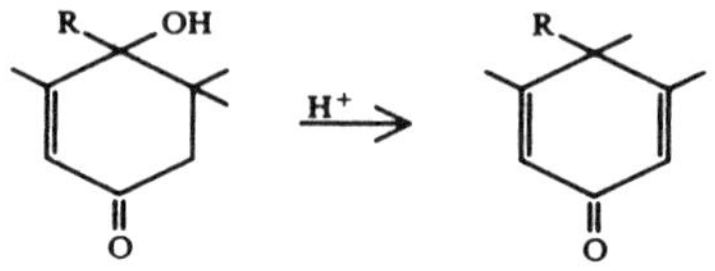

The photo-oxidation reaction is preferably carried out in an inert solvent such as methanol or a methanol-benzene mixture in the presence of a reaction sensitizer such as rose bengal and a base such as sodium hydroxide and potassium hydroxide. The photo-oxidation is carried out by bubbling oxygen through the reaction mass. The second of the above reaction sequences may be carried out using a mineral acid, e.g., dilute aqueous hydrochloric acid, sulfuric acid or phosphoric acid. Preferably the reaction is carried out in the presence of an inert solvent such as hexane or cyclohexane, at a temperature in the range of from 15° up to 70°C, most conveniently at room temperature.

At the end of each of the foregoing reactions, the reaction mass is distilled in the fractional distillation apparatus yielding the desired product. Organoleptic properties of described compounds are as follows:

Compound	Structure	Organoleptic Properties
2,4,4,6-tetra-methylcyclohexa-2,5-diene-1-one	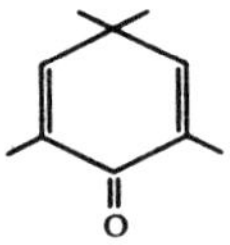	Flavor: medicinal Fragrance: warm, sweet, slightly minty, woody aroma with natural green and hay-like nuances Tobacco: Virginia tobacco-like
4-butyl-3,4,5-trimethyl-2,5-cyclohexa-diene-1-one	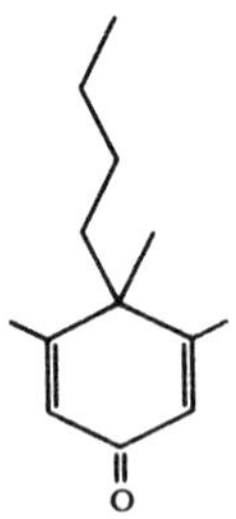	Flavor: sweet, hay-like, heliotropine-like, black pepper-like, woody, tobacco-like and "Oriental" aroma characteristics with sweet, hay-like, heliotropine-like, black pepper-like, woody, tobacco, "Oriental" flavor characteristics Fragrance: sweet, honey-like celery note with tobacco and floral nuances and a distinct Jasmine undertone Tobacco: dry black tobacco notes as well as pungent, spicey, celery, black pepper-like and sweet notes, both prior to and on smoking, in the main stream as well as the side stream

Examples 1 and 2 show typical preparations of the desired cyclohexadiene derivatives, and the further examples show some typical ways in which they can be utilized.

Example 1:

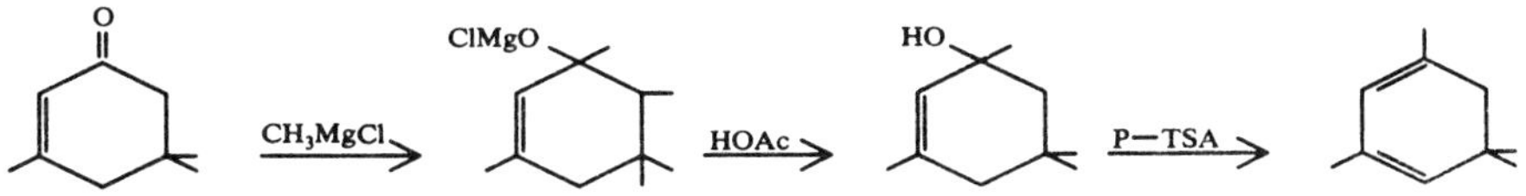

To a 5-liter flask fitted with stirrer, thermometer, dropping funnel, reflux condenser, drying tube and nitrogen purge is charged 2,100 ml (6.3 mols) of a 3 M solution of methyl magnesium chloride in tetrahydrofuran. The flask and contents are cooled to 0°C, and 730 g (5.3 mols) of isophorone is added at 0° to 5°C over a 2-hour period. The mixture is allowed to warm to room temperature and to stand overnight. The reaction mixture is poured over a mixture of 2,800 g of ice and 370 g of glacial acetic acid.

When all the ice has melted, the layers are separated and the organics are washed twice with 500 ml of water. The solvent is removed on a rotary evaporator, and the product is transferred to a 2-liter flask fitted with a condenser and a water separator. Benzene (300 ml) and p-toluenesulfonic acid (0.5 g) are added, and the mixture is refluxed until no additional water is separated. The product is distilled on a 12-inch Goodloe column using a reflux ratio of 9:1. A total of 494 g product is obtained (boiling point 66° to 67°C at 42 mm Hg) which is predominantly 1,3,5,5-tetramethylcyclohexa-1,3-diene.

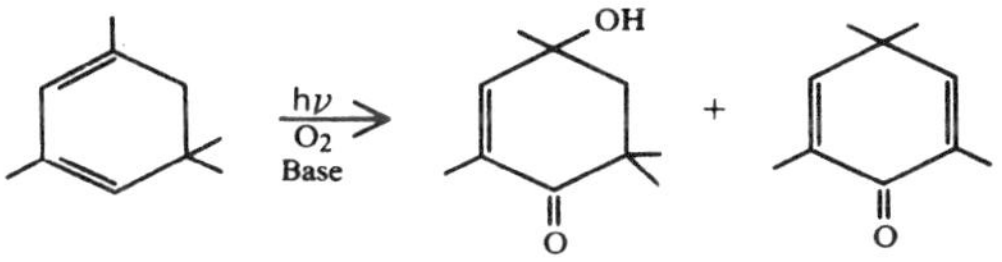

A solution of 40 g (0.3 mol) of the 1,3,5,5-tetramethylcyclohexa-1,3-diene (prepared as above), 13 g (0.23 mol) of potassium hydroxide, and 0.3 g of rose bengal in 600 ml of methanol is placed in a flask and irradiated at a temperature of 25° to 28°C with a 450 watt tungsten photo lamp over a period of two hours. A steady stream of oxygen is bubbled through the solution during irradation. At the end of the 2-hour irradiation period, analysis of a sample of the reaction mixture by GLC (10' x ⅛" 10% Carbowax 20M on Chromasorb W, programmed from 80° to 220°C at 8°/min shows that the starting material is completely consumed and two products have formed. Both products are trapped from the GLC column and analyzed. The first (about 20% of the reaction product) is found to be 2,4,4,6-tetramethylcyclohexa-2,5-diene-1-one (Compound A).

The second (about 80% of the product mixture) is found to be 4-hydroxy-2,4-6,6-tetramethylcyclohex-2-en-1-one (Compound B).

The reaction mixture is poured into 500 ml of water, and the organics are extracted with four 100 ml portions of ether. The ether is removed by means of a rotary evaporator, and the two products are separated by means of a careful distillation. Compound A is obtained first and has a boiling point of 60°C at 1.5 mm Hg. Compound B distills second with a boiling point of 100°C at 1.5 mm Hg.

Example 2: To a 2-liter flask fitted with a reflux condenser, a water separator and a stirrer is charged 265 g of 2,6,6-trimethylcyclohex-2-en-1,4-dione [prepared by the method described in *Helv. Chim. Acta*, 39:2041 (1956)], 266 g of propylene glycol, 500 ml of benzene and 1 g of p-toluenesulfonic acid. The reactants are stirred and refluxed until no additional water is collected in the water separator. The product is washed with water to a pH of 6, and the solvent is stripped off. The residue is vacuum distilled through a 9" Goodloe column to yield 257 g of the product, 2,7,9,9-tetramethyl-1,4-dioxaspiro-[4,5] dec-6-en-8-one, boiling point 92°C at 1 mm Hg (80% yield).

A solution of 20 g of this product in hexane is stirred while a solution of n-butyl lithium in hexane is added dropwise. When no additional exotherm occurs, the addition is stopped and the mixture is stirred for about 15 minutes. Water is added with stirring to hydrolyze the lithium salt. The organic layer is separated, and the hexane is allowed to evaporate leaving oily crystals which are recrystallized from cyclohexane yielding 10 g of 4-butyl-4-hydroxy-3,5,5-trimethyl-2-cyclohexyl-1-one, melting point 90° to 93°C.

A solution of 7 g of the 4-butyl-4-hydroxy-3,5,5-trimethyl-2-cyclohexen-1-one in hexane is stirred overnight with 10 ml of a 5% solution of HCl. The hexane is separated and evaporated, and the residue is distilled through a micro column using triethanolamine as a still base. The product obtained is 4-butyl-3,4,5-trimethyl-2,5-cyclohexadien-1-one, (Compound C), boiling point 112° to 119°C at 0.1 mm Hg.

Example 3: The following mixture is prepared:

Ingredients	Parts by Weight
Phenylacetic acid	70.0
Coumarin	20.0
Phenylethylphenyl acetate	100.0
Phenyl ethyl alcohol	5.0
Benzyl benzoate	100.0
Dimethylphenylethyl carbinol	10.0
Methyl anthranilate	5.0
Beta ionone	10.0
Compound A	30.0

The 2,4,4,6-tetramethyl-2,5-cyclohexadien-1-one (Compound A) imparts a warm, sweet, slightly minty, woody note to this formulation, and it enhances the herbaceous character thereof.

Example 4: The following mixture is prepared:

Ingredients	Parts by Weight
p-Cresol	3.0
Methyl N-acetylanthranilate	20.0
Farnesol	3.0

(continued)

Ingredients	Parts by Weight
Nerolidol	20.0
Cis-3-hexenyl benzoate	30.0
Indol	15.0
Eugenol	35.0
Benzyl alcohol	50.0
Methyl linoleate	100.0
Jasmine lactone	30.0
Dihydro methyl Jasmonate	20.0
Benzyl acetate	500.0
Cis-Jasmone	40.0
Linalool	200.0
Compound C	50.0

The 4-butyl-3,4,5-trimethyl-2,5-cyclohexadien-1-one (Compound C) imparts a green, herbaceous, floralcy of jasmine to this Jasmine formulation.

Example 5: *Preparation of a Detergent Composition* – A total of 100 g of a detergent powder (lysine salt of n-dodecylbenzene sulfonic acid as described in U.S. Patent 3,948,818) is mixed with 0.15 g of Compound A until a substantially homogeneous composition is obtained. This composition has a warm, sweet, slightly minty, woody aroma.

Example 6: *Preparation of a Cologne and Handkerchief Perfume* – The composition of Example 3 is incorporated in a cologne at a concentration of 2.5% in 85% aqueous ethanol; and into a handkerchief perfume at a concentration of 20% (in 95% aqueous ethanol). The use of 2,4,4,6-tetramethylcyclohexa-2,5-diene-1-one affords a distinct and definite strong warm, sweet, slightly minty, woody aroma with a definite honeylike fragrance having a herbaceous character.

Synthetic Lime Mix

R.J. Steltenkamp; U.S. Patent 4,093,565; June 6, 1978; assigned to Colgate-Palmolive Company has developed a process for producing a synthetic substitute for lime oil.

Lime oil is obtained commercially by the steam distillation of the macerated sour lime or Mexican lime. The essential oil of lime is in the peel of the fruit. Depending upon the condition and the size of the fruit, a barrel of limes, which contains 160 lb of limes, yields normally about 8 oz of distilled oil.

The synthetic lime mix is prepared by an acid catalyzed rearrangement of citral, and the reaction product mixture can be used directly without separation of components. Its fragrance closely resembles the fresh terpenic character of steam distilled lime oil.

The citral used in the reaction may be either natural or synthetic. Natural citral is obtained from the distillation of lemongrass oil which is produced in India and Guatemala. *Eucalyptus staigeriana* and *Litsea cubeba* are also natural sources for citral. Synthetic citral may be prepared by the process of U.S. Patent 2,795,617. It is not necessary to use pure citral since it would be possible to achieve the same reaction using oils with a significant citral content. The use of oils from lemongrass, *Eucalyptus staigeriana* or *Litsea cubeba* as sources of citral would result in a modification of the lime character of the reaction product to the extent of the additional notes from the impurities.

Example 1: *Preparation of Synthetic Lime Mix* – Citral (30.0 g) is added to 250 ml of a citric acid solution (pH 1.75) prepared by dissolving 36.0 g of citric acid (analyzed anhydrous powder) in distilled water to a 500-ml volume. This mixture is refluxed mildly for two hours in a single-neck 500-ml flask. After two hours the flask is equipped with a condenser and a lighter-than-water collection trap, and the product is water-distilled. The yield of distilled lime product is 22.0 g (73%). The remaining nondistillable product is highly resinous and does not possess the limelike fragrance. The distillable product mix contains about 60% terpene hydrocarbons, about 35% unidentified oxygen-containing compounds, and the remainder additional hydrocarbons. About 5% of the product mix is the terpene hydrocarbon 1-methyl-1,3-cyclohexadiene. Also present are trace amounts of 3-methylene-1-cyclohexene; 1-methyl-1,5-cyclohexadiene; isoprene; n-heptane; isopropyl formate; acetone; and isopropanol.

One of the important ingredients of the synthetic lime mix is the 1-methyl-1,3-cyclohexadiene. This material provides the characteristic "gassy" note of lime oil. Although it is difficult to obtain the 1-methyl-1,3-cyclohexadiene from the synthetic lime oil mix, several practical methods of preparation have been reported in the literature; and the method of A.J. Birch reported in the *Journal of the Chemical Society* of 1947 on page 1642 is a practical method for obtaining this isomer.

Example 2: A fragrant mixture suitable for incorporation into hand lotions, cream, and dishwashing formulations contains the following ingredients in the parts by weight indicated:

Ingredient	Percent
Benzaldehyde	55.0
Synthetic lime mix of Example 1*	10.0
Phenylethyl alcohol	10.0
Orange oil	3.0
Lemon oil	3.0
α-ionone	5.0
Citronellol	4.0
Coumarin	3.0
Linalool	2.0
Spike lavender	1.0
Cinnamic aldehyde	1.0
Benzyl acetate	2.0
Geranium bourbon	1.0

*or 1-methyl-1,3-cyclohexadiene

Example 3: A flavor mix containing the synthetic lime substitute suitable for cola-type flavors contains the following ingredients in the parts by weight indicated:

Ingredient	Percent
Synthetic lime mix of Example 1*	9.0
Vanilla extract	0.6
Solid extract Kola nuts	2.0
Kola flavor emulsion	17.4
Caramel	71.0

*or 1-methyl-1,3-cyclohexadiene

Derivatives of 1-Acetyl-3,3-Dimethylcyclohexane

M.A. Sprecker, M.H. Vock, F.L. Schmitt, J.B. Hall and J.M. Sanders; U.S. Patents 4,103,036; July 25, 1978; 4,107,093; August 15, 1978; 4,122,120; October 24, 1978; all assigned to International Flavors & Fragrances Inc. have found that one or more derivatives of 1-acetyl-3,3-dimethyl-cyclohexane having the generic formula:

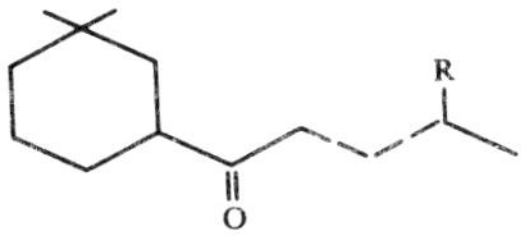

wherein one of the dashed lines may be a carbon-carbon single bond or a carbon-carbon double bond and the other of the dashed lines is a carbon-carbon single bond, and R is hydrogen or methyl, may be used for imparting and/or augmenting flavors and/or aromas to foods, chewing gums, toothpastes, tobacco, perfumes and perfumed toiletries.

These compounds may be produced by a process in which 1-acetyl-3,3-dimethylcyclohexane is reacted with an aldehyde (–RCHO) in the presence of an inorganic base, such as NaOH, KOH, or $Ba(OH)_2$ at a temperature of 20° to 160°C; or at a temperature of from 100° to 200°C when using a mixture of B_2O_3 and boric acid as the catalyst. The product may be used as is, or reduced with hydrogen.

Examples and organoleptic properties of materials produced according to the aforementioned process are as follows:

NAME	STRUCTURE	ORGANOLEPTIC PROPERTIES FLAVOR	FRAGRANCE
Mixture of 1-(3,3-dimethylcyclohexyl)-4-methyl-cis-2-penten-1-one, 1-(3,3-dimethylcyclohexyl)-4-methyl-trans-2-penten-1-one, and 1-(3,3-dimethylcyclohexyl)-4-methyl-3-penten-1-one	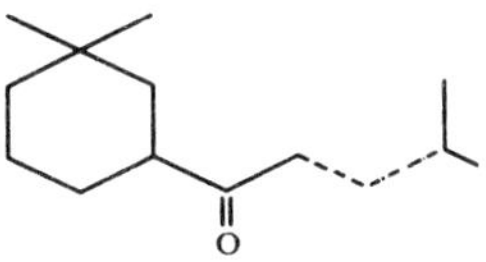	Sweet, woody, raspberry, lemon, piney aroma character with sweet, raspberry and lemon flavor character at 1.0 ppm.	Fruity woody, piney aroma with armoise, floral nuances and a slight chocolate undertone.
Mixture of 1-(3,3-dimethylcyclohexyl)-cis-2-penten-1-one, 1-(3,3-dimethylcyclohexyl)-trans-2-penten-1-one, and 1-(3,3-dimethylcyclohexyl) 3-penten-1-one	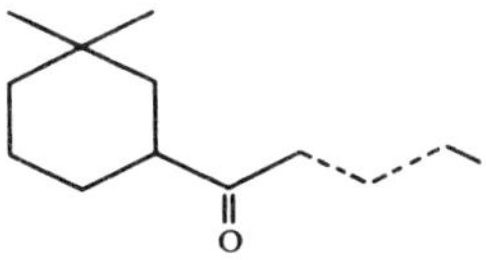	Sweet, raspberry, fruity, piney, floral, ionone-like aroma with sweet, sweet fruity, raspberry, red berry, piney, floral and ionone-like flavor characteristics.	Sweet, fruity, woody (pine) with pineapple-galbanum, hexalon notes.
3,3-dimethyl-1-(4-methylvaleryl)-cyclohexane	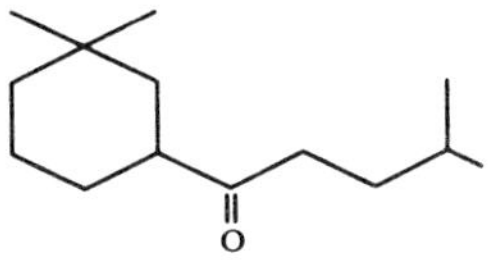	Sweet, piney, fruity, blueberry, woody aroma character with sweet, piney, fruity, nut meat characteristics.	Low-keyed, sweet, woody, somewhat chocolate like.

With respect to ultimate food compositions, chewing gum compositions, medicinal product compositions and toothpaste compositions, it is found that quantities of the 1-acetyl-3,3-dimethylcyclohexane derivative(s) ranging from a small but effective amount, e.g., 0.5 part per million up to about 100 parts per million based on total composition are suitable. Concentrations in excess of the maximum quantity stated are not normally recommended, since they fail to prove commensurate enhancement of organoleptic properties. In those instances, wherein a 1-acetyl-3,3-dimethylcyclohexane derivative is added to the foodstuff as an integral component of a flavoring composition, it is, of course, essential that the total quantity of flavoring composition employed be sufficient to yield an effective concentration in the foodstuff product.

The amount of 1-acetyl-3,3-dimethylcyclohexane derivative(s) which will be effective in perfume compositions as well as in perfumed articles and colognes depends on many factors, including the other ingredients, their amounts and the effects which are desired. It has been found that perfume compositions containing as little as 0.01% of a 1-acetyl-3,3-dimethylcyclohexane derivative(s) or even less (e.g., 0.005%) can be used to impart a fruity, woody, piney aroma with armoise and floral nuances and slight chocolate undertones to soaps, cosmetics or other products. The amount employed can range up to 70% of the fragrance components and will depend on considerations of cost, nature of the end product, etc.

Examples 1 and 2 show two of the possible methods for the preparation of 1-(3,3-dimethylcyclohexyl)-4-methyl-cis and trans-(2 and 3)-penten-1-ones (Compounds A and B). The other examples illustrate uses of these compounds.

Example 1: A solution of 474 g of 1-acetyl-3,3-dimethylcyclohexane, 500 ml methanol, 40 g of NaOH and 50 ml toluene is heated to reflux. Isobutyraldehyde (240 g) is added dropwise over a 45-minute period. The reaction mixture is heated to reflux for a further 90 minutes, at which time, the solution is cooled to room temperature and neutralized with hydrochloric acid. The methanol is removed by distillation, and the aqueous layer is removed from the resulting two phase mixture. The organic layer is washed once with water and distilled rapidly through a short column to afford a mixture of 1-(3,3-dimethylcyclohexyl)-4-methyl-cis and trans-(2 and 3)-penten-1-one (Flavor Composition) (360 g, 58%) (112° to 117°C, 2.8 mm) and higher boiling products.

Redistillation of this material through a 1½ x 12 inch Goodloe column gave the following fractions:

Fraction No.	Vapor Temperature (° C)	Liquid Temperature (° C)	Vacuum mm Hg	Weight (g)	Reflux Ratio
1	85–101	128	2.8	6.6	9:1
2	112	128	2.8	10.6	9:1
3	112	128	2.8	16.6	9:1
4	112	132	2.8	12.9	9:1
5	112	128	2.8	28.3	9:1
6	112	128	2.8	12.0	9:1
7	112	128	2.8	24.2	9:1
8	112	128	2.8	27.3	9:1
9	114	134	2.8	27.1	4:1
10	115	135	2.8	22.7	4:1
11	115	137	2.8	26.9	4:1
12	115	140	2.8	25.7	4:1
13	116	149	2.8	25.0	4:1
14	116	149	2.8	26.9	4:1
15	117	155	2.8	16.9	4:1
16	122	160	2.8	16.9	4:1
17	128	165	2.8	23.9	4:1
18	128	176	2.8	26.9	4:1
19	130	205	2.8	24.3	4:1
20	130	214	2.8	23.1	4:1
21	130	219	2.8	25.5	4:1
22	138	241	2.8	25.5	4:1
23	149	250	2.8	11.7	4:1

Fractions 5 through 13 contain > 99% desired product having the formulas:

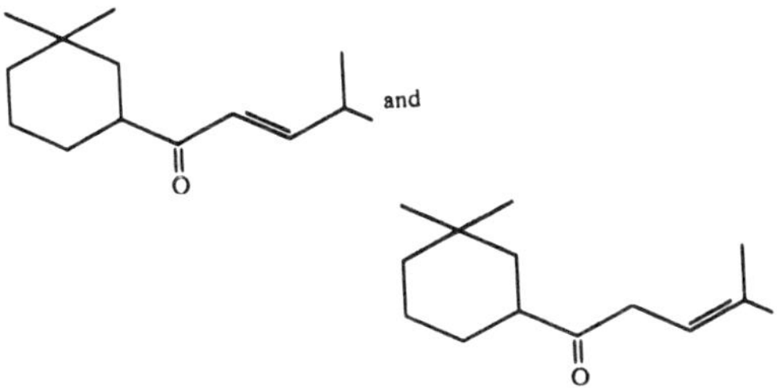

and

O

O

Fractions 5 through 13, from a fragrance standpoint, have fruity, woody, piney aromas with armoise, floral nuances and a slight chocolate undertone.

From a flavor standpoint, this material has a sweet, fruity, raspberry, woody, lemon, piney aroma with a sweet, raspberry and lemon flavor character.

Example 2: A mixture of 1-acetyl-3,3-dimethylcyclohexane (462 g), isobutyraldehyde (108 g), boric acid (6.2 g) and boron oxide (52 g) is heated and stirred in an autoclave at 150°C for 5 hours. The solution is filtered from inorganic salts and washed three times with 10% HCl. Distillation affords 319 g of 1-acetyl-3,3-dimethylcyclohexane and 138 g of 1-(3-3-dimethylcyclohexyl)-4-methy-cis- and trans-(2 and 3)-penten-1-one (22%, based on charged 1-acetyl-3,3-dimethylcyclohexane.

Example 3: *(A) Powder Flavor Composition* – 20 g of the flavor composition of Example 2 is emulsified in a solution containing 300 g gum acacia and 700 g water. The emulsion is spray-dried with a Bowen Lab Model Drier utilizing 260 cfm of air with an inlet temperature of 500°F, an outlet temperature of 200°F, and a wheel speed of 50,000 rpm.

(B) Sustained Release Flavor – The following mixture is prepared:

Ingredient	Parts by Weight
Liquid raspberry flavor Composition of Example 2	20
Propylene glycol	9
Cab-O-Sil M-5	5.00

The Cab-O-Sil is dispersed in the liquid raspberry flavor composition with vigorous stirring, thereby resulting in a viscous liquid. 71 parts by weight of the powder flavor composition of Part A is then blended into the viscous liquid, with stirring at 25°C for a period of 30 minutes resulting in a dry, free-flowing sustained release flavor powder.

Example 4: *Chewing Gum* – 100 parts by weight of chicle are mixed with 4 parts by weight of the flavor prepared in accordance with Example 3. 300 parts of sucrose and 100 parts of corn syrup are added. Mixing is effected in a ribbon blender with jacketed side walls of the type manufactured by Baker Perkins Co.

The resultant chewing gum blend is then manufactured into strips one inch in width and 0.1 inch in thickness. The strips are cut into lengths of 3 inches each. On chewing, the chewing gum has a pleasant, long lasting raspberry flavor.

Example 5: A perfume composition is prepared by admixing the following ingredients in the indicated proportions:

Ingredient		Amount, grams
n-Decyl aldehyde		1
n-Dodecyl aldehyde		2
Methyl nonyl acetaldehyde		0.5
Linalool		50
Linalyl acetate		70
Phenyl ethyl alcohol		100
Petigrain SA		20
Bergamot oil		30
α-Methyl ionone		25
Mixture produced according to Example 2		10
Cyclized bicyclo C-12 material produced according to the process of Example 4 of Canadian Patent 854,225		5
Isobornyl cyclohexyl alcohol		10
Benzyl acetate		25
2-n-Heptylcyclopentanone		5
	Total	353.5

The foregoing blend is evaluated and found to have a high degree of richness and persistence in its natural amber quality. This base composition can be admixed with aqueous ethanol, chilled and filtered to produce a finished cologne. The cologne so prepared has an amber aroma leaning towards a woody amber note with an excellent armoise nuance. The base composition can also be used to scent soap or other toilet goods such as lotion, aerosol, sprays, etc.

1-Acetyl-3,3-Dimethyl-(2-Propenyl)-Cyclohexane

J.B. Hall, M.A. Sprecker, F.L. Schmitt and M.H. Vock; U.S. Patents 4,073,751; February 14, 1978 and 4,117,015; September 26, 1978; both assigned to International Flavors & Fragrances Inc. have found that 1-acetyl-3,3-dimethyl-(2-propenyl)-cyclohexane (Compound A) is capable of augmenting or enhancing pineapple, raspberry and berry-fruit flavors by providing thereto fruity, pineapple, allyl caproatelike, galbanum and woody aroma characteristics along with rosey, ionone, oriental and wood flavor characteristics.

Compound A is also capable of modifying or enhancing the aroma characteristics of perfumed compositions and perfumed articles by imparting thereto woody, earthy, amberlike, sweet, animal, tobacco/animal notes with pineapple/amber nuances.

This compound is produced by reacting acetyl-3,3-dimethylcyclohexane with an allyl halide in the presence an inert solvent, and an alkali metal hydroxide and in the presence of a phase transfer agent. The reaction is illustrated as follows:

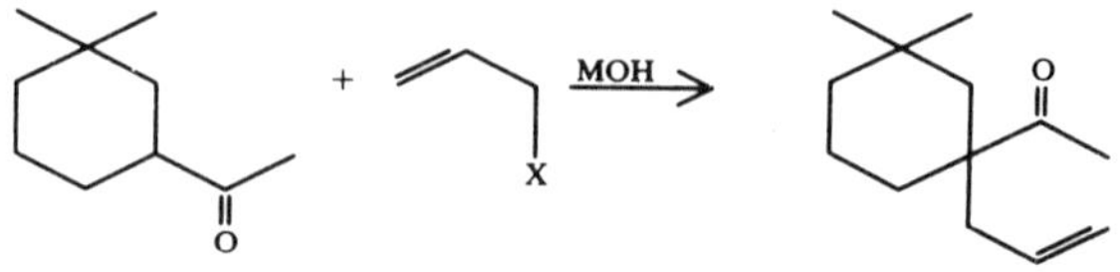

O + X MOH → O

wherein X is chloro or bromo and wherein M is alkali metal. Suitable phase transfer agents include the organic quaternary ammonium salts mentioned in previous patents in this chapter. Example 1 below describes the procedure followed for producing Compound A and the remaining examples show the way in which the compound may be used. All parts and percentages are by weight unless otherwise specified.

Example 1: Allyl chloride (222 g, 3.8 mols) is added over a period of 15 minutes to a stirred slurry of 1-acetyl-3,3-dimethylcyclohexane (462 g, 3 mols), granular sodium hydroxide (180 g, 4.5 mols), Aliquat 336 (tricapryl methyl ammonium chloride) (24 g) and toluene (200 g) at 70°C. The slurry is heated to reflux for 7.5 hours at the end of which period of time the temperature rises to 83°C. The slurry is then cooled and 1 liter of water is added to the reaction mass with stirring. The organic phase of the reaction mass is washed twice with water and distilled through a 12" x ½" packed Goodloe column fixed with an automatic reflux head set at a 19:1 reflux take-off ratio (85°C, 2.7 mm Hg pressure) yielding 275 g of Compound A (56% conversion), and 40 g of Compound B (8% yield); and 147 g of 1-acetyl-3,3-dimethylcyclohexane.

A B

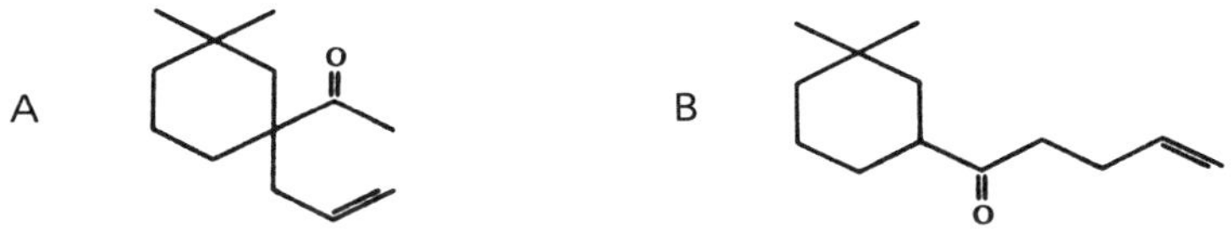

Example 2: A perfume composition is prepared by admixing the following ingredients in the indicated proportions:

Ingredient	Amount, grams
n-Decyl aldehyde	1
n-Dodecyl aldehyde	2
Methyl nonyl acetaldehyde	0.5
Linalool	50
Linalyl acetate	70
Phenyl ethyl alcohol	100
Petigrain SA	20
Bergamot oil	30
α-Methyl ionone	25
Compound A produced according to Example 1	10
Cyclized bicyclo C_{12} material produced according to the process of Example 4 of Canadian Patent 854,225, issued October 20, 1970	5
Isobornyl cyclohexyl alcohol	10
Benzyl acetate	25
2-n-Heptyl cyclopentanone	5
Total	353.5

The foregoing blend is evaluated and found to have a high degree of richness and persistence in its natural amber quality. This base composition can be admixed with aqueous ethanol, chilled and filtered to produce a finished cologne.

The cologne so prepared has an amber aroma leaning towards a woody amber note. The base compositions can also be used to scent soap or other toilet goods such as lotion, aerosol, sprays, etc.

Example 3: *Preparation of a Cologne and Handkerchief Perfume* – Compound A prepared according to the process of Example 1 is incorporated in a cologne at a concentration of 2.5% in 85% aqueous ethanol; and into a handkerchief perfume at a concentration of 20% (in 95% aqueous ethanol). A distinct and definite woody, earthy, amber aroma is imparted to the cologne and to the handkerchief perfume.

Example 4: *Raspberry Flavor Formulation* – The following basic raspberry flavor formulation is produced:

Ingredient	Parts by Weight
Vanillin	2.0
Maltol	5.0
p-Hydroxybenzylacetone	5.0
α-ionone (10% in propylene glycol)	2.0
Ethyl butyrate	6.0
Ethyl acetate	16.0
Dimethyl sulfide	1.0
Isobutyl acetate	13.0
Acetic acid	10.0
Acetaldehyde	10.0
Propylene glycol	930.0

Compound A is added to half of the above formulation at the rate of 2.0%. The formulation with the Compound A is compared with the formulation without Compound A at the rate of 0.01% (100 ppm) in water and evaluated by a bench panel.

The flavor containing Compound A is found to have substantially sweeter aroma notes and a sweet raspberry, raspberry kernel-like and sweet aftertaste and mouthfeel missing in the basic raspberry formulation. It is the unanimous opinion of the bench panel that Compound A rounds the flavor out and contributes to a very natural fresh aroma and taste as found in full ripe raspberries.

Example 5: 10 parts by weight of 50 Bloom pigskin gelatin is added to 90 parts by weight of water at a temperature of 150°F. The mixture is agitated until the gelatin is completely dissolved and the solution is cooled to 120°F. 20 parts by weight of the liquid flavor composition of Example 4 is added to the solution which is then homogenized to form an emulsion having particle size typically in the range of 2 to 5 microns. This material is kept at 120°F under which conditions the gelatin will not jell.

Coacervation is induced by adding, slowly and uniformly, 40 parts by weight of a 20% aqueous solution of sodium sulfate. During coacervation, the gelatin molecules are deposited uniformly about each oil droplet as a nucleus.

Gelation is effected by pouring the heated coacervate mixture into 1,000 parts by weight of 7% aqueous solution of sodium sulfate at 65°F. The resulting jelled coacervate may be filtered and washed with water at temperatures below the melting point of gelatin, to remove the salt.

Hardening of the filtered cake, in this example, is effected by washing with 200 parts by weight of 37% solution of formaldehyde in water. The cake is then washed to remove residual formaldehyde.

Substituted 1-Acetyl-3,3-Dimethylcyclohexanes

In another patent assigned to *International Flavors & Fragrances Inc., J.B. Hall, M.A. Sprecker, F.L. Schmitt and M.H. Vock; U.S. Patent 4,081,479; March 28, 1978* describe 1-acetyl-3,3-dimethylcyclohexane derivatives having the generic structure:

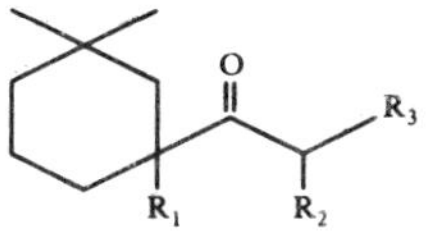

wherein R_1 or one or both of R_2 and/or R_3 is methallyl and the other of R_1 or R_2 and/or R_3 is hydrogen and an economical process of synthesizing such 1-acetyl-3,3-dimethylcyclohexane derivatives by reaction of methallyl halides with acetyl-3,3-dimethylcyclohexane, and utilization of such 1-acetyl-3,3-dimethylcyclohexane derivatives for their organoleptic properties in perfumes, perfumed articles, foodstuffs, medicinal products, tobaccos and tobacco articles.

Example 1: *Preparation of 1-Acetyl-1-Methallyl-3,3-dimethylcyclohexane (Compound C) and 1-(2-Methallyl-5-methyl-4-pentenoyl)-3,3-dimethylcyclohexane (Compound D)* – A slurry of 1-acetyl-3,3-dimethylcyclohexane (624 g, 4 mols), granular sodium hydroxide (240 g, 6 mols), Aliquat 336 (35 g), methallyl chloride (432 g, 4.8 mols), and 400 ml of toluene is heated at reflux for 4¼ hours. At the end of this time, one liter of water is added to the cooled reaction mass. The aqueous layer is discarded, and the organic layer is distilled rapidly through a short column to afford 203 g of recovered 1-acetyl-3,3-dimethylcyclohexane, 500 g of Compound C and 73 g of Compound D.

A part of the latter compound is isolated by GLC chromatography and exhibits fragrance properties having green, herbaceous, melony, galbanumlike, gingerlike and citrusy notes and fatty, sweet, fruity, citrus, green and waxy flavor notes.

Example 2: A tobacco blend is made up by mixing the following materials:

Ingredient	Parts by Weight
Bright	40.1
Burley	24.9
Maryland	1.1
Turkish	11.6
Stem (flue cured)	14.2
Glycerin	2.8
Water	5.3

The above tobacco is used in producing cigarettes and the following formulation is compounded and incorporated into each of these cigarettes.

Ingredient	Parts by Weight
Ethyl butyrate	00.05
Ethyl valerate	00.05
Maltol	2.00
Cocoa extract	26.00
Coffee extract	10.00
Ethyl alcohol	20.00
Water	41.90

The above flavor is incorporated into model "filter" cigarettes at the rate of 0.1%. One-third of these model cigarettes are treated in the tobacco section with the mixture of Compounds C and D produced according to Example 1 at 100 ppm per cigarette. Another third of these model cigarettes are treated in the filter with the mixture of the compounds at the rate of 2×10^{-5} g and 3×10^{-5} g. When evaluated by paired comparison, the cigarettes treated both in the tobacco and in the filter with the mixture of Compounds C and D are found, in smoke flavor, to have a green, sweet, fruity, floral and Virginia tobacco-like flavor and aroma nuance in the mainstream and in the sidestream on smoking and, in addition, prior to smoking.

Example 3: A perfume formulation is prepared by admixing:

Ingredients	Parts
Linalool	30
Linalyl acetate	10
Terpineol coeur	5
Nerol coeur	10
Terpinyl acetate	2
Geranyl acetate	2
Neryl acetate	2
Methyl anthranilate	1
Citral	10
n-Decyl alcohol	1
n-Dodecyl alcohol	5
n-Dodecanal	15
n-Decanal	30
n-Nonanol	3
n-Nonanal	5
n-Decyl acetate	5
n-Dodecyl acetate	3
Compound C	5

Compound C imparts a natural, tart, orange character to this terpeneless orange perfume formulation.

Hydroxy Cyclohexenone Derivatives

K.K. Light, B.M. Spencer, J.F. Vinals, J. Kiwala, M.H. Vock and E.J. Shuster; U.S. Patents 4,119,574; October 10, 1978 and 4,084,009; April 11, 1978; both assigned to International Flavors & Fragrances Inc. describe hydroxy cyclohexenone derivatives having the generic formula

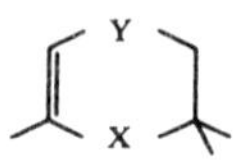

Y

X

where either X or Y is a keto moiety and the other is a carbinol moiety. These compounds give to foodstuffs, chewing gums, toothpaste, medicinal products and tobacco citrusy, sweet labdanum, and woody aromas and flavors. To perfumes and perfumed articles they provide green, minty, herbaceous, strong fruity aromas with earthy and mossy notes; and to tobaccos and tobacco flavoring compositions they give sweet, tobaccolike, floral and green aromas prior to smoking and bright tobaccolike notes on smoking.

These hydroxy cyclohexenone derivatives may be produced by one or two processes:

(1) subjecting 1,3,5,5-tetramethylcyclohexa-1,3-diene to an oxidation step comprising continuously insufflating air or oxygen through the diene at a temperature of between 15° and 30°C, while agitating the diene and irradiating the diene with ultraviolet light in the presence of an alkali metal hydroxide and Rose Bengal according to the following reaction sequence:

$$+ O_2 \xrightarrow[KOH]{\text{Rose bengal},\ h\nu}$$

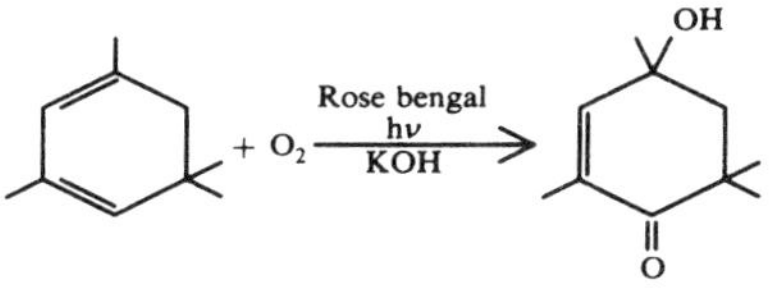

(2) reacting 2,7,9,9-tetramethyl-1,4-dioxaspiro[4,5]-dec-6-en-8-one with an alkyl lithium to form a lithium salt and then hydrolyzing the salt while simultaneously deketalizing the resulting compound according to the following reaction sequence:

$$+ R_1Li \longrightarrow \xrightarrow{H_2O}$$

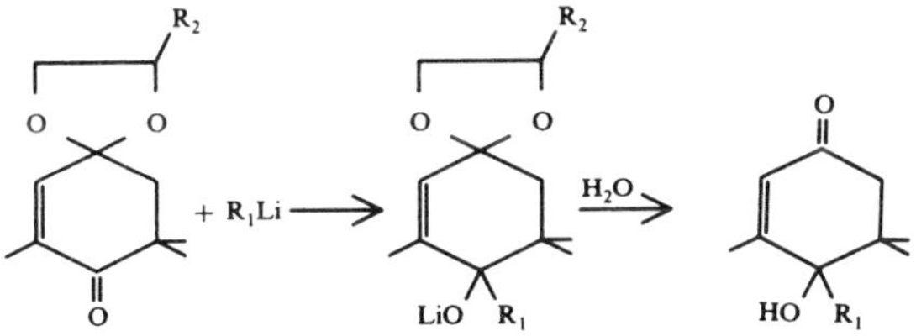

wherein R_1 is C_{1-4} alkyl and R_2 is lower alkyl or H.

Examples and organoleptic properties of materials produced according to these processes are as follows:

NAME	STRUCTURE	ORGANOLEPTIC PROPERTIES FLAVOR	FRAGRANCE	TOBACCO AROMA AND FLAVOR
4-hydroxy-2,4,6,6-tetramethylcyclohex-2-en-1-one	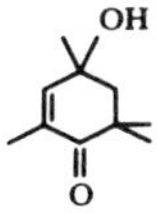	Citrus/labdanum, sweet woody aroma characteristics with sweet, woody, earthy, labdanum/citrus flavor characteristics at 10 ppm.	Minty, green, herbaceous aroma.	Sweet tobacco-like, floral, green aroma with slight cooling effect prior to smoking and bright tobacco-like aroma and taste with sweet and floral nuances in the mainstream and in the sidestream.

(continued)

NAME	STRUCTURE	ORGANOLEPTIC PROPERTIES FLAVOR	FRAGRANCE	TOBACCO AROMA AND FLAVOR
4-butyl-4-hydroxy-3,5,5-trimethyl-2-cyclohexen-1-one	O, HO	Bitter chemical.	Strong fruity and earthy mossy aroma with herbaceous, green, sweet, woody, minty and slightly earthy nuances.	Sweet, hay-like, Virginia tobacco, flue-cured-like aroma with floral and sweet hay nuances prior to and on smoking in the main and side-streams.

Example 1: *Preparation of 1,3,5,5-Tetramethylcyclohexa-1,3-Diene –*

O —CH_3MgCl→ ClMgO —H+→ OH —p-TSA→

To a 5-liter flask fitted with a stirrer, thermometer, dropping funnel, reflux condenser, drying tube and nitrogen purge is charged 2,100 ml (6.3 mols) of a 3 M solution of methyl magnesium chloride in tetrahydrofuran. The flask and contents are cooled to 0°C and 730 g (5.3 mols) of isophorone is added at 0° to 5°C over a 2-hour period. The mixture is allowed to warm to room temperature and to stand overnight. The reaction mixture is poured over a mixture of 2,800 g of ice and 370 g of glacial acetic acid. When all the ice melts, the layers are separated and the organics are washed twice with 500 ml of water. The solvent is removed on a rotary evaporator and the product is transferred to a 2-liter flask fitted with a condenser and a water separator. Benzene (300 ml) and p-toluene-sulfonic acid (0.5 g) are added and the mixture is refluxed until no additional water separated. The product is distilled on a 12-inch Goodloe column using a reflux ratio of 9:1. A total of 494 g of product is obtained (BP 66° to 67°C at 42 mm Hg) which is predominantly 1,3,5,5-tetramethylcyclohexa-1,3-diene.

Example 2: *Preparation of 4-Hydroxy-2,4,6,6-Tetramethylcyclohex-2-en-1-one and 2,4,4,6-Tetramethylcyclohexa-2,5-Diene-1-one –*

—hν, O_2, Base, Rose Bengal→ OH, O + O

A solution of 40 g (0.3 mol) of 1,3,5,5-tetramethylcyclohexa-1,3-diene (prepared as in Example 1), 13 g (0.23 mol) of potassium hydroxide and 0.3 g of Rose Bengal in 600 ml of methanol is placed in a flask and irradiated with a 450 watt tungsten photo lamp over a period of two hours. A steady stream of oxygen is bubbled through the solution during irradiation. At the end of the

2-hour irradiation period analysis of a sample of the reaction mixture by GLC (10' x 1/8" 10% Carbowax 20M on Chromasorb W, programmed from 80° to 220°C at 8°C/min) shows that the starting material is completely consumed and two products had formed. Both products are trapped from the GLC column and analyzed. The first (about 20% of the reaction product) is found to be 2,4,4,6-tetramethylcyclohexa-2,5-diene-1-one. The second (about 80% of the product mixture) is found to be 4-hydroxy-2,4,6,6-tetramethylcyclohex-2-en-1-one.

Example 3: *Geranium Perfume Formulation* – The following mixture is prepared:

Ingredients	Parts by Weight
Geraniol coeur	100.0
Citronellol coeur	130.0
Linalool	5.0
Citronellyl formate	25.0
Geranyl acetate	10.0
Benzyl butyrate	2.0
Guaicwood oil	10.0
4-Hydroxy-2,4,6,6-tetramethyl-cyclohex-2-en-1-one, produced according to Example 2	30.0

The 4-hydroxy-2,4,6,6-tetramethylcyclohex-2-en-1-one imparts the green, minty, herbaceous tone of geranium to this synthetic geranium formulation.

Example 4: A standard tobacco blend is made up and a flavor formulation consisting of:

Ingredient	Parts by Weight
Ethyl butyrate	00.05
Ethyl valerate	00.05
Maltol	2.00
Cocoa extract	26.00
Coffee extract	10.00
Ethyl alcohol	20.00
Water	41.90

The above formulation is incorporated into model filter cigarettes at the rate of 0.1%. One-third of these model cigarettes are treated in the tobacco section with the 4-hydroxy-2,4,6,6-tetramethylcyclohex-2-en-2-one produced according to Example 2 at 200 ppm per cigarette. Another third of these model cigarettes are treated in the filter with the 4-hydroxy-2,4,6,6-tetramethylcyclohex-2-en-1-one at the rate of 2×10^{-5} g and 3×10^{-5} g. When evaluated by paired comparison, the cigarettes treated both in the tobacco and in the filter with the 4-hydroxy-2,4,6,6-tetramethylcyclohex-2-en-1-one are found, in smoke flavor, to be more aromatic, sweeter, cooling (sensation in the mouth) and containing sweet tobacco-like, floral and green nuances.

Example 5: A standard IF&F raspberry flavor formulation is made up. When added at the rate of 1% to the formulation, the 4-hydroxy-2,4,6,6-tetramethyl-cyclohex-2-en-1-one adds a more natural character thereto. A 5-member bench panel unanimously agrees that the formulation containing 1% of 4-hydroxy-2,4,6,6-tetramethylcyclohex-2-en-1-one is more raspberry-kernel-like; more piney;

has a wild raspberry or herbaceous taste and has a natural berry character. The flavor formulation containing the 1% of 4-hydroxy-2,4,6,6-tetramethylcyclohex-2-en-1-one is unanimously preferred over the flavor formulation not containing any 4-hydroxy-2,4,6,6-tetramethylcyclohex-2-en-1-one. The flavor formulations are compared side-by-side at the rate of 40 ppm in water.

Example 6: The following mixture is prepared:

Ingredients	Parts by Weight, %
Natural Raspberry	
Concentrate juice	2.50
Water	85
Sugar Syrup	
(37.5° Baume)	12.50

The wild raspberry, herbaceous and seedy, raspberry kernel notes of this raspberry juice is imparted in increased strength by addition of 4-hydroxy-2,4,6,6-tetramethylcyclohex-2-en-1-one at the rate of from 20 ppm up to 50 ppm.

1-Butanoyl-3,3-Dimethylcyclohexane

M.A. Sprecker, M.H. Vock, F.L. Schmitt, J.B. Hall and J.M. Sanders; U.S. Patents 4,081,481; March 28, 1978; 4,119,576; October 10, 1978; 4,147,727; April 3, 1979; all assigned to International Flavors & Fragrances Inc. have discovered that 1-butanoyl-3,3-dimethylcyclohexane (Compound I) may be used to provide fruity, berrylike and herbaceous aromas to perfumes and perfumed articles and sweet rootbeerlike, cedarwood flavors to foodstuffs.

Compound I and one or more auxiliary perfume ingredients, including, e.g., alcohols, aldehydes, nitriles, esters, cyclic esters (or lactones) and natural essential oils, may be admixed so that the combined odors of the individual components produce a pleasant and desired fragrance, particularly and preferably in amber fragrances.

The 1-butanoyl-3,3-dimethylcyclohexane can be used to alter, modify or enhance the aroma characteristics of a perfume composition, e.g., by utilizing or moderating the olfactory reaction contributed by another ingredient in the composition. All parts and percentages given in the examples are by weight unless otherwise specified.

Example 1: *Synthesis of 1-(3,3-Dimethylcyclohexyl)-Cis and Trans-(2 and 3)-Buten-1-one* –

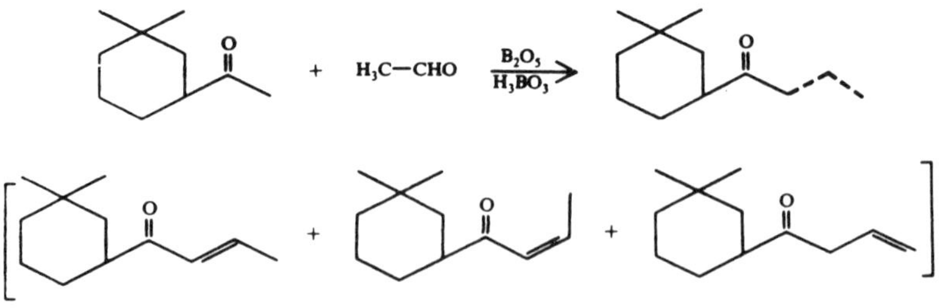

$$+ \quad H_3C\text{—}CHO \xrightarrow[H_3BO_3]{B_2O_3}$$

[+ +]

A mixture of 616 g of 1-acetyl-3,3-dimethylcyclohexane (4 mols), 132 g of acetaldehyde (3 mols), 105 g boron oxide (1.5 mols) and 12 g of boric acid (0.2 mol) is heated at 150°C in an autoclave for 3 hours. After cooling to room temperature the solution is filtered from inorganic salts, washed once with aqueous sodium carbonate and twice with saturated salt solution. Distillation through a short column afforded 127.2 g of 1-acetyl-3,3-dimethylcyclohexane and 127.2 g of a mixture of 1-(3,3-dimethylcyclohexyl)-cis and trans-2-buten-1-one and 1-(3,-3-dimethylcyclohexyl)-3-buten-1-one.

A portion of the distillate rich in the desired product(s) was distilled through a 1" x 12" Goodloe column giving twelve fractions. Bulked fractions 6 through 10 at 2 ppm had cedarwood aroma and a sweet, rootbeerlike, cedarwood flavor.

Example 2: *Hydrogenation of 1-(3,3,Dimethylcyclohexyl)-Cis and Trans-(2 and 3)-Buten-1-one —*

H_2/Pd

Compound I

Into a 250 cc Parr shaker is placed 107 g (0.6 mol) of 1-(3-3-dimethylcyclohexyl)-cis and trans-(2 and 3)-buten-1-one prepared according to Example 1, 1 g of 5% palladium-on-carbon catalyst and 125 g of isopropyl alcohol. The Parr shaker is sealed and heated to a temperature of 125°C at a pressure of 50 to 150 psig of hydrogen and maintained at that pressure range and temperature range for a period of 1.5 hours. At the end of the 1.5 hour period the reaction product is filtered, the solvent is removed in vacuo, and the residual oil is distilled on an 8 plate Vigreaux column, producing 9 fractions.

From a perfumery standpoint fraction 7 at 10% in food grade alcohol has a fruity, berry, herbaceous aroma with green, tobacco nuances.

Example 3: A perfume composition is prepared by admixing the following ingredients in the indicated proportions:

Ingredient	Amount, grams
n-Decyl aldehyde	1
n-Dodecyl aldehyde	2
Methyl nonyl acetaldehyde	0.5
Linalool	50
Linalyl acetate	70
Phenyl ethyl alcohol	100
Petigrain SA	20
Bergamot oil	30
Alpha methyl ionone	25
Compound I produced according to Example 2	10
Cyclized bicyclo C_{12} material produced according to the process of Example 4 of Canadian Patent 854,225	5

(continued)

Ingredient		Amount, grams
Isobornyl cyclohexyl alcohol		10
Benzyl acetate		25
2-n-Heptylcyclopentanone		5
	Total	353.5

The foregoing blend is evaluated and found to have a high degree of richness and persistence in its natural amber quality. This base composition can be admixed with aqueous ethanol, chilled and filtered to produce a finished cologne. The cologne so prepared has an amber aroma leaning towards a woody amber note with excellent fruity and herbaceous nuances. The base composition can also be used to scent soap or other toilet goods such as lotion, aerosol, sprays, etc.

Example 4: *Preparation of a Cosmetic-Powder Composition* – A cosmetic powder is prepared by mixing in a ball mill, 100 g of talcum powder with 0.25 g of Compound I prepared according to Example 2. It has an excellent fruity, berry, herbaceous aroma with green and tobacco nuances.

Example 5: *Preparation of a Detergent Composition* – A total of 100 g of a detergent powder (lysine salt of n-dodecyl benzene sulfonic acid as more specifically described in U.S. Patent 3,948,818 issued April 6, 1976) is mixed with 0.15 g of the 1-(3,3-dimethylcyclohexyl)-butan-1-one of Example 2 until a substantially homogeneous composition is obtained. This composition has an excellent fruity, berry, herbaceous aroma with green and tobacco nuances.

Certain Tricyclic Compounds

B. Maurer; U.S. Patent 4,142,997; March 6, 1979; assigned to Firmenich S.A., Switzerland has developed processes for preparing certain tricyclic compounds for use as perfumery or flavor ingredients. These compounds have the formula:

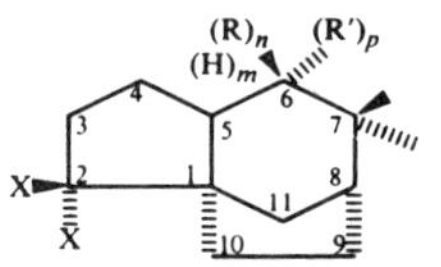

containing a single or a double bond in the position indicated by the dotted line at position 5 and wherein the indexes m, n and p represent the integers 0 or 1, the symbols X represent when taken together an oxygen atom, or when taken separately one of them represents an acyl, a hydroxyl, an O-acyl, a CH_2OH, a CH_2O-acyl or a COO-alkyl group and the other represents a hydrogen atom, and wherein: (1) the symbol R represents a lower alkyl group when both m and p are identical and equal to 0 and n is 1; or (2) the symbol R represents a lower alkylidene group when m and n are identical and equal to 1 and p is 0; or (3) one of the symbols R and R' represents a lower alkyl group and the other is a hydrogen atom when each of the indexes m, n and p is equal to 1.

Among the tricyclic compounds which can be prepared by the described process, two of them, 7,7-dimethyl-6-methylene-tricyclo[6.2.1.$0^{1,5}$]undecan-2-one (Compound A) and 7,7-dimethyl-6-methylene-tricyclo[6.2.1.$0^{1,5}$]undec-2-yl methyl alcohol (Compound B), are naturally occurring compounds. Both were in fact isolated from vetiver oil by means of an extremely complex and expensive process.

Owing to its particular olfactive properties Compound A can be widely used in the art of perfumery. Its use is broader than that of vetiver oil itself and it enables the perfumer to create totally original woody notes. Furthermore, the ketone possesses a much desired reinforcing and fixative effect.

One of the described processes consists in preparing Compound A by using a relatively cheap starting material, known in the art as vetivenic or zizanoic acid (7,7-dimethyl-6-methylene-tricyclo[6.2.1.$0^{1,5}$]undec-2-yl carboxylic acid). This compound may be obtained in large amounts and at a low price, as a by-product resulting from the purification of vetiver essential oil.

By means of this process of synthesis, industrially and economically more advantageous than the isolation of the tricyclic ketone from the natural vetiver oil, it is now possible to place at the disposal of perfumers and flavorists a pure 7,7-dimethyl-6-methylene-tricyclol[6.2.1.$0^{1,5}$]undecan-2-one. The difficulties of supply, storage and purification of the natural essential oil are thus avoided.

Compound B may be easily obtained from zizanoic acid by reducing this latter compound by means of an alkali metal aluminum hydride.

Example 1: *Preparation of 7,7-Dimethyl-6-Methylene-Tricyclo[6.2.1.$0^{1,5}$]undecan-2-one (Compound A)* – (a) A mixture of 4.68 g (20 mmol) vetivenic acid and 16.4 g of anhydrous sodium acetate dissolved in 45 ml of acetic acid was heated to 75°C. A single portion of 12.0 g of lead tetraacetate was then added to the reaction mixture. An evolution of CO_2 started immediately. The reaction mixture was kept at 75°C for 60 minutes, then cooled, diluted with 300 ml of water and finally extracted with 6 portions of 200 ml of n-pentane. After the usual treatments of washing and drying, the organic extracts were evaporated to afford 4.61 g (92%) of crude material containing 80% of the desired acetate. This material was used without any further purification for the following step.

(b) A solution of 9.7 g (39 mmol) of the acetate prepared in part (a) (purity 80%) in 50 ml of anhydrous ether was added dropwise to a suspension of 1.85 g (48 mmol) of $LiAlH_4$ in 100 ml of anhydrous ether. The reaction mixture was then heated with an equal portion of water and finally extracted, washed and dried according to the usual techniques. There were thus obtained 8.4 g of a crude material containing 8.0% of the desired alcohol.

(c) A solution of chromic acid was first prepared as indicated below. A solution of 100 g (330 mmol) of $Na_2Cr_2O_7 \cdot 2H_2O$ in 300 ml of water was added to 136 ml of concentrated sulfuric acid and finally diluted with water to 500 ml.

52 g of the above solution were then added dropwise under vigorous stirring, to a cold (0°C) solution of 5.2 g of the alcohol prepared in part (b) in 120 ml of ether. After 15 minutes of stirring, the reaction mixture was poured onto crushed ice, extracted with ether, the organic layer being successively washed with a 10% aqueous solution of $NaHCO_3$ and a saturated solution of NaCl. After the usual treatments of drying and evaporation, there were obtained 3.8 g of a crude material containing 80% of the desired ketone.

Pure 7,7-dimethyl-6-methylene-tricyclo[6.2.1.$0^{1,5}$]undecan-2-one (Compound A) was obtained after column chromatography. The thus obtained compound was identical with a pure sample prepared according to a known method.

Example 2: *Preparation of 6,7,7-Trimethyl-tricyclo[6.2.1.$0^{1,5}$]undec-5-en-2-one (Compound C)* – 350 mg of 7,7-dimethyl-6-methylene-tricyclo[6.2.1.$0^{1,5}$]undecan-2-one (see Example 1) were heated at 40°C for 5 days in the presence of 20 ml of 98% formic acid. After the usual treatments of extraction, washing, drying and evaporation, there were isolated 330 mg of crude material (purity 90%). The pure compound was obtained after crystallization in n-pentane.

Example 3: A base perfume composition for a masculine eau de cologne was prepared by admixing the following:

Ingredients	Parts by Weight
Sage oil	20
Lavender oil	150
Synthetic bergamot	200
Lemon oil	140
Sweet orange oil	40
Synthetic galbanum 10%*	20
Muscone 10%*	50
Methyl 2-pentyl-3-oxo-cyclopentyl-acetate	10
1,1-Dimethyl-6-tert-butyl-4-acetylindane	10
α-Isomethylionone	50
Synthetic ylang	80
Synthetic jasmine	25
Synthetic geranium	50
Synthetic neroli	100
Coriander oil	5
Total	950

*in diethyl phthalate

By adding to 95 g of the above base composition 5 g of a 10% solution of Compound A in diethyl phthalate, there was obtained a new perfume composition possessing a very elegant and original woody character having moreover a very natural richness.

By adding Compound C in the same proportions to the above base, there was obtained a perfume composition possessing a very rich woody and amberlike note.

Example 4: A commercial cranberry jam was flavored with a 10% ethanolic solution of Compound C in the proportions of 10 ml of the ethanolic solution per 100 kg of flavored material. The thus flavored foodstuff was then compared by a panel of flavor experts with an unflavored jam containing ethyl alcohol in the above given proportions. It was declared that the flavored jam possessed a woody, slightly balsamic taste reminiscent of that of fresh cranberrries.

Cyclic C_6 Ketones

K.-H. Schulte-Elte, B. Willhalm and F. Gautschi; U.S. Patent 4,147,672; April 3, 1979; assigned to Firmenich SA, Switzerland have formulated a number of oxygenated alicyclic derivatives useful as perfuming and odor modifying agents in the manufacture of perfumes or perfumed products, and as flavoring and taste-modifying agents in the aromatization of foodstuffs in general and imitation flavors for foodstuffs, beverages, animal feeds, pharmaceutical preparations and tobacco products.

The compounds have the formula:

(1)

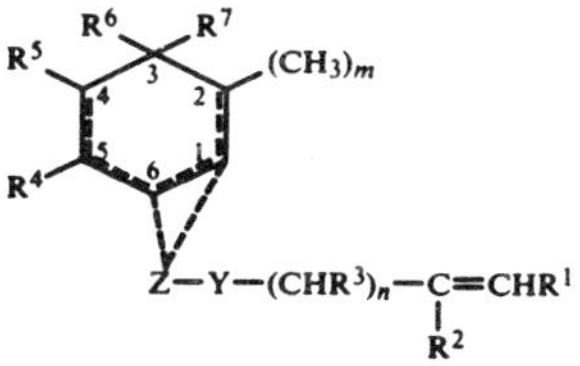

R^6, R^7, R^5, $(CH_3)_m$, R^4

$Z{-}Y{-}(CHR^3)_n{-}\underset{R^2}{\underset{|}{C}}{=}CHR^1$

having a saturated ring or an isolated double bond in position 1, 4 or 6 or two double bonds in position 1 and 4, 1 and 5, or 4 and 6 of the ring as indicated by the dotted lines, and wherein: m stands for integers 0 or 1; n stands for integers 0, 1 or 2; Z is bound to the ring carbon atoms in position 1 or 6 and represents the group –CO, $-COR^8$ or

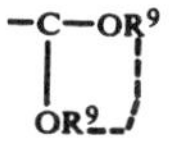

$-C-OR^9$, OR^9

(wherein symbol R^8 represents an acyl group, and R^9 represents, when taken individually, an alkyl group having from 1 to 6 carbon atoms, or, when taken together, an alkylenyl group having from 2 to 6 carbon atoms); Y represents an oxygen atom or a methylene group; each of the symbols R^1, R^2, R^3, R^4 and R^5 designates a hydrogen atom or one of them represents a methyl radical and each of symbols R^6 and R^7 represents an alkyl radical having from 1 to 3 carbon atoms or one of them represents an alkyl radical as defined above and the other a hydrogen atom.

Typically, the compounds of formula (1) develop various flavor notes such as fruity, green, woody and oily notes. These flavoring characters are reminiscent of those developed by galbanum or Juniper oil.

Further, the compounds of formula (1) possess useful olfactive properties and, accordingly, can be used in the art of perfumery. They impart a variety of fragrance notes such as green, herbal, and fruity notes. These aromatic properties are reminiscent of the odor developed in particular by galbanum oil, and enable harmonious matches with a great variety of compositions such as floral, woody, green, chypre or animal type compositions. The incorporation of compounds (1) into perfume compositions brings about a distinct and lifting green character of great richness and power.

Compounds (1) are also particularly useful for the reconstruction of certain essential oils as well as for the manufacture of perfumed articles such as toilet soaps, cosmetics, detergents, household materials, waxes or air-refresheners.

The names of thirty-seven specific examples of Compound (I) and a number of possible preparations for these compounds are given in the patent. An example of the synthesis and one of the applications of two of the most useful compounds are as follows.

Example 1: *Preparation of 1-(3,3-Dimethyl-Cyclohex-6-en-1-yl)pent-4-en-1-one [Compound A] and 1-(3,3-Dimethyl-Cyclohex-1-en-1-yl)pent-4-en-1-one [Compound B]* — (a) 100 g of α-dehydrolinalool were heated at 70°-75°C in the presence of 30 g of $B(OH)_3$ at 12 torrs until no further water formation was observed. The bath temperature was then slowly increased to 160° to 170°C and at that temperature a mixture comprising 3-methylene-7-methyl-oct-1-yn-7-ene and 3,7-dimethyl-oct-1-yn-3,7-diene was collected by distillation in 47% yield. By using β-dehydrolinalool, instead of the corresponding α-isomer, the mixture of unsaturated olefins was obtained in 66% yield. The obtained olefins could be separated one from the other by means of fractional distillation followed by gas chromatography.

(b) 2 g respectively of the olefin mixtures obtained as indicated above in 5 ml isopropanol were added dropwise to a cooled mixture of 1.1 g of KOH, 0.15 g of K_2CO_3, 0.1 g of CuCl in 5 ml methanol. After 30 minutes, to the thus-obtained reaction mixture, 1.5 g of allyl chloride were slowly added. After having been left overnight under stirring, the mixture was diluted with water and petrol ether and the separated organic phase was subjected to washing, drying and evaporation. By distilling the residue in a bulb apparatus 6-methylene-10-methyl-undeca-1,10-dien-4-yne and 6-methylene-10-methyl-undeca-1,9-dien-4-yne were respectively obtained in about 80% yield in admixture with 6,10-dimethyl-undeca-1,6,10-trien-4-yne and 6,10-dimethyl-undeca-1,6,9-trien-4-yne, respectively.

(c) 100 g of the mixture of olefins obtained as indicated above were heated at 90° to 100°C in the presence of 200 ml of 80% formic acid. After 1½ hours two-thirds of the starting material was converted into 1-(3,3-dimethylcyclohex-1-en-1-yl)-pent-4-en-1-yne and 1-(3,3-dimethylcyclohex-6-en-1-yl)-pent-4-en-1-yne. Their separation from the reaction mixture was possible by dilution of the mixture with water and petrol-ether, separation of the organic phase followed by the usual treatments of washing and drying, evaporation and distillation. The olefin mixture (77 g) was then separated into its two components by gas chromatography.

(d) 3 g of the olefin mixture obtained above were then heated to about 90°C in the presence of 80% formic acid for 2½ hours. After cooling, the reaction mixture was diluted with water and petrol-ether and, upon separation, the organic phase was subjected to the usual treatments.

On distillation 2.8 g of a fraction having BP 65° to 75°C/12 torrs was obtained. This fraction comprised 75% of a 3:2 mixture of Compound A and Compound B.

Example 2: A base perfume composition of the chypre-type was prepared by admixing the following:

Ingredient	Parts by Weight
Synthetic Jasmin	150
Vetiveryl acetate	60
Synthetic rose of may	10

(continued)

Ingredients		Parts by Weight
Synthetic bulgarian rose		50
Synthetic bergamot		150
Synthetic lemon		50
Angelica roots oil 10%*		20
α-Isomethyl-ionone		80
Cyclopentadecanolide 10%*		50
Muscone 10%*		50
γ-Undecalactone 10%*		50
Undecylenic aldehyde 10%*		50
Absolute oak moss 50%*		50
Dodecanal 10%*		10
Synthetic civet		50
Ylang extra		20
Sandal wood oil Mysore		20
Musk ketone		20
1,1-Dimethyl-4-acetyl-6-tert-butyl-indane		10
Synthetic lily-of-the-valley		50
	Total	1,000

*in diethyl phthalate

By adding to 95 g of the above indicated base, 5 g of a 10% solution of Compound A, a composition was obtained which possessed an original, harmonious tonality which proved to be distinctly more sophisticated than that shown by the base composition. It possessed, moreover, a green, herbal and lifting note which matches particularly well to the base odoriferous character.

By substituting Compound B for Compound A in the above example, analogous results were observed. The effects achieved were, however, less marked.

Substituted Dimethyl Dihydroxy Benzene and Cyclohexadiene Compounds

J.B. Hall, M.A. Sprecker, E.J. Shuster, F.L. Schmitt and J.F. Vinals; U.S. Patents 4,155,431; September 19, 1978; 4,155,867; May 22, 1979; and 4,169,072; Sept. 25, 1979; all assigned to International Flavors & Fragrances Inc. describe processes for preparing substituted dimethyl dihydroxy benzene and cyclohexadiene compounds and compositions using these compounds in perfumes and perfumed articles.

These substituted dimethyl dihydroxy benzene and cyclohexadiene compounds are represented by the generic structures:

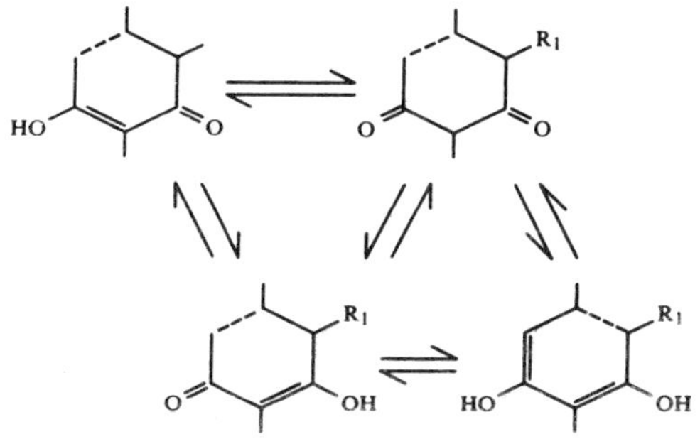

(and other tautomers) wherein the dashed line represents a carbon-carbon single bond or a carbon-carbon double bond and R_1 is either an acetyl or nitrile grouping with the proviso that when the dashed line is a carbon-carbon single bond, R_1 is nitrile.

When one or more of the compounds: 3,6-dimethyl-2,4-dioxocyclohexanecarbonitrile; 2,4-dihydroxy-3,6-dimethylbenzonitrile; and 2,4-dihydroxy-3,6-dimethyl-1-acetophenone are combined with other perfume compositions they serve as substitute(s) for "Oak Moss" in the combination of components which determine the odor note of various perfume compositions.

In addition, each of these chemicals, taken alone or together, not only simulates with great fidelity the characteristic odor note of Oak Moss, but each of them, taken alone or in combination, are less expensive than natural Oak Moss or its prior art synthetic simulations. Each of the compounds is prepared from readily available materials in commercial synthesis.

The following Examples 1 through 3 give preparations of these compounds and Examples 4 through 8 show compositions using the compounds for fragrance or flavor.

Example 1: *Preparation of 3,6-Dimethyl-2,4-Dioxocyclohexanecarbonitrile* – To a solution of sodium methoxide (162 g, 3 mols) in 400 ml of anhydrous methanol is added, dropwise, 276 g (3.2 mols) of methylcyanoacetate over a period of 1 hour at 60°C with stirring. The resulting slurry is stirred at 60°C for 10 minutes, whereupon 288 g (3.0 mols) of 2-hexen-4-one is added over a 20-minute period. The resulting reaction mixture is stirred and heated at reflux for 6 hours, whereupon a distillation take-off condenser is affixed to the reaction flask and 240 ml of methanol are removed under reduced pressure (65 mm Hg). The reaction mixture is then cooled to room temperature and poured into 2 liters of water. Toluene (500 ml) is added and the mixture is transferred to a separatory funnel.

The aqueous layer is removed and again extracted with toluene (500 ml). To the aqueous solution, 500 ml of 18% HCl are added with external cooling at 25°C, whereupon a solid material precipitates. The solid is filtered, dried and recrystallized from aqueous methanol to afford colorless needles of 3,6-dimethyl-2,4-dioxocyclohexanecarbonitrile (388 g, 68.3% yield based on 2-hexene-4-one); MP 178° to 179°C. The resulting product is represented by the following equilibrium mixture:

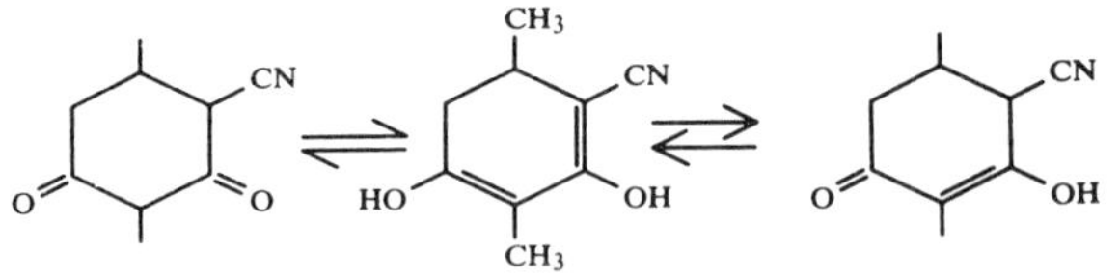

CN ⇌ CH_3 CN HO OH CH_3 ⇄ CN O OH

Example 2: *Preparation of 2,4-Dihydroxy-3,6-Dimethylbenzonitrile* – 3,6-Dimethyl-2,4-dioxocyclohexanecarbonitrile (165 g, prepared according to Example 1) is dissolved in 300 ml of glacial acetic acid. The resulting solution is cooled to 5°C, whereupon 64 g of chlorine is bubbled through the reaction mass using a gas dispersion tube. The reaction mixture is heated at 50°C for 1 hour subsequent to the chlorine addition and 3 liters of water are added thereto. The resulting aqueous solution is extracted three times with 250 ml of toluene (the toluene extracts are discarded) and evaporated under reduced pressure to afford a final volume of 800 ml. Upon standing, 95 g of crystals are recovered in two crops. Recrystallization from aqueous methanol yields light-tan needles, MP 162° to 163°C, consisting of 60% 2,4-dihydroxy-3,6-dimethylbenzonitrile and 40% 3,6-dimethyl-2,4-dioxocyclohexanecarbonitrile (% composition determined by NMR). NMR, IR and Mass Spectral analyses confirm the structure:

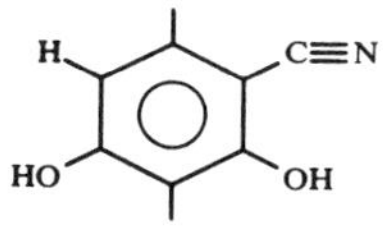

H C≡N

HO OH

Example 3: *Preparation of 2,4-Dihydroxy-3,6-Dimethylacetophenone* – To a 12-liter reaction flask equipped with stirrer, thermometer and reflux condenser are added 300 g of potassium hydroxide dissolved in 3 liters of methanol. To the methanol/potassium hydroxide mixture is added 3 mols (588 g) of 2,4-dihydroxy-3,6-dimethyl-carbomethoxybenzene ("Veramoss"). The reaction mixture is heated to 68°C and refluxing is commenced. At this point, an additional amount of KOH solution is added over a period of 45 minutes while maintaining the reaction temperature at 68° to 75°C. The reaction mass continues to be heated and refluxed for a period of 4 hours, at the end of which time 3,600 ml of 12% hydrochloric acid solution is added over a period of 30 minutes. At the end of the addition of the hydrochloric acid the methanol is removed and the reaction mass is allowed to cool down, yielding crystals of β-Orcenol having the structure:

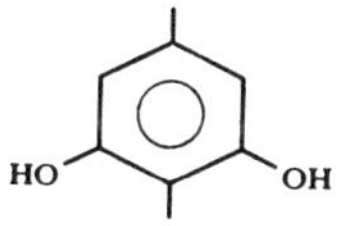

HO OH

The resulting β-Orcenol (20 g/0.143 mol) is then placed in a 250 ml reaction flask equipped with gas inlet tube and gas outlet tube and cooling bath. Acetonitrile (0.28 mol/12 g), zinc chloride (4 g) and anhydrous diethyl ether (100 ml) are added to the reaction mass and hydrogen chloride gas is passed through the reaction mass for a period of 2 hours at a temperature of 0°C. At the end of the 2-hour period, the hydrogen chloride gas addition is complete and the reaction mass is permitted to warm up to room temperature with stirring. 22 g of crystals are filtered.

The resulting material is then dissolved in 100 ml water in a 250 ml flask equipped with stirrer and reflux condenser. The reaction mass is heated to mild reflux for a period of 1.2 hours and then cooled to room temperature. The resulting crystals are filtered. NMR, IR and Mass Spectral analyses yield the information that the structure of the resulting material is as shown on the following page.

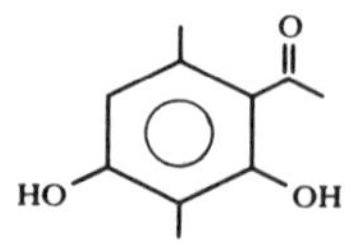

Example 4: A tobacco mixture and flavor formulation (see Example 6, Patent 4,068,012, as given in this book) are made up, and the flavor formulation is applied at the rate of 0.1% to all of the cigarettes produced using the tobacco formulation. Half of the cigarettes are then treated with 500 or 1,000 ppm of 2,4-dihydroxy-3,6-dimethyl acetophenone prepared according to Example 3. The control cigarettes not containing the acetophenone prepared according to the process of Example 3 and the experimental cigarettes which do contain the acetophenone are evaluated by paired comparison and the results are as follows.

The experimental cigarettes are found to have a sweet, slightly smokey, slightly woody, oakmosslike aroma prior to smoking and a woody, oakmosslike aroma on smoking in both the mainstream and in the sidestream.

The tobacco of the experimental cigarettes, compared with the control cigarettes, is more aromatic and more tobaccolike. In addition, the control cigarettes have none of the interesting and pleasant oakmoss aroma in the mainstream or in the sidestream on smoking.

Example 5: *Chypre Perfume Compositions* – The following example illustrates the use in perfumery compositions of 2,4-dihydroxy-3,6-dimethyl acetophenone produced according to Example 3. This material can replace partially or completely natural "Oak Moss." By adding any quantity of the 2,4-dihydroxy-3,6-dimethyl acetophenone, the odor quality of natural Oak Moss may be achieved with great fidelity but with great reduction in cost:

Ingredient	Parts by Weight
Oak Moss extract (Evernial)	2.5
2,4-Dihydroxy-3,6-dimethyl acetophenone produced according to Example 3	2.5
Oil of bergamot	22.5
Oil vetiver bourbon	7.5
Oil of lavender	5.0
Oil of sandalwood	7.0
Oil patchouli	1.0
Oil of cloves	3.5
Extract of jasmine	10.0
Oil of rose	8.0
Isobutyl salicylate	7.0
Cinnamyl alcohol	5.0
Heliotropin	10.0
Coumarin	5.0
Oleoresin tonka beans	2.0
Methyl nonyl acetaldehyde	1.5

Example 6: *Preparation of a Cosmetic-Powder Composition* – A cosmetic powder is prepared by mixing in a ball mill, 100 g of talcum powder with 0.25 g of 2,4-dihydroxy-3,6-dimethyl acetophenone prepared according to Example 3. It has an excellent "Mousse de Chene" aroma.

Example 7: *Preparation of a Cologne and Handkerchief Perfume* — 3,6-Dimethyl-2,4-dioxocyclohexanecarbonitrile prepared according to Example 1 or 2,4-dihydroxy-3,6-dimethylbenzonitrile prepared according to Example 2 or 2,4-dihydroxy-3,6-dimethyl acetophenone prepared according to Example 3 are each separately incorporated into a cologne at a concentration of 2.0%, 2.5%, 3.0% and 3.5% in 85% aqueous ethanol; and into a handkerchief perfume at concentrations of 15%, 20%, 25%, 30% and 40% (in 95% aqueous ethanol) in each of the cases with each of the compounds prepared according to Examples 1, 2 and 3, a distinct and definite Mousse de Chene aroma is imparted to the cologne and to the handkerchief perfume, the intensity increasing with increasing concentrations of substituted dimethyl dihydroxy benzene or cyclohexadiene compound.

Example 8: *Preparation of Soap Composition* — 100 g of soap chips are mixed with 1 g each of the following materials: (1) 3,6-dimethyl-2,4-dioxocyclohexanecarbonitrile (prepared according to the process of Example 1); (2) 2,4-dihydroxy-3,6-dimethylbenzonitrile (prepared according to the process of Example 2); and (3) 2,4-dihydroxy-3,6-dimethyl acetophenone (prepared according to the process of Example 3), until a substantially homogeneous composition is obtained. The perfumed soap composition manifests an excellent mossy, Mousse de Chene aroma.

DERIVATIVES OF NORBORNANE AND RELATED COMPOUNDS

Norbornane Derivatives

In a long series of patents, all assigned to *International Flavors & Fragrances, Inc., K.K. Light, J.M. Sanders, M.H. Vock, E.J. Shuster, J. Vinals, W.L. Schreiber, J.B. Hall, D.E. Hruza, Sr., V. Kamath, B.D. Mookherjee, C.Y. Tseng, and M.A. Sprecker* describe the preparation of a number of norbornane derivatives and their use as (1) flavoring in foodstuffs and related products such as chewing gums and medicinal products; (2) fragrances and perfume aroma enhancers in perfumes, colognes, soaps, detergents and toiletries; and (3) aroma enhancers and flavor or "taste" improvers for tobacco products and/or in cigarette filters.

U.S. Patent 4,076,853; February 28, 1978 discusses the use of Compounds C, D, and S (see the table which follows for names, abbreviations and structures of compounds) as flavorings for foodstuffs. Other norbornane derivatives used for flavorings are described in *U.S. Patents 4,136,208; January 23, 1979* (Compounds J and L); and *4,151,309; April 24, 1979* (Compounds A, B, F, I, M, and P). *U.S. Patents 4,119,577; October 10, 1978* and *4,123,393; October 31, 1978* address the subject of perfume compositions, using the substituted norbornane derivatives:

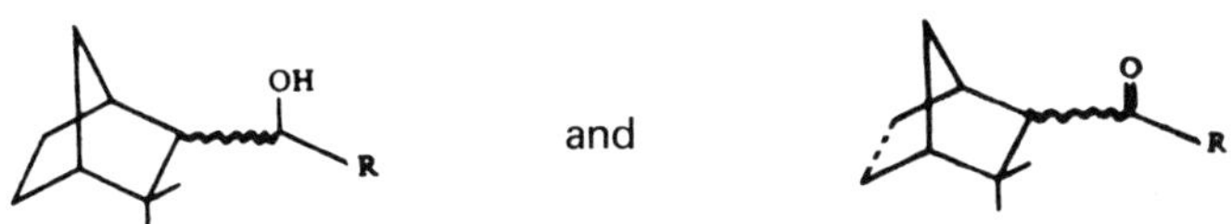

U.S. Patents 4,126,569; 4,126,570 and 4,126,571; all granted on November 21, 1978 and *4,153,568; May 8, 1979* take up the subject of enhancing the aromas of soaps and detergents by the use of Compounds A, B, S, T, and related compounds. *U.S. Patent 4,143,074; March 6, 1979* and *4,153,811; May 8, 1979* describe the organoleptic properties of carbocyclic ketones and hydroxy alkenyl derivatives having the structures, respectively, of:

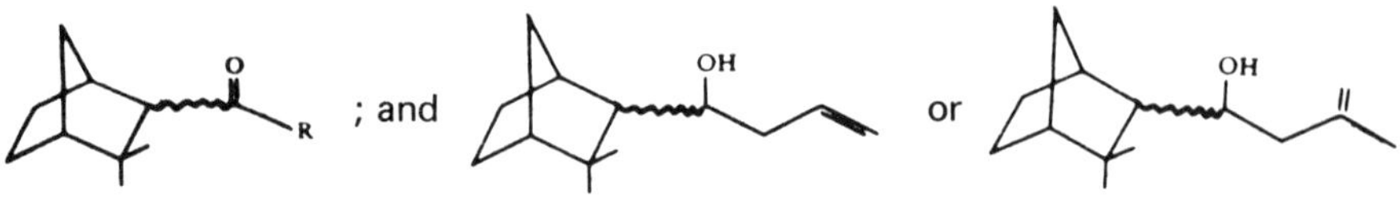

; and or

where R is ethyl, n-propyl, or n-butyl.

This group of substituted norbornane derivatives has the generic structure:

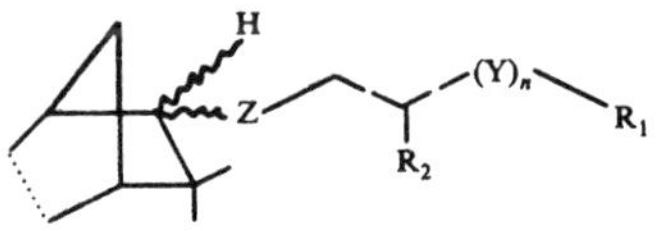

wherein each of the dashed lines represents a carbon-carbon single bond or a carbon-carbon double bond with the proviso that at least one of the dashed lines is a carbon-carbon single bond; wherein n is 0 or 1 with the proviso that n is 1 when both dashed lines are single bonds and n is 0 when one of the dashed lines is a double bond; wherein R_1 and R_2 are each the same or different hydrogen or lower alkyl; wherein Y is:

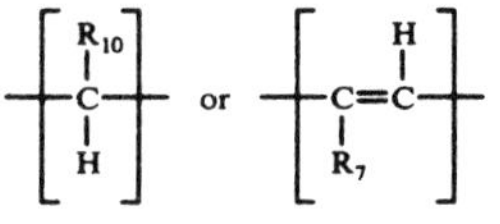

or

("cis and/or "trans")

wherein Z is one of the moieties:

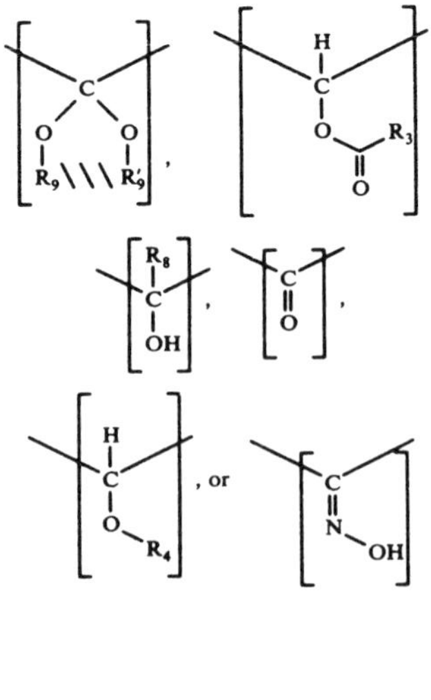

, or

wherein R_3 and R_4 are each alkyl; wherein R_7, R_8 and R_{10} are each the same or different hydrogen or lower alkyl; wherein R_9 and R_9' taken separately are the same or different lower alkyl, or taken together is lower alkylene; wherein the dotted line represents a carbon-carbon single bond or a carbon-carbon double bond; and wherein each of the wavy lines represents, in the alternative, exo or endo isomers. These compounds may be made by a variety of processes, e.g.:

(1) Reaction of an aldehyde with 2-acetyl-3,3-dimethylnorbornane in the presence of a base or a mixed boron oxide-boric acid catalyst. The product can be used as is, or can be reacted with an alkyl Grignard reagent or organolithium compound, followed by acid hydrolysis. It can then be further hydrogenated, if desired.

(2) Oxidation of camphene to form camphene oxide, rearranging this compound to an aldehyde, reacting the aldehyde as above with a Grignard reagent or an organolithium compound, followed by hydrolysis to form the alcohol. The alcohol can be used as is, or further oxidized or reduced.

(3) Reaction of 3,3-dimethylnorbornane-2-carboxaldehyde with a Grignard or organolithium reagent and using the resulting alcohol; or oxidizing the alcohol to form the ketone and then forming an ester with it.

(4) Reaction of 2-acetyl-3,3-dimethyl-5-norbornene with acetaldehyde.

(5) Reaction of 2-acetyl-3,3-dimethylnorbornane with an allylic halide in the presence of a phase transfer agent.

(6) Reaction of 2-acetyl-3,3-dimethylnorbornane (or the related 5-norbornene) with a dialkyl carbonate, treating the resulting compound with a base, reacting the resulting enolate with an allyl halide and saponifying the resulting ketocarboxylic acid ester to form a salt, acidifying the salt and then decarboxylating the ketocarboxylic acid to obtain the ketone.

Several examples of product syntheses will be given later.

The following table shows the structure of selected norbornane derivatives and the evaluations of their organoleptic properties for use as flavors, perfume constituents, or tobacco additives.

STRUCTURE OF COMPOUND	NAME OF COMPOUND	PERFUMERY EVALUATION	FLAVOR EVALUATION	TOBACCO EVALUATION
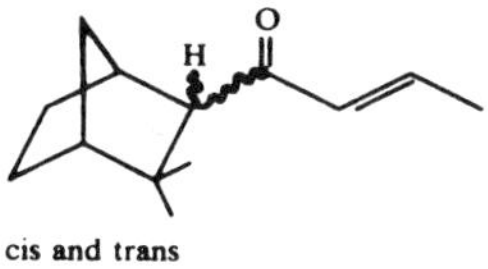 H, O; cis and trans	2(2-butenoyl)-3,3-dimethylnorbornane (A)	Sweet, woody, fruity, spicey (nutmeg, pepper), herbaceous aroma with pine needle nuances.	Fruity, pine needle/green, winey aroma with piney, eugenol/clove, spicey, fruity, winey flavor.	
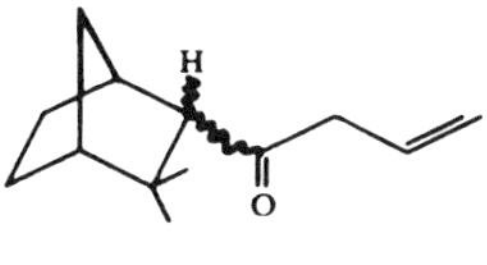 H, O	2-(3′-butenoyl)-3,3-dimethylnorbornane (B)	Fruity, herbaceous, piney aroma with armoise character and fir-balsam characteristics.	Fruity-estery, cedarwood, floral, woody, balsam tree resin-like aroma with sweet, fruity, banana-like, estery, balsam tree resin-like, piney flavor character.	

(continued)

STRUCTURE OF COMPOUND	NAME OF COMPOUND	PERFUMERY EVALUATION	FLAVOR EVALUATION	TOBACCO EVALUATION
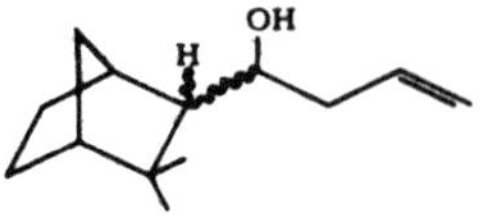	alpha-allyl-3,3-dimethyl-2-norbornanemethanol (C)	Sweet, woody, thujone-like aroma with armoise, cedar leaf, piney and camphoraceous nuances.	A sweet, fruity berry aroma with woody, incense, warm tea-like, floral nuances and an incense and cedarwood flavor with rosey, berry and tea nuances.	A woody, sweet fruity, cooling aroma and taste before and on smoking.
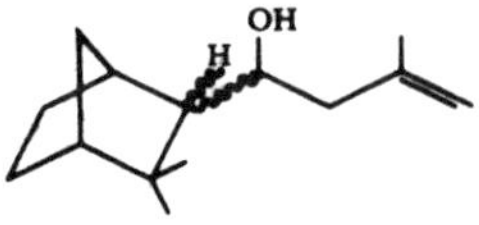	alpha-methallyl-3,3-dimethyl-2-norbornanemethanol (D)	Fruity, musty, sweet, woody, minty aroma with fresh cut pine/spruce nuances.	Incense and woody aroma and an incense and woody flavour with astringent and bitter notes.	
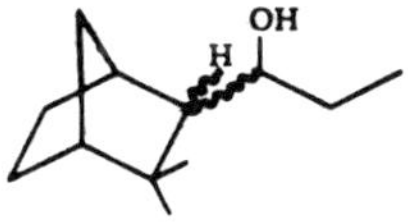	alpha-ethyl-3,3-dimethyl-2-norbornanemethanol (E)	Musty, sweet fruity, woody aroma with artemesia characteristics.	A sweet, woody, minty aroma with camphoraceous green/earthy nuances and a woody, green and minty flavour with spicey, earthy, red beet-like and camphoraceous nuances as well as a cooling effect.	
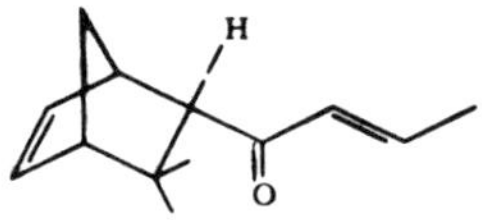	endo-2(2′-butenoyl)-3,3-dimethyl-5-norbornene (F)	Woody, piney, natural-like aroma with fruity, spicey, cresylic, borneol nuances, as well as a definitive berry top note.	Balsam, blueberry, piney, green fruity, woody aroma with balsam resin-like, piney, woody and blueberry flavour characteristics.	
cis and trans				
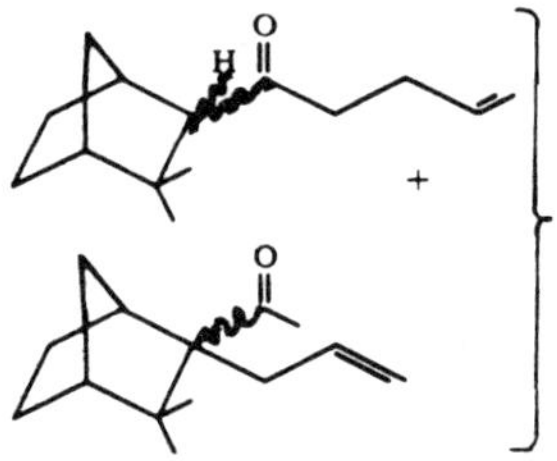	Mixture of 2-acetyl-2-allyl-3,3-dimethylnorbornane and 3,3-dimethyl-2-(4′-pentenoyl)-norbornane (G)	Camphoraceous, woody, fruity (orange) aroma with green, piney undertones.		
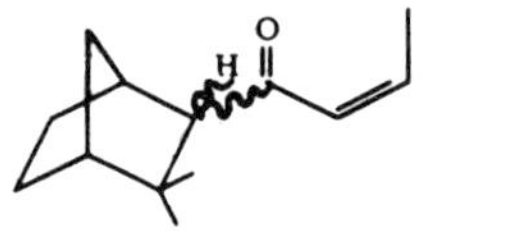	2-(cis-2′-buten-1′-oyl)-3,3-dimethyl)-norbornane (H)	Intense, piney, fruity, with floral, berry and cologne nuances.		
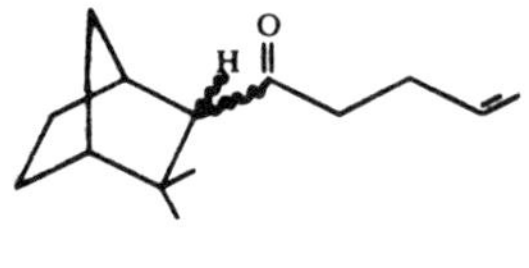	3,3-dimethyl-2-(4′-pentenoyl)-norbornane (I)	A sweet, fruity, fir-balsam, floral aroma with woody and pineapple notes.	Sweet, fruity, pineapple, green, woody, apple juice-like, piney aroma character with berry, piney, oriental-like and woody flavour character.	

(continued)

STRUCTURE OF COMPOUND	NAME OF COMPOUND	PERFUMERY EVALUATION	FLAVOR EVALUATION	TOBACCO EVALUATION
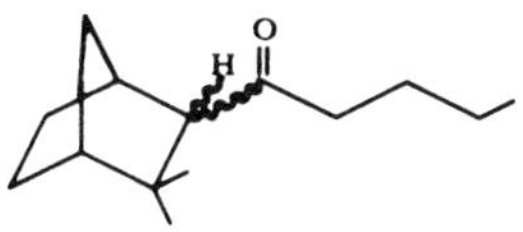	3,3-dimethyl-2-pentanoylnorbornane (J)	Sweet, fruity, fir-balsam aroma with floral (ylang, jasmin) and berry notes.	A sweet, fruity/berry, raspberry, woody and plum aroma with sweet, fruity/berry, raspberry, valerian oil-like, woody and plum flavour characteristics.	
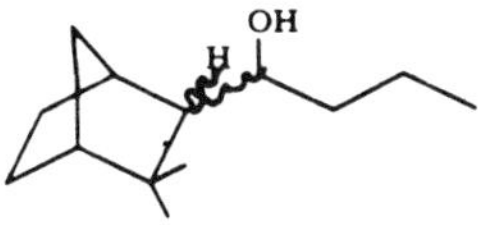	3,3-dimethyl-alpha-propyl-2-norbornanomethanol (K)	Camphoraceous, woody, green, aroma with minty nuances.	A sweet, delicate piney, fruity, berry-like, valerian oil-like, woody aroma with sweet, delicate piney, evergreen-like, valerian oil-like, fruity, berry and astringent flavor characteristics.	
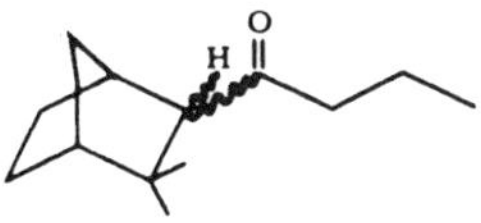	2-butyryl-3,3-dimethylnorbornane (L)	A sweet, piney needle-like, fruity camphoraceous aroma.	A sweet, delicate, piney, fruity, fir balsam needle oil-like, blueberry, woody, and strawberry-like aroma with sweet, delicate piney, blueberry, juniper-like, strawberry and astringent nuances.	
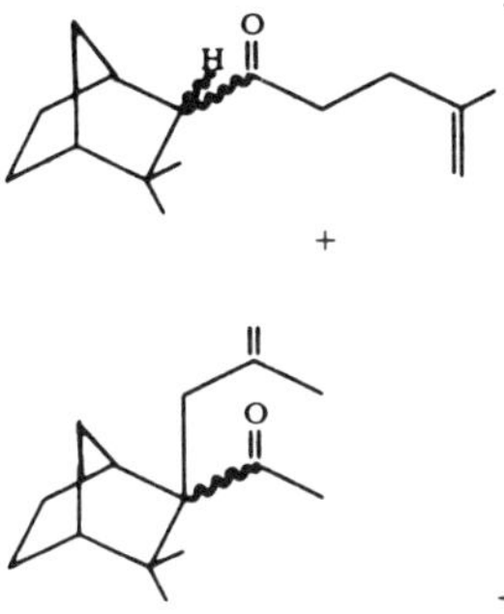 +	Mixture of 3,3-dimethyl-2(4'-methyl-4'-pentenoyl)-norbornane and 3,3-dimethyl-2-acetyl-2(2'-methyl-2'-propenyl)-norbornane (M)	A fruity, woody, low-keyed but long lasting aroma.	A camphoraceous, woody and minty aroma with a strong cooling nuance and a strong, woody and minty flavor.	A woody aroma, prior to and on smoking with sweet, spicey, floral nuances.
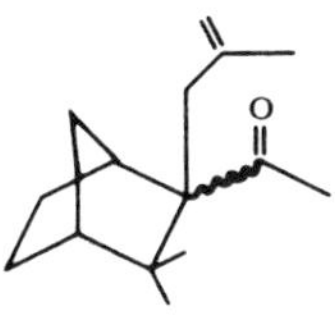	2-acetyl-2-methallyl-3,3-dimethylnorbornane (N)	A low-keyed, sweet champhoraceous, woody (cedarwood) aroma with green, herbaceous and piney nuances.		
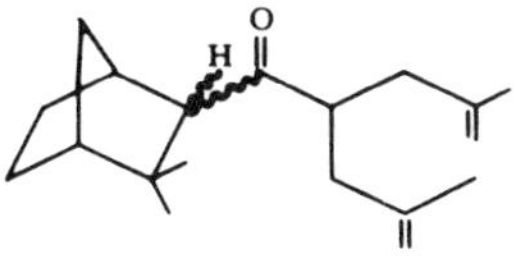	3,3-dimethyl-2-(2-methallyl-4'-methyl-4'-pentenoyl)-norbornane (O)	A low-keyed, sweet green, woody, herbaceous aroma with a waxy nuance.		

(continued)

STRUCTURE OF COMPOUND	NAME OF COMPOUND	PERFUMERY EVALUATION	FLAVOR EVALUATION	TOBACCO EVALUATION
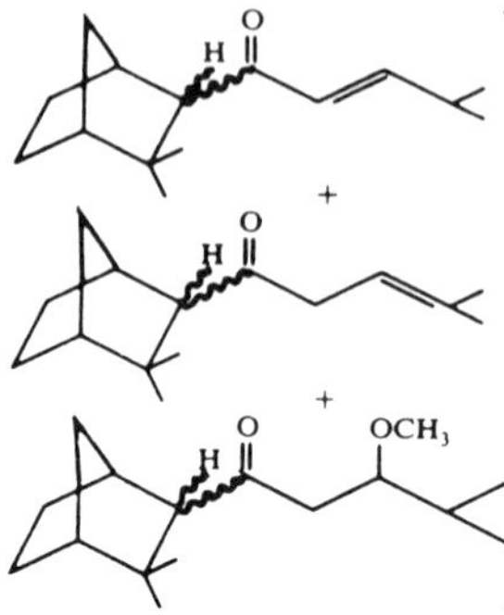 O, H + O, H + O, OCH_3, H	Mixture of 3,3-dimethyl-2-(4'-methyl-2'-pentenoyl)-norbornane; 3,3-dimethyl-2-(4'-methyl-3'-pentenoyl)-norbornane; and 3,3-dimethyl-2-(4'-methyl-3-methoxypentenoyl)-norbornane (P)	A long lasting excellent, fruity and berry aroma.	A fruity, raspberry and sweet aroma with a green nuance and a fruity, raspberry, green flavor with a bitter nuance.	A sweet, fruity-raspberry-like aroma prior to smoking and a sweet, fruity, raspberry aroma with green nuances on smoking, when used in both the tobacco and in the filter.
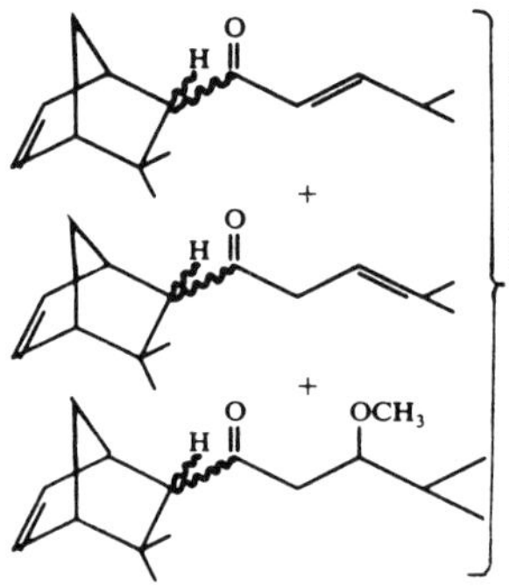 O, H + O, H + O, OCH_3, H	Mixture of 3,3-dimethyl-2-(4'-methyl-2'-pentenoyl)-5-norbornene; 3,3-dimethyl-2-(4'-methyl-3-pentenoyl)-5-norbornene; and 3,3-dimethyl-2-(4'-methyl-3-methoxypentenoyl)-5-norbornene (Q)	A fruity, berry-like aroma.	A sweet, red berry jam-like, fruity aroma with citrus, ionone-like, berry-like, pine needle-like and pungent nuances and a fruity, red berry-like berry jam-like flavor with green, fresh raspberry and piney nuances.	A sweet, floral and green aroma prior to smoking, and a sweet, fruity, floral aroma on smoking in the mainstream and in the side-stream.
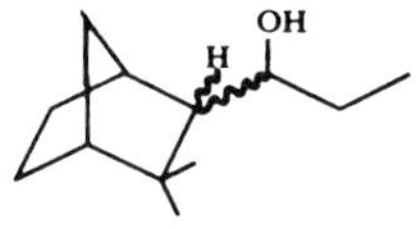 OH, H	alpha-ethyl-3,3-dimethyl-2-norbornanemethanol (R)	A minty, sweet fruity, woody aroma with some artemesia character and, in addition, herbaceous and pine needle nuances.	A sweet, woody, minty aroma with green and earthy nuances and a woody, green, minty flavor with spicey, earthy, cooling and red beet nuances.	A sweet, woody, piney aroma and a strong, cooling effect when used in the filter, both prior to, and, on smoking. Also reduces Harshness and gives rise to a "fresh" effect.
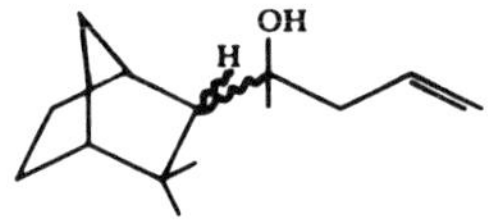 OH, H	alpha-allyl-alpha-3,3-trimethyl-2-norbornanemethanol (S)	Low-keyed, sweet, woody (cut pine) note.	Sweet, earthy, piney, cedarwood aroma with sweet, cedarwood, fruity, blueberry-like flavor with patchouli and walnut nuances.	
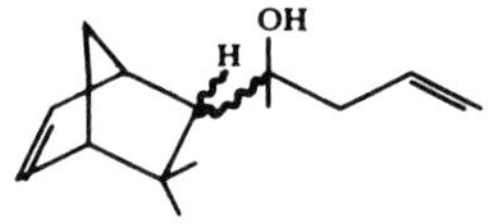 OH, H	alpha-allyl-alpha-3,3-trimethyl-5-norbornene-2-methanol (T)	A sweet, minty, fruity, slight melony, woody aroma.	A piney, fruity, woody, patchouli-like, carrot, walnut aroma with a sweet, fruity, woody, camphoraceous, carrot, dried apple, patchouli-like, walnut and astringent flavor.	

Example 1: *Preparation of 3,3-Dimethyl-2-Norbornanecarboxaldehyde* –

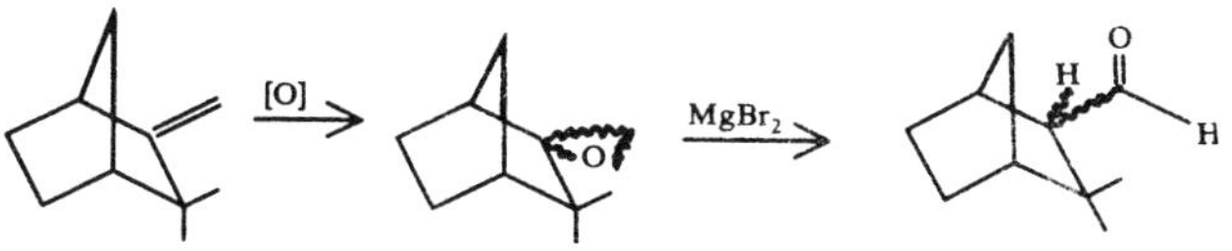

Into a 3-liter flask fitted with a stirrer, addition funnel, reflux condenser and thermometer are placed the following materials: 500 g camphene (80%), 500 cc methylene chloride, and 300 g sodium carbonate. The slurry is stirred while 600 cc of peracetic acid (40%) is added over a 6-hour period. The temperature rises until the mixture is at reflux (44°C in the beginning, 48°C at the end of the addition). The course of the reaction is monitored by GLC. The mixture is allowed to stir for 2 hours after the completion of the addition. The salts are then removed by filtration. The methylene chloride is stripped off and 500 cc of benzene is added. The mixture is azeotroped using a water separation trap to remove traces of water. After all water is removed, the mixture (670 g) is ready for use in the next step. A GLC indicated 26% benzene, 9.6% tricyclene and 61.2% camphene oxide, corresponding to a yield of 408 g of camphene oxide, or 92% of theory.

50 g of magnesium bromide is added and the mixture is refluxed for 4 hours. At the end of this time a GLC analysis shows complete conversion of the camphene oxide to 3,3-dimethyl-2-norbornanecarboxaldehyde.

Example 2: *Preparation of α-Allyl-3,3-Dimethyl-2-Norbornanemethanol (Compound C)* –

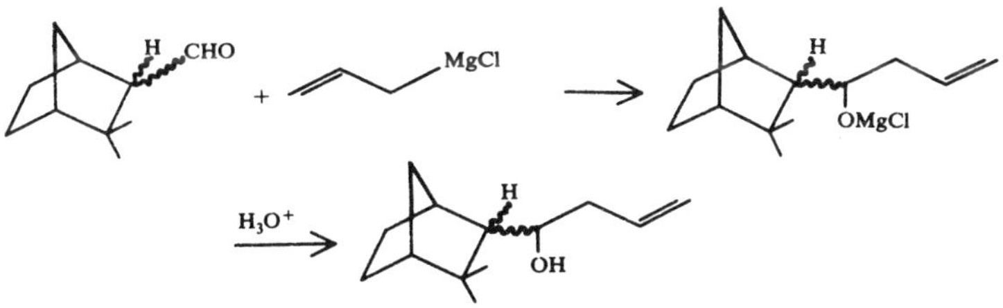

Into a 3-liter reaction flask fitted with a stirrer, reflux condenser, thermometer, dropping funnel and a N_2 bleed, is added 55 g of magnesium turnings (2.26 mols) and 1 liter of anhydrous tetrahydrofuran. The apparatus is blanketed with dry N_2, and 2 mols of allyl chloride (152 g, 162 ml) is placed in the addition funnel; about 10 cc is added at once with stirring. When the reaction begins (exotherm to 30°C), the flask and contents are cooled to 0°C (dry ice/isopropyl alcohol bath) and the remainder of the allyl chloride is added over a 3-hour period while the temperature is kept at 0° to 5°C. When addition is complete the contents of the flask are stirred for an additional hour while warming to 22°C at room temperature. The Grignard solution is decanted from the excess magnesium into another reaction flask and 330 g of a solution of 3,3-dimethyl-2-norbornanecarboxaldehyde in benzene prepared as in Example 1 is added over one hour. The reaction is exothermic and the temperature rises until the solution refluxes (65° to 70°C). At the end of the addition, the mixture is heated and refluxed for an additional one-half hour. The Grignard complex is hydrolyzed by the addition of about 400 cc of 15% HCl (to a pH of 1). The aqueous layer is separated and extracted with 200 cc of benzene. The combined organic layers are stripped on a rotary evaporator and the residue is distilled through a 1' stone packed column after adding thereto 30 g of Primol, at 70° to 77°C at 0.2 to 0.25 mm Hg pressure, yielding 206 g product. This corresponds to a yield of 92% theory based on 3,3-dimethyl-2-norbornanecarboxaldehyde added.

Example 3: *Preparation of 2-(3-Butenoyl)-3,3-Dimethylnorbornane (Compound B) Using a Sodium Dichromate-Sulfuric Acid-Benzene Oxidizing System* –

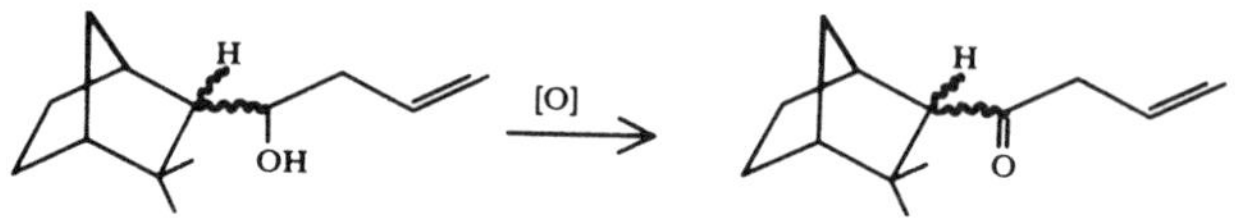

Sulfuric acid (60 cc, 110 g) is added slowly to 300 ml water with stirring. When the solution cools to room temperature, 300 cc (263 g) of benzene and 50 g of α-allyl-3,3-dimethyl-2-norbornanemethanol is added. The mixture is stirred at room temperature and 20 g of sodium dichromate is added. After one hour a second 20 g portion is added. After an additional one hour of stirring the layers are separated. The benzene layer is washed with saturated $NaHCO_3$ solution and then with water. The solvent is stripped and the residue is distilled rapidly to yield two distillation fractions. The first distillation fraction has a boiling point range of 65° to 97°C at 1.2 mm and weighs 2.5 g. The second distillation fraction has a boiling point range of 97° to 99°C at 1.0 mm Hg and weighs 39.4 g, corresponding to a yield of 80% of theory.

The second distillation fraction is primarily a compound having the desired structure, which is purified by means of column chromatography on a column of silicic acid (deactivated by the addition of 5% water). The compound is eluted with a mixture of 2% diethyl ether in isopentane. Thus, from 1.5 g of this second distillation fraction was obtained, after solvent stripping, 1.0 g of an oil which was distilled on a microdistillation apparatus, yielding pure Compound B.

Example 4: *Preparation of 2-(2-Butenoyl)-3,3-Dimethylnorbornane (Compound A) (Primarily the Trans Isomer)* –

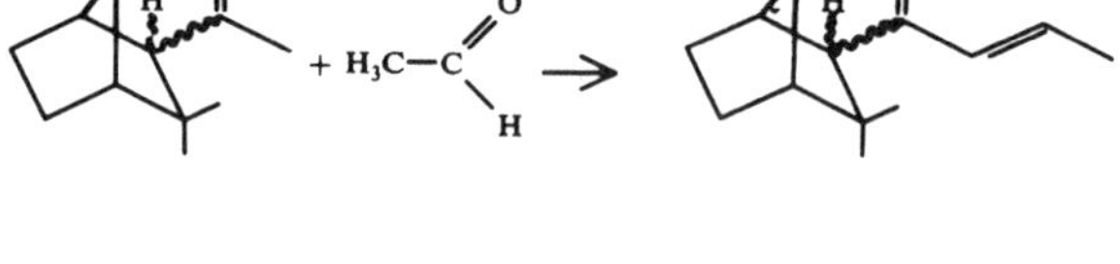

Into a 2-liter autoclave the following materials are placed:

Ingredients	Quantity
2-Acetyl-3,3-dimethylnorbornane	644 g (4 mols)
Acetaldehyde	132 g (3 mols)
Boron oxide	105 g (1.5 mols)
Boric acid	12 g (0.2 mol)

The autoclave is sealed and the contents are heated at 150°C for a period of 3 hours at autogeneous pressure. At the end of the 3-hour period the contents of the autoclave are cooled to room temperature and filtered. The filter cake is washed with 250 ml cyclohexane and the filtrate is distilled rapidly through a 2-inch column, after adding thereto 20 g Primol and 0.1 g Ionox to give 649 g of oil collected at 77° to 129°C and a pressure of 2.7 to 4.0 mm Hg. This material is then fractionally distilled on a 12" x 1" Goodloe column, after adding thereto Primol and Ionox, to give 490 g of oil collected at 73° to 78°C and a pressure of 3.3 to 4 mm Hg. The resulting material, insofar as its fragrance properties are concerned, has a sweet, woody, fruity, spicy and herbaceous aroma with pine needle nuances. Insofar as flavor characteristics are concerned, this material has a fruity, pine needle/green, winey aroma with a piney, eugenol/clove, spicy, fruity and winey flavor.

NMR, Infrared and Mass Spectral analyses confirm the structure of Compound A.

Example 5: *Blueberry Flavor Formulation* – The following formulation is prepared:

Ingredients	Parts by Weight
Heliotropin	3.0
Terpinenol-4 (10% in 95% aqueous food grade ethanol)	0.2
Benzaldehyde	1.5
Anisaldehyde	0.2
Phenyl acetaldehyde	0.4
Benzyl formate	0.5
Benzyl acetate	2.0
Cis-3-hexenyl benzoate (10% in 95% aqueous food grade ethanol)	0.5
Methyl hexanoate	2.0
Hexanal	1.0
Eucalyptol (1% in 95% aqueous food grade ethanol)	0.5
Eugenol	0.2
Acetaldehyde	3.0
Ethyl acetate	21.0
Ethyl butyrate	26.0
Propylene glycol	38.0

The above formulation is split into 2 portions. To the first portion is added 1% of Compound A. The second portion contains nothing added thereto. Both formulations, with and without Compound A (produced according to Example 4), are combined with water at the rate of 100 ppm. The flavor with the Compound A has a more winey, fruity, piney character and is closely similar to the flavor of wild blueberries. It is therefore preferable to the basic blueberry formulation which does not contain Compound A.

Example 6: *Pine Needle Oil Formulation* – The following formulation is prepared:

Ingredients	Parts by Weight
Turpentine gum oil	100
Limonene	70
Gum camphor	10

(continued)

Ingredients	Parts by Weight
Isobornyl acetate	50
Borneol	30
Compound A (produced according to Example 4)	40
Compound B (produced according to Example 3)	100
Compound C (produced according to Example 2)	70

The 2-(2-butenoyl)-3,3-dimethylnorbornane (Compound A) imparts a green, melony, herbal, ozoney and twiggy character of pine needle to the middle portion of the aroma profile and dry out.

The mixture of 2-(2-butenoyl)-3,3-dimethylnorbornane and 2-(3-butenoyl)-3,3-dimethylnorbornane (Compound B) produced according to Example 3 imparts the sweet, bright, melony, herbal lift so necessary to this pine needle oil formulation.

The α-allyl-3,3-dimethyl-2-norbornanemethanol (Compound C) produced according to Example 2 imparts the sweet, woody, melony, cut wood note to the dry out aroma of this pine needle oil formulation.

Example 7: *Perfumed Liquid Detergent* – Concentrated liquid detergents (lysine salt of n-dodecyl benzene sulfonic acid as more specifically described in U.S. Patent 3,948,818) with intense pine needles aromas are prepared containing 0.10, 0.15 and 0.20% of the perfume composition produced according to Example 6. The detergents are prepared by adding and homogeneously mixing the appropriate quantity of perfume oil composition. The detergents all possess intense pine needle aromas, with green, melony, herbal, twiggy and cut wood characteristics, with the intensity increasing with greater concentrations of the formulations of Example 6.

Example 8: *Preparation of α-Ethyl-3,3-Dimethyl-2-Norbornanemethanol (Compound E)* –

H CHO C_2H_5MgCl → H OMgCl $\xrightarrow{H_3O^+}$ H OH

Into a 1-liter reaction flask equipped with a stirrer, thermometer, condenser, nitrogen inlet and dropping funnel is placed 129 cc of a 2.0 M solution of ethyl magnesium chloride in tetrahydrofuran. Over a 20-minute period 19.0 g of 3,3-dimethyl-2-norbornanecarboxaldehyde is added with stirring. The reaction is exothermic. The reaction mass is then refluxed for a period of 3 hours. At the end of the 3-hour period the reaction mass is cooled to room temperature, and 100 cc of saturated ammonium chloride solution is added thereto. The organic layer is then extracted with 500 cc diethyl ether, and the extract is concentrated and distilled through a micro Vigreux column after adding thereto 3 g of Primol.

Mass spectral, NMR and infrared analysis of the distilled product, BP 85°C at 1.5 mm Hg, indicates that it is compound E.

Example 9: *Tobacco Filter* – Into a 20 mm cellulose acetate filter is added α-ethyl-3,3-dimethyl-2-norbornanemethanol (Compound E) at the rate of 1,000 ppm (10 microliter of a 10% solution of norbornanemethanol is added to the filter).

The filter is then attached to a full flavor cigarette on the market, e.g., (1) Marlboro, (2) Winston or (3) Viceroy, as well as on a Kentucky 1A3 reference cigarette (produced by the University of Kentucky), yielding the following results:

(1) Both cigarettes containing the norbornanemethanol additive, when compared to a cigarette having a filter without the norbornanemethanol additive, give rise to sweet, woody, piney aromas on smoking, with a pleasant, cooling effect and rather noticeable reduced harshness.

(2) Both cigarettes containing the norbornanemethanol additive have a lesser degree of "hotness" and give rise to a "fresh" taste on smoking.

1-(3,3-Dimethyl-2-Norbornyl)-2-Propanone

W.L. Schreiber, J.N. Siano, M.H. Vock and E.J. Shuster; U.S. Patent 4,090,985; May 23, 1978; assigned to International Flavors & Fragrances Inc. describe for use in foods, chewing gums, toothpastes and medicinal products flavor and aroma and perfume aroma augmenting, enhancing, modifying and imparting, compositions of 1-(3,3-dimethyl-2-norbornyl)-2-propanone (hereinafter referred to as the "norbornylpropanone derivative") having the structure:

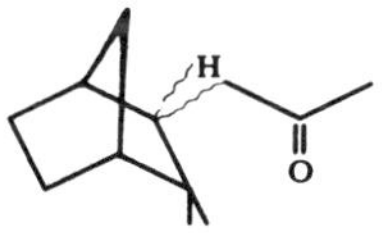

H
O

(which structure is intended to cover both the "endo" and the "exo" isomers thereof). Addition of this norbornylpropanone derivative produces:

(a) In foodstuffs, foodstuff flavorings, chewing gums, toothpastes and medicinal products, fruity/berry, pine needlelike aromas and sweet fruity/berry, blueberrylike, pine needlelike, pine balsam flavor characteristics; and

(b) In perfumes, colognes and perfumed articles sweet, berry, woody, piney, mint aroma notes with a damascenonelike character and buttery and minty nuances.

Example 1 shows how the norbornylpropanone derivative is prepared. The other examples give specific formulations for products using the compound for flavor and perfume enhancers. All parts and percentages are by weight unless otherwise specified.

Example 1: A 1-liter water-jacketed photochemical reaction vessel equipped with magnetic stirrer and pyrex immersion well is charged with 418 g camphene and 372 g acetaldehyde and 30 g diacetyl (photochemical hydrogen abstractor). The mixture is irradiated with a 450 medium pressure mercury lamp (Hanovia) for 55 hours, during which time five 30-g portions of diacetyl are added at varying intervals. The reaction mixture is combined with 150 g of 1,2,4-trimethyl benzene and washed with water. The material is distilled without fractionation and then redistilled through a 5' Vigreux column to yield 224 g colorless product, BP 70° to 72°C (55 mm Hg).

Example 2: The following mixture is prepared:

Ingredients	Parts by Weight
p-Hydroxy benzyl acetone	5
Vanillin	2
Maltol	3
α-Ionone (1% solution in propylene glycol)	15
Isobutyl acetate	15
Ethyl butyrate	5
Ethyl acetate	5
Dimethyl sulfide (10% solution in propylene glycol)	5
Acetic acid	15
Acetaldehyde	20
Propylene glycol	910
1-(3,3-Dimethyl-2-norbornyl)-2-propanone produced in Ex. 1	3

When 1-(3,3-dimethyl-2-norbornyl)-2-propanone is added to the above formulation it adds an excellent raspberry kernel character causing the formulation to be much more natural in taste and aroma. The formulation is preferred over the same formulation not containing the 1-(3,3-dimethyl-2-norbornyl)-2-propanone.

Example 3: A perfume composition is prepared by admixing the following ingredients in the indicated proportions:

Ingredients	Amount, grams
n-Decyl aldehyde	1
n-Dodecyl aldehyde	2
Methyl nonyl acetaldehyde	0.5
Linalool	50
Linalyl acetate	70
Phenyl ethyl alcohol	100
Petigrain SA	20
Bergamot oil	30
Alpha methyl ionone	25
1-(3,3-Dimethyl-2-norbornyl)-2-propanone produced according to Example 1	10
Cyclized bicyclo C_{12} material produced according to the process of Example 4 of Canadian Patent 854,225	5

(continued)

Ingredients		Amount, grams
Isobornyl cyclohexyl alcohol		10
Benzyl acetate		25
2-n-Heptyl cyclopentanone		5
	Total	353.5

The foregoing blend is evaluated and found to have a high degree of richness and persistence in its natural piney quality. This base composition can be admixed with aqueous ethanol, chilled and filtered to produce a finished cologne. The cologne so prepared has a sweet, berry, woody, piney, mint aroma note with a damascenonelike character and buttery and minty nuances. The base composition can also be used to scent soap or other toilet goods such as lotions, aerosols, sprays, etc.

Norbornanol Derivatives

The process developed by *K.K. Light, M.H. Vock, E.J. Shuster and F.L. Schmitt; U.S. Patents 4,089,986; May 16, 1978; and 4,107,094; August 15, 1978; both assigned to International Flavors & Fragrances Inc.* relates to the production of 3,3-dimethyl-2(3-butenyl) norbornanol-2 (hereinafter referred to as the "norbornanol derivative") having the structure:

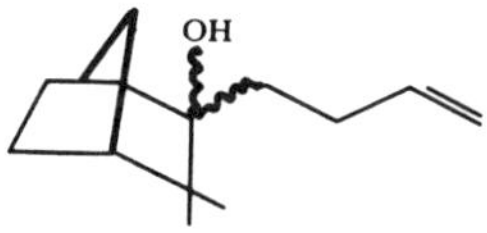

OH

(which structure is intended to cover both the "endo" and the "exo" isomers thereof) produced by a process which comprises reacting an allylic Grignard reagent having the structure: $CH_2{=}CHCH_2MgX$, where X is chloro, bromo or iodo, with camphene oxide having the structure:

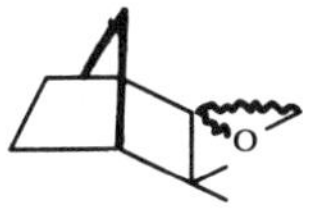

O

(either the exo or endo form or a mixture). Addition of the norbornanol derivative to perfumes and colognes produces sweet, woody, thujonelike aromas with strong fir-balsam, armoise, cedarleaf, piney, earthy and camphoraceous nuances.

When used as a flavoring agent, the norbornanol derivative produces sweet, fruity, berrylike aromas with woody, incenselike, cedarwoodlike, tealike and floral nuances and incenselike and cedarwoodlike flavors with rosy, berry and tealike nuances.

The reaction with the allylic Grignard reagent takes place at a temperature of between 25° and 50°C in the presence of a solvent inert to any of the reactants, diethyl ether or tetrahydrofuran. The Grignard reagent is first formed in the presence of the same solvent by reacting magnesium with allyl chloride, allyl bromide or allyl iodide. The camphene oxide is previously premixed with additional allyl chloride and the resulting mixture is added into the preformed Grignard reagent. The reaction time in forming the organometallic salt, at a temperature of 40°C, is between 2 and 10 hours. At higher temperatures the reaction time is somewhat shorter.

The hydrolysis of the organometallic salt to the tertiary alcohol norbornanol derivative takes place at a temperature in the range of between 20° to 60°C using a mineral acid such as hydrochloric acid, sulfuric acid, phosphoric acid or p-toluene sulfonic acid. The reaction takes place in an excess of water, which is one of the reactants.

At the end of the reaction, the reaction mass is concentrated and the desired product is fractionally distilled at a low pressure (of between 1 and 10 mm Hg).

Example 1: *Production of 3,3-Dimethyl-2-(3-Butenyl) Norbornanol-2 –*

Into a 12-liter reaction vessel is placed 164 g of magnesium and 985 g of tetrahydrofuran. While stirring the magnesium/tetrahydrofuran mixture, 20 g of allyl chloride is added and the temperature of the reaction mass rises to 40°C. A mixture of 2,410 g of camphene oxide (43% in benzene solution) and 520 g of allyl chloride is slowly added to the reaction mass and the reaction mass is maintained at a temperature of 35° to 40°C with external cooling. The allyl chloride and camphene oxide are added over a period of 7 hours. 10 g of additional magnesium is added and stirring is continued for an extra hour while maintaining the reaction mass at 40°C with external cooling. The reaction mass is then cooled to room temperature and maintained at room temperature for a period of 15 hours.

6.8 liters of water is then admixed with 170 g of concentrated 93% sulfuric acid. The 3.8 liters of reaction product is then added to the sulfuric acid solution whereupon the temperature rises from 20° up to 40°C. The resulting mixture is an emulsion and, in order to separate the phases of the emulsion, a solution containing 178 g of concentrated sulfuric acid and 1,200 ml water is added. The water and the organic layers are separated and the organic layer is washed with one liter of water containing 5 g of concentrated sulfuric acid solution then with two 100 ml portions of water. The organic layer is then stripped on a Rotovap yielding 1,249 g of product, which is then fractionally distilled on an 8" Goodloe column after adding thereto 41 g of Primol. The fractions shown on the following page are produced. Fractions 12-23 are substantially all product and are combined yielding 633 g of good material.

Fraction	Vapor Temperature (°C)	Liquid Temperature (°C)	Vacuum mm Hg	Weight (g)
1	43-60	12-122	3	–
2	60	111	2	–
3	65	108	1	–
4	76	115	1	–
5	80	118	1	–
6	91	122	1	–
7	91	121	1	18.9
8	90-95	128-126	3	3.7
9	101	123	3	26.4
10	103	123	3	28.3
11	103	123	3	37.8
12	101	124	3	23.4
13	86	124	1	41.5
14	88	124	1	38.4
15	88	124	1	31.6
16	91	125	1	36
17	93	126	1.5	76.5
18	86	122	1	81.3
19	102	125	2	59.3
20	88	125	2	75.9
21	88	125	2	64.1
22	88	127	2	60.3
23	88	134	2	44.5
24	93	161	2	40.1
25	91	196	2	40.2

Example 2: The following basic cinnamon flavor is formulated:

Ingredient	Parts by Weight
Cinnamic aldehyde	70
Eugenol	15
Cuminic aldehyde	1.0
Furfural	0.5
Methyl cinnamate	5.0
Cassia oil	8.5

Fraction 5 is incorporated at the rate of 5% into the abovementioned flavor formulation, and so incorporated, adds a more cinnamon barklike character in aroma and taste and a more woody, sweet aroma. It is preferred over the same formulation not containing the 3,3-dimethyl-2(3-butenyl) norbornanol-2.

Example 3: The composition of Example 1 is incorporated in a cologne at a concentration of 2.5% in 85% aqueous ethanol; and into a handkerchief perfume at a concentration of 20% (in 95% aqueous ethanol). The use of 3,3-dimethyl-2(3-butenyl) norbornanol-2 in the composition of Example 1 affords a distinct and definite sweet, woody, thujonelike aroma with strong, fir-balsam, armoise, cedarleaf, piney, earthy and camphoraceous notes to the handkerchief perfume and cologne.

Oxygenated Tricyclic Norbornene Derivatives

W. Skorianetz and G. Ohloff; U.S. Patent 4,118,343; October 3, 1978; assigned to Firmenich, SA, Switzerland report that the compound of the formula, tricyclo[6.2.1.$0^{2,7}$]undec-9-en-3-one, (Compound 1):

(1)

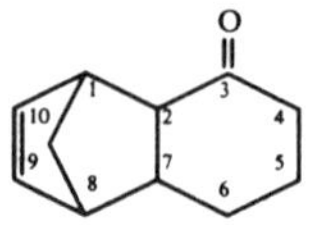

possesses useful organoleptic properties and consequently it can be advantageously used in a variety of perfume compositions to confer or improve aromatic odoriferous notes. Its odor is reminiscent of that developed by plants such as wormwood (*Artemisia Absinthium*) or *liatris*; it is very powerful and it confers an overall effect of naturalness to the composition to which it is added.

Compound 1 can be prepared according to the known process described in the cited literature [see: *J. Org. Chem.*, 39, 3063, (1974)].

Example 1: *Perfume Composition* – A base perfume composition destined to be incorporated into a medical shampoo was prepared by mixing together the following ingredients (parts by weight):

Ingredient		Parts
Trimethylhexyl acetate		200
Lavandin oil		100
Linalol		100
p-tert-Butylcyclohexyl acetate		80
Synthetic bergamot oil		80
Benzyl benzoate		70
Discolorized oak-moss absolute (50%)*		60
Methoxyphenethylol		40
Citrate mint oil		40
trans-Epoxy-ocimene 10%*		40
Cyclopentadecanolide 10%*		30
Isobornyl acetate		20
Methyl-nonyl acetaldehyde 10%*		20
Methyl salicylate 1%*		10
Discolorized mate absolute 10%*		10
Aspic oil		40
Caryolan**		10
	Total	950

*in diethyl phthalate
**See U.K. Patent 1,418,600

The above indicated perfume base possessed a rather pleasant lavender scent. By adding 5 g of Compound 1 to 95 g of this perfume base there was obtained a composition which had distinct "medical" note.

Example 2: *Tobacco Aromatization* – 0.5 g of a 1% solution of Compound 1 in 95% ethanol were sprayed onto 100 g of a tobacco mixture of the "American blend" type. The tobacco thus flavored was then used for the manufacture of test cigarettes, the smoke of which was subjected to an organoleptic evaluation by a panel of flavor experts. These latter declared that the smoke of the test cigarettes presented a more herbal, slightly fruity and floral flavor note.

Tricyclic Derivatives

G. Ohloff and W. Skorianetz; U.S. Patent 4,159,258; June 26, 1979; assigned to Firmenich, SA, Switzerland describe a process for preparing compounds of formula

(1)(a, b, c)

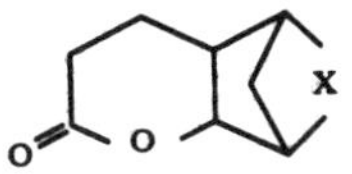

in which X may be $-CH_2-\overset{|}{C}=CH$(a), $-CH_2-CH_2-$(b), or $-CH_2-\overset{|}{C}H-CH_2-CH_3$(c). The process comprises:

(A) Reacting acrolein with a compound of formula

(2)

where X represents the radical (a) or (b) above;

(B) Reacting the product of (A) with water to give a hydroxy

(3) and

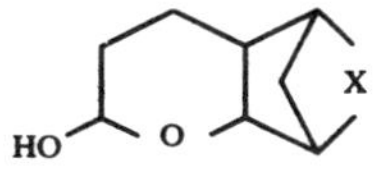

(C) Oxidizing the Compound 3.

To produce Compound 1(c), Compound 3 is catalytically hydrogenated. Specific compounds useful for perfumery are: 3-oxa-9-ethyl-tricyclo[6.2.1.$0^{2,7}$]undecan-4-one; 3-oxa-10-ethyl-tricyclo[6.2.1.$0^{2,7}$]undecan-4-one; 3-oxa-9-ethylidene-tricyclo[6.2.1.$0^{2,7}$]undecan-4-one; and 3-oxa-10-ethylidene-tricyclo[6.2.1.$0^{2,7}$]undecan-4-one.

Compounds 1(a, b, c) develop an original odor note which could be defined as very natural, green, fresh, herbaceous and slightly fatty or even spicy in some instances. Depending upon the nature of the products to which they are added, Compounds 1(a, b, c) can develop extremely varied odorous notes, such as green, fruity, aromatic notes, reminiscent in certain instances of tonka beans.

Compounds 1(a, b, c) develop also interesting gustative notes ranging from fatty, cuminic and carvonelike notes, reminiscent in certain instances of the taste of coconuts. 3-Oxa-tricyclo[6.2.1.$0^{2,7}$]undecan-4-one presents moreover a slightly fruity and animal character. Consequently, these compounds can be used for manufacturing artificial flavor compositions and for aromatizing foodstuffs and beverages in general. Typical proportions are in this case of the order of 1 to 100 ppm, preferably of about 1 to 5 ppm by weight based on the weight of the flavored foodstuffs.

Example 1: *3-Oxa-10-Ethylidene-Tricyclo[6.2.1.$0^{2,7}$]undecan-4-one* – Two autoclaves have been charged with two fractions, each containing 288 g (2.4 mols) of ethylidene norbornene, 44.8 g (0.8 mol) of acrolein and 1 g of hydroquinone. The reaction is carried out by heating the mixture at 190°C for 15 hours. The contents of the two vessels are then combined and fractionally distilled by means of a spinning band column to give a substance at BP 45° to 55°C/0.1 torr. 150 g of a mixture of 3-oxa-9- and 3-oxa-10-ethylidene-tricyclo[6.2.1.$0^{2,7}$] undec-4-ene were thus obtained. 430 g of starting ethylidene norbornene could be recovered and used for an operation.

A mixture of 176 g (1 mol) of the obtained mixture of undec-4-ene derivative, 250 g of water, 8 g of concentrated sulfuric acid and 1,500 ml of acetone was kept at 55°C for 2 hours, whereupon it was diluted with 1,000 ml of a saturated NaCl aqueous solution and extracted with 3 fractions of 1,000 ml each of ether. The combined organic extracts were subjected to the usual treatments of neutralization, washing and drying over Na_2SO_4. Subsequent evaporation of the volatile components over $NaHCO_3$ gave a mixture of 3-oxa-4-hydroxy-9- and 3-oxa-4-hydroxy-10-ethylidene-tricyclo[6.2.1.$0^{2,7}$] undecane (186 g; yield 95%).

30 g of the mixture obtained according to the above were heated under nitrogen at 220°C in the presence of 1.5 g of copper chromite. The reaction is over in approximately 2 hours. After cooling and filtration, the mixture was distilled over a short Vigreux column to give 16 g (yield ca 59%) of the desired mixture of 3-oxa-9- and 3-oxa-10-ethylidene-tricyclo[6.2.1.$0^{2,7}$] undecan-4-one; BP 128° to 134°C/0.1 torr.

Example 2: A base perfuming composition was obtained by admixing the following ingredients (parts by weight):

Benzyl salicylate	200
Phenylpropanol	100
Hydratropic alcohol	80
Brazil rose wood oil	60
Methyl-nonyl-acetaldehyde 10%*	60
Terpineol	60
p-tert-Butyl-cyclohexyl acetate	50
Lavandin oil	50
Galbanum oil 10%*	50
α-Ionone	40
Cyclopentadecanolide 10%*	40
Pine balsam absolute 10%*	30
Hydroxy-citronellal	20
1,1-Dimethyl-6-tert-butyl-4-acetyl-indane	10
Oak moss concrete 50%*	20
Trimethyl-hexanal 10%*	20
α-Damascone 10%*	10
β-Damascone 10%*	5
Ethyl acetyl-acetate	5
Isobornyl acetate	20
Diethyl phthalate	20
Total	950

*in diethyl phthalate

The above base possesses a pleasant odor of herbaceous type and is particularly suitable to perfume shampoos or lotions.

HETEROCYCLIC COMPOUNDS

2,4,6-Triisobutyl-1,3,5-Trioxane

D.A. Withycombe, B.D. Mookherjee, M.H. Vock and J.F. Vinals; U.S. Patent 4,093,752; June 6, 1978; assigned to International Flavors & Fragrances Inc. describe the organoleptic properties of the compounds in the table below and procedures for their synthesis.

Name of Compound	Structure	Organoleptic Properties as Food Flavor	Organoleptic Properties as Tobacco Flavor
2,4,6-triisobutyl-1,3,5-trioxane	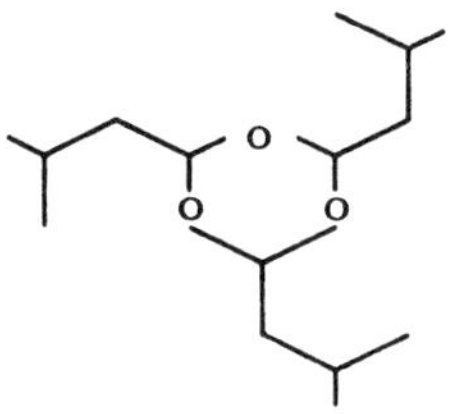	Sweet, dairy-like, creamy, vanilla, berry-like, milk chocolate-like and cocoa butter-like aroma and a sweet, dairy-like, creamy, vanilla, berry, milk chocolate-like and cocoa butter-like flavor at 15 ppm	A chocolate-like, cocoa-like and sweet, creamy aroma and flavor prior to smoking and a chocolate-like aroma and more tobacco-like nuances on smoking in the main stream and side stream.
5-ethyl-2-isobutyl-4-methyl-3-oxazoline	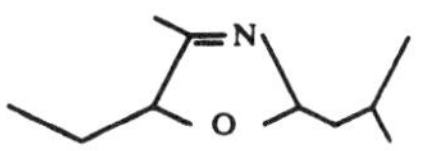	Cooked green pepper, spicey, peppery and meaty aroma with cooked green pepper, spicey, peppery and meaty flavor characteristics at 10 ppm.	Green pepper-like and slight vegetable-like aroma characteristics prior to and on smoking in the main stream and the side stream
4-ethyl-2-isobutyl-5-methyl-3-oxazoline	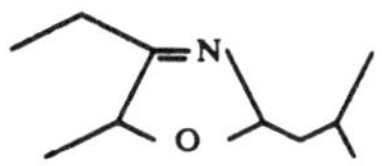	Green pepper, chocolate and roasted aroma characteristic with green pepper, hydrolyzed vegetable protein mouthfeel, and chocolate, nutty, and roasted notes at 6ppm.	Chocolate-like, nutty, green, burnt roasted-like aroma prior to smoking and on smoking in the main stream and in the side stream.

A preferred process for producing 2,4,6-triisobutyl-1,3,5-trioxane comprises the trimerization of isovaleraldehyde in the presence of an acidic catalyst such as concentrated sulfuric acid or concentrated phosphoric acid at a temperature between 0° and 25°C and in an inert solvent such as benzene, toluene or xylene.

A reaction product mixture containing all three of the compounds in the table is obtained by a process which involves first reacting isovaleraldehyde with aqueous ammonia at a temperature in the range of 0° to 10°C to form isovaleraldehyde imine, then reacting the imine with either 3-hydroxy-2-pentanone or 2-hydroxy-3-pentanone or a mixture of the two. This reaction is carried out at a temperature in the range of 30° to 50°C. The resulting reaction product is extracted with an inert solvent such as diethyl ether. The organic extract is evaporated and the residue is distilled. Preparative GLC may then be used to separate the oxazoline derivatives and the trioxane derivative from each other and from other reaction products.

Example 1: *Preparation of 2,4,6-Triisobutyl-1,3,5-Trioxane* — Into a one-liter, three-necked reaction flask equipped with mechanical stirrer, condenser, immersion thermometer, Y adapter and cold bath is placed 200 ml of isovaleraldehyde and 100 ml of benzene. The resulting mixture is cooled to about 5°C using an ice water bath. 2 g of concentrated sulfuric acid is then added dropwise to the reaction mixture while it is being agitated using the mechanical stirrer. The reaction mass heats up to a temperature of about 12°C and is then stirred for a period of 19 hours at a temperature of about 22°C. The resulting reaction product is made basic by washing with five 20 ml portions of 5% aqueous sodium hydroxide. The resulting product is then washed to neutral with two 50 ml portions of water. The resulting product is then distilled at 95°C and 1 mm Hg yielding 2,4,6-triisobutyl-1,3,5-trioxane.

The resulting material, when added to tobacco, gives rise to a chocolate-like, cocoa-like, and sweet, creamy flavor and aroma prior to smoking; and a chocolate-like aroma nuance with more tobacco-like nuances on smoking in the main stream and the side stream. In foodstuffs, the resultant material gives rise to sweet, dairy-like, creamy, vanilla, berry-like, milk chocolate-like and cocoa butter-like aromas and sweet, dairy-like, creamy, vanilla, berry-like, milk chocolate and cocoa butter-like flavor characteristics.

The resulting 2,4,6-triisobutyl-1,3,5-trioxane is then compared with isovaleraldehyde at about the same intensity insofar as their taste characteristics are concerned: at the rate of 10 ppm for 2,4,6-triisobutyl-1,3,5-trioxane; and at the rate of 0.1 ppm for isovaleraldehyde.

The 2,4,6-triisobutyl-1,3,5-trioxane has a cocoa/floral aroma and taste as found in dark chocolate. The isovaleraldehyde is substantially weaker in aroma notes at about the same flavor strength for both chemicals. Its aroma and taste characteristics are more cocoa powder-like. Therefore, both chemicals are completely different in their aroma and taste characteristics.

Example 2: *Preparation of Isobutyl Substituted Heterocyclic Compounds* — Into a 500 ml reaction flask equipped with mechanical stirrer, 250 ml addition funnel, thermometer, Y adapter, Freidrich condenser and isopropanol/dry ice bath is placed 86.2 ml of aqueous ammonia (0.5 mol). The aqueous ammonia is cooled to 3°C. Dropwise, with stirring, is added 43 g of isovaleraldehyde. On completion of the addition of the isovaleraldehyde, the resulting imine forms a solid mass in the aqueous medium which is dispersed at a temperature of about 35°C. The addition time of the isovaleraldehyde is 1.5 hours.

The resulting reaction mass is stirred for an additional 1.5 hours before the dropwise addition of 51.9 g of 2(3)-hydroxy-3(2)-pentanone. The reaction between the 2(3)-hydroxy-3(2)-pentanone and the imine is carried out for a period of 2 hours. The resulting reaction mass now exists in two phases. The aqueous phase is extracted with three 100 ml portions of diethyl ether and combined with the organic phase. The combined organic phases are then washed with three 100 ml portions of saturated sodium chloride solution.

The resulting product is fractionally distilled at 110°C and 10 mm Hg to yield a large number of products, including all three of the compounds in the above table.

Example 3: *Tobacco Formulation* — A tobacco mixture is produced by admixing the following ingredients:

Ingredient	Parts by Weight
Bright	40.1

(continued)

Ingredient	Parts by Weight
Burley	24.9
Maryland	1.1
Turkish	11.6
Stem (flue-cured)	14.2
Glyerin	2.8
Water	5.3

Cigarettes are prepared from this tobacco. The following flavor formulation is prepared:

Ingredient	Parts by Weight
Ethyl butyrate	0.05
Ethyl valerate	0.05
Maltol	2.00
Cocoa extract	26.00
Coffee extract	10.00
Ethyl alcohol	20.00
Water	41.90

The abovestated tobacco flavor formulation is applied at the rate of 0.1% to all of the cigarettes produced using the above tobacco formulation. Half of the cigarettes are then treated with 500 or 1,000 ppm of 2,4,6-triisobutyl-1,3,5-trioxane produced according to the process of Example 1. The control cigarettes not containing the 2,4,6-triisobutyl-1,3,5-trioxane and the experimental cigarettes which contain the 2,4,6-triisobutyl-1,3,5-trioxane produced according to the process of Example 1 are evaluated by paired comparison, and the results are as follows: the experimental cigarettes are found, on smoking, to have very pleasant chocolate-like aroma notes and to be more tobacco-like; the tobacco of the experimental cigarettes, prior to smoking, has chocolate-like, cocoa-like and sweet, creamy notes. All cigarettes are evaluated for smoke flavor with a 20 mm cellulose acetate filter.

Spirane Derivatives

W. Renold, W. Skorianetz, K.-H. Schulte-Elte and G. Ohloff; U.S. Patent 4,114,630; September 19, 1978; assigned to Firmenich SA, Switzerland have found that spirane derivatives of the formula:

(1)

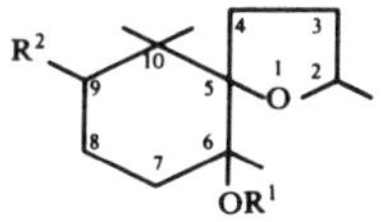

will reproduce certain typical nuances of natural patchouli oil odor and can, at times, be used as a replacement for the more expensive natural product.

In the above formula R^1 represents hydrogen or an acyl radical with 1 to 6 carbons, and R^2 represents hydrogen or methyl.

The compounds of formula (1) are characterized by their woody-type odor. Ester derivatives differ from the corresponding alcohols by a greater diffuseness associated with an amber-like, balsamic, flowery and even herbaceous notes.

Some of these compounds, particularly 2,6,9,10,10-pentamethyl-1-oxaspiro[4.5]-decan-6-yl acetate, can also develop fruity sulfury organoleptic notes which are reminiscent of the character developed by black currant fruits.

Owing to their particular organoleptic properties, these compounds can also be used in the flavor industry as ingredients for the preparation of artificial flavors or for the aromatization of foodstuffs, animal feeds, beverages, pharmaceutical preparations and tobacco products.

Depending on the nature of the products in which they are incorporated, these compounds can enhance or develop various gustative notes such as woody, amber-like, earthy and in certain cases slightly flowery notes, or even notes reminiscent of that of cedar wood oil. They are, therefore, particularly appreciated for the preparation of artificial flavors such as citrus fruits or even mushroom flavors wherein the woody and earthy gustative note is often requested.

Spirane derivatives of formula (1) can be produced by a process which comprises epoxidizing a compound of formula (2) to produce a compound of formula (3), subjecting compound (3) to a catalytic hydrogenation to produce a compound of formula (4), then reducing the epoxide of formula (4) and, if necessary, esterifying the obtained reaction product.

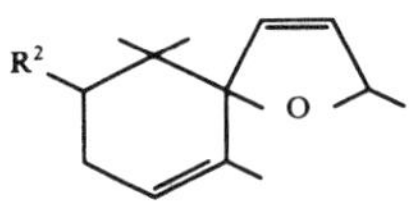

R^2 O (2) R^2 O O (3) R^2 O O (4)

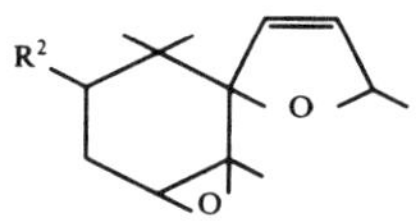

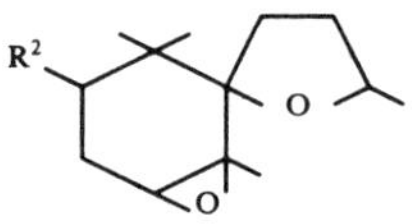

The patent contains detailed instructions for preparation of Isomers A and B of formula (1). The following examples illustrate uses of this compound.

Example 1: A black tea infusion possessing a relatively bland taste was prepared from 6 g of commercial black tea leaves and 1 liter of boiling water. After a few minutes, the above infusion was poured into small cups, at a rate of 30 ml per cup. Test samples were then prepared by adding to each cup of hot tea 0.06 ml of a 0.001% solution of Formula 1 in 95% ethyl alcohol. A control sample was obtained by adding 95% ethyl alcohol to the hot tea, in the same proportions.

Test and control samples were then subjected to the evaluation of a panel of flavor experts who unanimously declared that the test samples possessed a particularly pleasant woody character when compared to the unflavored material.

Analogous organoleptic effects were obtained, however less pronounced, by replacing the above alcohol by its B isomer. Similar effects were observed by using in the same proportion 2,6,9,10,10-pentamethyl-1-oxaspiro[4.5]decan-6-ol instead of the tetramethyl derivative.

Example 2: 100 mg of a 0.1% ethanolic solution of formula (1) were sprayed onto 100 g of an American blend tobacco mixture. The tobacco thus flavored was used for the manufacture of test cigarettes, the smoke of which was then subjected to organoleptic evaluation by comparison with unflavored control cigarettes. The tobacco used to prepared the control was preliminarily treated with a corresponding amount of 95% ethyl alcohol.

A panel of flavor experts defined the taste of the smoke of the test cigarettes as possessing a typical woody and earthy character.

Example 3: A base perfume composition for shampoo was prepared by admixing the following ingredients:

	Parts by Weight
Phenylethyl alcohol	130
Ethylene brassylate	100
Aspic oil, 10%*	80
Synthetic geraniol	80
Synthetic bergamot	80
Linalyl acetate	60
Synthetic geranium	60
Oak moss absolute, 10%*	60
Patchouli oil	40
Labdanum resinoid, 10%*	40
p-tert-Butylcyclohexyl acetate	30
Musk ketone	30
Coumarin	30
Terpineol	30
Methylisoeugenol	30
Rosemary oil	20
Galbanum oil, 10%*	20
Isocamphyl-cyclohexanol	20
Synthetic civet, 10%*	20
Lavandin absolute, discolorized	10
Synthetic neroli	15
Lilial (Givaudan)	10
Myrtle oil	5

*In diethyl phthalate

The above base composition possesses an intense and typical woody note, due to the presence of patchouli oil.

By replacing, in the above base, the 40 parts of patchouli oil by 80 parts of formula (1), there was obtained a perfume composition similar to the above base, the woody and balsamic notes thereof being more pronounced.

In *W. Renold, W. Skorianetz, K.-H. Schulte-Elte, and G. Ohloff; U.S. Patent 4,120, 830; October 17, 1978; assigned to Firmenich SA, Switzerland*, the spirane derivatives are used for additional formulations.

Example 1: A base perfume composition for a masculine Eau de Toilette was prepared by mixing the following ingredients:

	Parts by Weight
Bergamot oil dist.	160
p-tert-Butylcyclohexyl acetate	100
Galbanum resinoid, 50%*	80
Absolute oak moss decolorized, 50%*	80
Pentadecanolide, 10%*	80
Cedryl acetate	60
Thibetine, 10%*	40

(continued)

	Parts by Weight
Cedrene	40
Lemon oil	40
Vetiveryl acetate	40
Florida orange oil	40
α-Isomethylionone	40
Synthetic civette, 10%*	20
Lavender oil	20
Opopanax oil, 10%*	10
Eugenol	10
Coriander oil	10
Nutmeg oil	10
4-Isopropyl-cyclohexylmethanol	10
α-Ionone	10

*In diethyl phthalate

By adding to 90 g of the above base 2 g of 2,6,9,10,10-pentamethyl-1-oxaspiro-[4.5] deca-3,6-diene, a new perfume composition with an odor possessing a better defined lifting character was obtained. The odor, which is moreover richer than that of the above base, presents an original citrus-like note.

Example 2: Two syrups of raspberry and black currant-type, respectively, were prepared by diluting 1 part by weight of commercial syrup with 4 and 9 parts by weight, respectively, of water. The beverages thus obtained were flavored with a proportion of 0.3 and 0.5 ppm, respectively, of 2,6,9,10,10-pentamethyl-1-oxaspiro[4.5] deca-3,6-diene.

The flavored beverages were subjected to organoleptic evaluation by a panel of experienced tasters whose judgment was expressed as follows: the flavored raspberry syrup possessed an improved top note and an overall aroma which was fuller and fresher than that of the unflavored syrup; and the flavored black currant syrup showed a fuller and a more natural taste than the unflavored one. It possessed moreover a better defined herbal and fruity note.

2-Alkyl-4-Phenyl-Dihydropyrans

J.F. Vinals, J. Kiwala, D.E. Hruza, Sr., J.B. Hall, and M.H. Vock; U.S. Patents 4,070,491; January 24, 1978; 4,071,535; January 31, 1978; and 4,115,406; September 19, 1978; all assigned to International Flavors & Fragrances Inc. describe the 2-alkyl-4-phenyl-dihydropyrans having the generic structure:

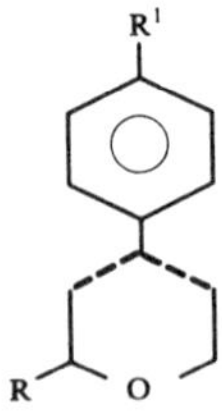

R^1

R O

where one of the dashed lines is a carbon-carbon double bond and the other of the dashed lines is a carbon-carbon single bond, R^1 is one of C_{1-3} alkyl, and R is one of C_{2-4} alkyl.

They have found that these pyran derivatives produce green, floral, neroli-like, jasmine, and peach-like tastes when added to foods, toothpastes and chewing gums and they give to tobacco products sweet, spicy, coumarin-like and vanilla-like flavors and aromas.

When these pyran derivatives are used as food flavor adjuvants, their primary purpose is the enhancement of flavors of their coingredients. It is, therefore, preferable that the pyran derivative, in quantities ranging from 0.2 to about 200 ppm based on the total composition, be combined with at least one of the following flavoring compounds: natural orange oil; methional; rum ether; diethyl malonate; ethyl propionate; amyl isovalerate; acetic acid; ethyl nonanoate; phenyl ethyl acetate; isobutyl formate; methyl anthranilate; ethyl heptanonate; cinnamyl propionate; ethyl acetate; ethyl butyrate; cinnamyl alcohol; cinnamyl isovalerate; cinnamyl propionate; citral; ethyl benzoate; ethyl caproate; ethyl methylphenylglycidate; ethyl oenanthate; ethyl pelargonate; hydroxycitronellal; petitgrain oil; terpinenyl acetate; tolualdehyde; methyl naphthyl ketone; benzylidene acetone; alpha-ionone; lemon essential oil; and lime essential oil.

Example 1: *Preparation of 2-Isopropyl-4-Phenyl-Dihydropyran —*

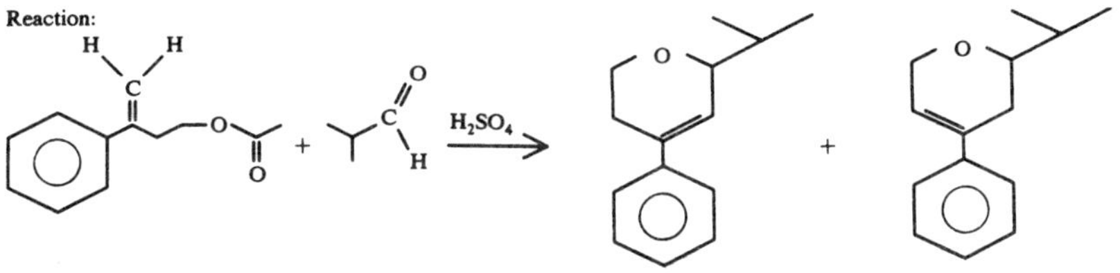

Reaction:

$\xrightarrow{H_2SO_4}$

In a 500 ml, 3-neck reaction flask equipped with stirrer, thermometer, reflux condenser and addition funnel is placed 72 g (1 mol) of isobutyl aldehyde. The isobutyl aldehyde is heated to 66°C (reflux) and from the addition funnel, over a period of 1 hour, is added a mixture of 204 g (1 mol) of 3-phenyl-3-buten-1-yl acetate and 1 g concentrated sulfuric acid. During addition of the 3-phenyl-3-buten-1-yl acetate/sulfuric acid mixture, the reaction mass temperature increases to 80° to 85°C. At the end of the addition of the mixture, the reaction mass is stirred for a period of 3 hours at 80° to 85°C. The reaction mass is then cooled to room temperature and 100 ml of a 10% aqueous sodium hydroxide solution is added thereto with stirring, followed by 100 ml cyclohexane.

The resulting mixture is transferred to a separatory funnel and the aqueous layer is removed. The organic layer is then washed with one 100 ml portion of 10% aqueous sodium chloride. The organic layer is then stripped of solvent and rushed over. To the rushed-over material 4.5 g Primol, 0.5 g triethanolamine and 0.1 g of Ionox is added. The resulting material is then distilled in an 8-plate Vigreux column yielding 13 fractions. Fractions 7-10 (BP 155° to 195°C at 3.5 mm Hg) are bulked and these fractions have a sweet, spicy, coumarin-like aroma, which are useful in augmenting or enhancing the flavor of tobacco.

Infra-red, NMR and Mass Spectral analyses yield the information that the resulting material is a mixture of compound having the structures shown as the product in the equation above.

Example 2: *Use of 2-Isopropyl-4-Phenyl-Dihydropyran in Tobacco* – The following tobacco flavor formulation (A) is prepared:

Ingredients	Parts by Weight
Ethyl butyrate	0.05
Ethyl valerate	0.05
Maltol	2.00
Cocoa extract	26.00
Coffee extract	10.00
Ethanol (95% aqueous)	20.00
Water	41.90

A tobacco formulation (B) is prepared as follows:

Ingredients	Parts by Weight
Bright tobacco	40.1
Burley tobacco	24.9
Maryland tobacco	1.1
Turkish tobacco	11.6
Stem (flue-cured) tobacco	14.2
Glycerin	2.8
Water	5.3

The flavor formulation (A) is added to a portion of the smoking tobacco formulation (B) at the rate of 0.1% by weight of the tobacco. The flavored and nonflavored tobacco formulations are then formulated into cigarettes by the usual manufacturing procedures:

(1) At the rate of 100 ppm to one-fourth of the cigarettes in each group is added 2-isopropyl-4-phenyl-dihydropyran (bulked Fractions 7–10, produced according to Example 1). The use of the 2-isopropyl-4-phenyl-dihydropyran in the cigarettes causes the cigarettes prior to smoking to have a sweet, spicy, coumarin-like aroma. In smoke flavor, in the main stream, the use of the 2-isopropyl-4-phenyl-dihydro-2H-pyran promotes salivation, decreases harshness and has a sweet, creamy, vanilla and coumarin-like effect; and also enhances the natural tobacco-like character. In the side stream, the same effects are observed.

(2) At the rate of 200 ppm to one-fourth of the cigarettes in each group is added 2-isopropyl-4-phenyl-dihydropyran (bulked Fractions 7–10, produced according to Example 1). The results are the same as in (1).

One-fourth of the cigarettes remains totally without added flavorants. One-fourth of the cigarettes contain only formulation (A) added thereto; but no 2-isopropyl-4-phenyl-dihydro-2H-pyran.

In general, on smoking, the cigarettes containing the 2-isopropyl-4-phenyl-dihydro-2H-pyran are more desirable than those cigarettes not containing this pyran whether or not the ingredients of formulation (A) are included in the tobacco.

Example 3: *Use in Foodstuff Flavors of 2-Isopropyl-4-Phenyl-2H-Dihydropyran* – The following basic Concord grape flavor formulation is prepared:

Ingredients	Parts by Weight
Methional (10% in 95% food grade aqueous ethanol)	10
Rum ether	10
Diethyl malonate	15
Ethyl propionate	40
Amyl isovalerate	40
Glacial acetic acid	40
Fusel oil	35
Ethyl nonanoate	40
Phenyl ethyl acetate	50
Isobutyl formate	50
Methyl anthranilate	100
Ethyl heptanoate	60
Cinnamyl propionate	120
Ethyl acetate	200
Ethyl butyrate	190

The foregoing formulation is split into two parts. To the first part, at the rate of 2% by weight is added 2-isopropyl-4-phenyl-2H-dihydropyran. To the second portion of the formulation nothing additional is added thereto. Both formulations are compared at the rate of 50 ppm in water. The flavor formulation containing the 2-isopropyl-4-phenyl-2H-dihydropyran has more of the delicate aroma and taste characteristics of fresh Concord grapes and a fuller, more natural Concord grape juice taste. The 2-isopropyl-4-phenyl-2H-dihydropyran improves the grape flavor and it is, therefore, preferred over the grape flavor that does not contain the 2-isopropyl-4-phenyl-2H-dihydropyran.

Example 4: *Use of 2-Isopropyl-4-Phenyl-2H-Dihydropyran in Orange Juice* – 2-isopropyl-4-phenyl-2H-dihydropyran, produced according to the process of Example 1, is added to canned, unsweetened orange juice at the rate of 5 ppm. This compound adds a pleasant orange flavor-like aroma, improves the cooked orange juice taste and makes the juice fresher, fruitier and is, therefore, more preferred.

Certain Substituted Tetrahydrofurans

M. Baumann, W. Hoffmann and K. von Fraunberg; U.S. Patent 4,116,982; September 26, 1978; assigned to BASF AG, Germany describe a process for making compound (1),

(1)

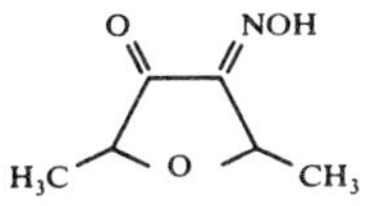

O NOH

H_3C O CH_3

3-oximino-4-oxo-2,5-dimethyl-tetrahydrofuran, which has a fine caramel-like fragrance, similar to, but slightly weaker than, that of the sought-after scent and flavor material 4-oxo-3-hydroxy-2,5-dimethyl-4,5-dihydrofuran. Because of its interesting olfactory properties, the compound is useful for the manufacture of scents and, preferably, flavor material for a great diversity of applications.

The compound is also an interesting intermediate for the manufacture of the 4-oxo-3-hydroxy-2,5-dimethyl-4,5-dihydrofuran.

Example 1: (a) 40 g of 4-oxo-2,5-dimethyl-3-carboethoxytetrahydrofuran in 200 ml of benzyl alcohol are heated at 180° to 200°C. The methanol formed is distilled off. The excess benzyl alcohol is distilled off under reduced pressure and the residue is fractionated. 45 g (corresponding to 78% of theory) of 4-oxo-2,5-dimethyl-3-carbobenzyloxytetrahydrofuran of boiling point 102°C/0.02 mm Hg and refractive index n_D^{25} = 1.5070 are obtained.

(b) 20 g of the benzyl ester obtained as described in (a) are hydrogenated in 100 ml ethyl acetate with 1 g of 10% strength Pd on charcoal, under normal pressure. 2,280 ml of H_2 are taken up. After distilling off the solvent under reduced pressure, 3-carboxy-4-oxo-2,5-dimethyl-tetrahydrofuran is obtained as a colorless oil which solidifies to crystals of melting point 55° to 57°C, and the spectroscopic data of which confirm the structure. The yield is quantitative.

(c) 60 g of 4-oxo-3-carboxy-2,5-dimethyl-tetrahydrofuran are dissolved in 100 ml of water. A solution of 26 g of $NaNO_2$ in about 50 ml of water is slowly added dropwise thereto at from 0° to 10°C, while cooling with ice. Vigorous elimination of CO_2 occurs and the contents of the flask turn yellowish.

After adding about half the solution, the evolution of gas ceases. When all the $NaNO_2$ has been added, about 40 ml of 50% strength H_2SO_4 are slowly added dropwise at from 0° to 10°C. CO_2 is again eliminated. Toward the end of the addition, the evolution of gas subsides and the pH is then about 2–3. The reaction mixture is stirred for a further hour and is then neutralized with NaOH solution and extracted repeatedly with ether, and the ether phase is dried and concentrated. 42 g of 4-oxo-3-oximino-2,5-dimethyl-tetrahydrofuran remain in the form of a pale yellow oil, the spectroscopic data and analysis of which confirm the stated structure. The crude yield is 78% of theory. A sample was distilled under reduced pressure; at the boiling point of 90°C/0.5 mm Hg, partial decomposition takes place.

Example 2: 15.8 g of 4-oxo-3-carboxy-2,5-dimethyl-tetrahydrofuran are dissolved in 80 ml of ethanol and 15 g of ethyl nitrate are added. A slow stream of HCl is passed in while cooling with ice. On warming to 20°C, evolution of gas occurs, accompanied by a slight exothermicity. When evolution of gas has ceased (which is the case after about 1 hour), the mixture is concentrated, the residue is taken up in ether and the ether solution is washed neutral with aqueous $NaHCO_3$ solution, dried and concentrated under reduced pressure. 13.5 g of 4-oxo-3-oximino-2,5-dimethyl-tetrahydrofuran (corresponding to a crude yield of 94% of theory) remain.

Example 3: *Scission of the Oxime with Aqueous Acids* – 10 g of 4-oxo-3-oximino-2,5-dimethyl-tetrahydrofuran and 50 ml of 10% strength H_2SO_4 are heated at 80°C and the hydrolysis is followed by thin layer chromatography. The oxime has been converted after 3 hours. In its place, 4-oxo-3-hydroxy-2,5-dimethyl-4,5-dihydrofuran has been formed, as established by isolating the product and comparing the chromatographic and spectroscopic data with those of a comparative substance.

Example 4: *Scission of the Oxime by Transoximation with Reactive Carbonyl Compounds* – 10 g of 4-oxo-3-oximino-2,5-dimethyl-tetrahydrofuran are stirred with 50 ml of a 40% strength formaldehyde solution and 30 ml of 20% strength sulfuric acid for 24 hours at room temperature. Subsequent examination by thin layer chromatography shows that the oxime has been completely converted to 4-oxo-3-hydroxy-2,5-dimethyl-4,5-dihydrofuran.

Example 5: *Oxidative Scission of the Oxime* – 40 g of the crude 4-oxo-3-oximino-2,5-dimethyl-tetrahydrofuran obtained as described in Example 1 are suspended in 100 ml of 30% strength H_2SO_4 and a solution of 20 g of $NaNO_2$ in 50 ml of water is slowly added dropwise to this suspension at room temperature. A vigorous evolution of gas occurs and the temperature rises; it is kept at 40°C by cooling with ice water. After completion of the dropwise addition, the reaction mixture is allowed to continue reacting for 60 minutes at room temperature and the solution is then neutralized with aqueous NaOH. Thereafter, it is extracted with ether and the extract is dried and concentrated. 19.8 g of 4-oxo-3-hydroxy-2,5-dimethyl-4,5-dihydrofuran are left in the form of a yellow oil which, according to analysis by thin layer chromatography, still contains traces of impurities. The yield is 55% of theory.

The further purification of the end product can be carried out by distillation, or by recrystallizing the viscous oil from ether. In the latter case, a colorless substance of melting point 79° to 80°C is obtained.

SULFUR-CONTAINING COMPOUNDS

Allylmercaptans

G. Frater, T. Sigg-Grütter, and J. Wild; U.S. Patent 4,067,910; January 10, 1978; assigned to Givaudan Corporation have developed a process for making allylmercaptans which comprises heating an allyl xanthogenate of the general formula:

$$(1)\quad \begin{array}{ccccccccc} R^2 & & R^3 & & R^4 & & & & S \\ | & & | & & | & & & & \| \\ C & = & C & - & C & - O - & C & - & S^- \ Cat^+ \\ | & & & & | & & & & \\ R^1 & & & & R^5 & & & & \end{array}$$

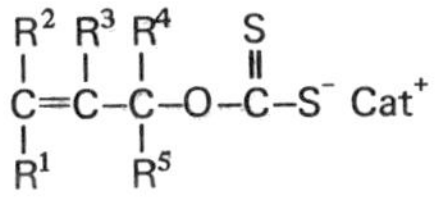

wherein R^1, R^2, R^3 and R^5 each individually represent a hydrogen atom or a lower alkyl, lower alkenyl, lower alkynyl, lower cycloalkyl or lower cycloalkenyl group which may carry an oxygen-containing substituent, or R^2 and R^3 or R^2 and R^4 or R^3 and R^4 can be joined together to form a ring, and Cat^+ denotes the cation of a base, in an aqueous medium to give a mixture of an allylmercaptan of the general formula:

$$(2)\quad \begin{array}{ccccc} R^2 & & R^3 & & R^4 \\ | & & | & & | \\ C & = & C & - & C-SH \\ | & & & & | \\ R^1 & & & & R^5 \end{array}$$

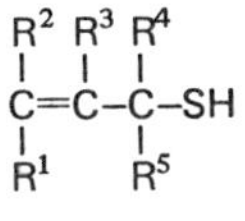

and an allylic rearrangement product thereof of the general formula:

$$(3)\quad \begin{array}{ccccc} & R^2 & & R^3 & R^4 \\ & | & & | & | \\ HS- & C & - & C & =C \\ & | & & & | \\ & R^1 & & & R^5 \end{array}$$

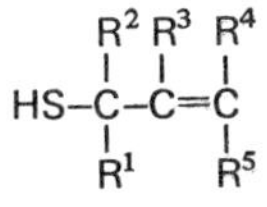

wherein R^1, R^2, R^3, R^4 and R^5 have the significance given earlier.

Examples of groups denoted by R^1, R^2, R^3, R^4 and R^5 are especially straight-chain or branched-chain alkyl, alkenyl or alkynyl groups containing up to 6 carbon atoms.

Especially valuable compounds of the formulas (2) and (3), which can be used in odorant or flavoring compositions, especially those having berry, fruit and/or flower notes are 2-methyl-hept-2-ene-1-thiol, 2-methyl-hept-1-ene-3-thiol, 1-[1,1,3-trimethyl-3-cyclohexen-2-yl]-but-2-ene-1-thiol, 1-[1,1,3-trimethyl-3-cyclohexen-2-yl]-but-1-ene-3-thiol, 2-isopropyl-5-methyl-cyclohex-2-ene-1-thiol, 3,7,11-trimethyl-dodeca-1,6,10-triene-3-thiol, 3,7,11-trimethyl-dodeca-2,6,10-triene-1-thiol, oct-1-ene-3-thiol and oct-2-ene-1-thiol.

These compounds can be used in the manufacture of high grade fruit aromas for foods (e.g., milk products, yoghurt, etc.), luxury goods (e.g., confectionary products such as bonbons) and drinks (e.g., table-water, mineral water, etc.). Their pronounced aromatic qualities enable them to be used in low concentrations (e.g., in the range of 0.1 to 10 ppm, preferably in the range of 1 to 10 ppm). They can, however, also be used in higher concentrations (e.g., 50 ppm). Thus, for example, a mixture of 3,7,11-trimethyl-dodeca-1,6,10-triene-1-thiol and its allylic rearrangement products 3,7,11-trimethyl-dodeca-1,6,10-triene-3-thiol in the ratio of 1:1 can be used for the improvement of flavor properties (e.g., of citrus aromas such as grapefruit aromas) of fruit juices, mineral waters, etc. having such prevailing flavor. In this case, in particular, the character of the grapefruit peel is emphasized in a desirable manner and the fruit character becomes balanced.

These compounds also show advantageous properties in odorant compositions (e.g., in compositions having flowery notes) in that they are capable of modifying the odor of such compositions in a desirable manner. They can accordingly also be used as odorants for the production of perfumes, especially those having wood and flower notes, in an amount of from about 0.001 to 5 wt %, preferably from about 0.01 to 1 wt %.

3,7,11-Trimethyl-dodeca-1,6,10-triene-1-thiol or the allylic rearrangement product thereof is an especially interesting odorant. In particular, it is used because of its sulfur-like, fruity, metallic and green notes, for example, in odorant compositions having a flowery character (e.g., honeysuckle, orange blossom, tuberose, etc.)

Substituted Thiatetrahydrofuranones

B.D. Mookherjee, W.J. Evers, and A.E. Goossens; U.S. Patent 4,092,333; May 30, 1978; assigned to International Flavors & Fragrances Inc. have found that foodstuffs, chewing gum, medicinal compounds and toothpaste having coffee-like roasted, caramel-like, licorice-like and roasted almond aromas and flavors; tobacco flavoring compositions having sweet, caramel, coffee, nutty, and roasted aromas and tastes; and perfumes and perfumed articles having pleasant woody aromas may be produced by the use of one or more 2-acyl-5-substituted thiatetrahydrofuran-4-ones (cis- or trans-stereoisomers or a mixture). These compounds have the generic structure shown on the following page.

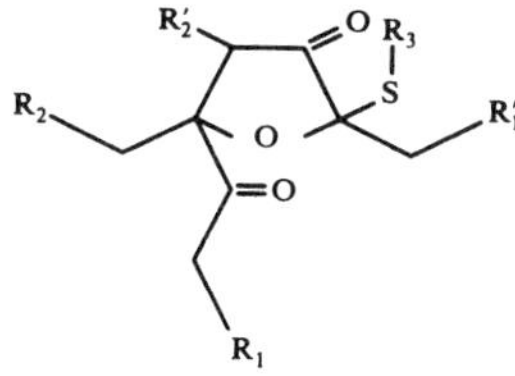

In the above formula, R_1, R_2, R_1' and R_2' are the same or different and each represents hydrogen or methyl and R_3 represents one of the group C_{1-9} alkyl, benzyl, phenyl, substituted or unsubstituted allyl; or a substituted or unsubstituted 3-furyl having the structure:

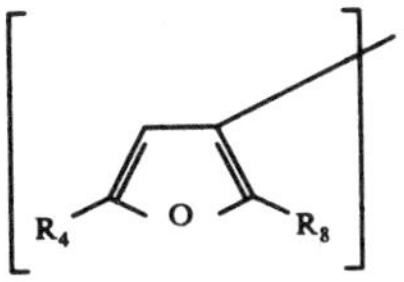

wherein R_4 and R_8 are the same or different and each represents hydrogen or methyl; hydroxyalkyl, oxoalkyl, hydroxycycloalkyl or oxocycloalkyl having the structure:

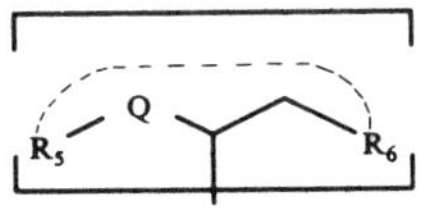

wherein R_5 and R_6, taken separately, represent hydrogen or C_{1-3} alkyl or R_5 and R_6, taken together, complete a cycloalkyl group and wherein Q is one of the moieties >CO or >CHOH.

An example will illustrate the general process of reacting one or more dimers of C_{4-6} alpha, beta diketones with a mercaptan. All parts and percentages are by weight unless otherwise specified.

Example: The preparation of 5-acetyl-2(2-methyl-3-furylthio)-dihydro-2,5-dimethyl-3,2H-furanone according to the reaction of 2-methyl-3-furanthiol with the diacetyl dimer is as follows:

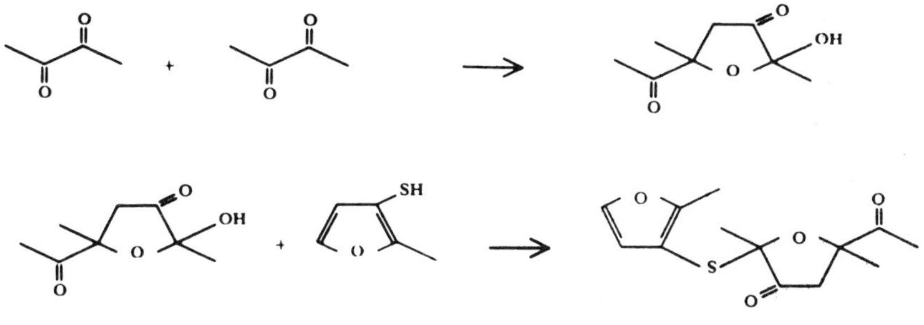

Into a 5-liter reactor equipped with reflux condenser, stirrer and cooling bath are placed 320 g of diacetyl and 3,200 ml of water. The reaction mass is cooled to 3°C at which point a solution of 54.4 g of potassium hydroxide and 1,600 ml of water is added over a period of 20 minutes. 85 g of 50% sulfuric acid is then added to the reaction mixture bringing the pH to 5.

The reaction mass is then saturated with 1,600 g of sodium chloride and then extracted with four 200 ml portions of methylene dichloride. The reaction mass is then dried over anhydrous sodium sulfate and stripped of solvent. The resulting product is then distilled through a 6" Vigreaux column followed by redistillation yielding the diacetyl dimer.

Into a 250 ml reaction flask equipped with reflux condenser, stirrer, thermometer and heating mantle is placed 20 g (17.7 ml) of 2-methyl-3-furanthiol. 10 ml tetrahydrofuran and 0.5 ml concentrated HCl is then added. The reaction mass is heated to 50°C and, over a 10 minute period, 30 g of diacetyl dimer prepared as above in 20 g of tetrahydrofuran is added. 20 ml tetrahydrofuran additional used to rinse the addition flask is then added.

The reaction mass is then stirred for 5 hours at 50°C with GC samples taken each hour.

The reaction mass is stirred for 5 hours and then 20 ml saturated sodium bicarbonate is added. The resulting mixture is transferred to a separatory funnel containing 200 ml of water and in the separatory funnel the oil phase is separated from the aqueous phase. The aqueous phase is extracted with one portion (30 ml) of methylene dichloride and the extract is combined with the oil layer. The resulting material is dried over anhydrous sodium sulfate and filtered and then stripped at atmospheric pressure to 80°C. The resulting portion is then distilled through a micro Vigreaux column.

The resulting compound is confirmed by IR, GLC, NMR and Mass Spectral analyses to be two isomers having the structures:

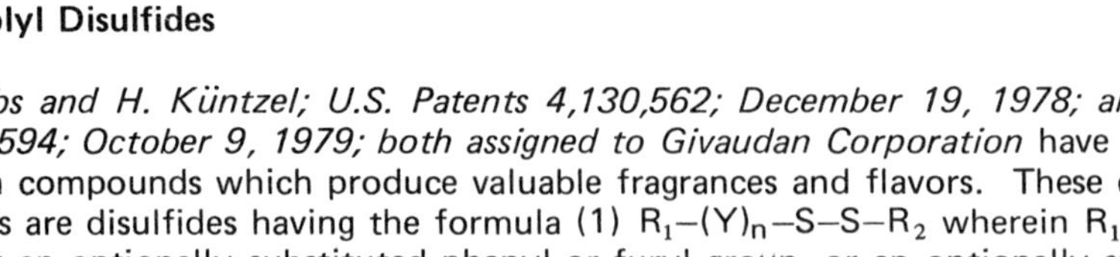

and

Thiazolyl Disulfides

P. Dubs and H. Küntzel; U.S. Patents 4,130,562; December 19, 1978; and 4,170,594; October 9, 1979; both assigned to Givaudan Corporation have found certain compounds which produce valuable fragrances and flavors. These compounds are disulfides having the formula (1) R_1–$(Y)_n$–S–S–R_2 wherein R_1 represents an optionally substituted phenyl or furyl group, or an optionally substituted monocyclic group containing nitrogen and/or sulfur and, if desired, also oxygen as the hetero atom, R_2 represents a lower alkyl or lower alkenyl group, Y represents an optionally mono- or di-substituted methylene group and n stands for 0–5 with the proviso that n is 2–5 when R_1 represents a phenyl or furyl group.

The disulfides of formula (1) can be manufactured by reacting one of the compounds of the general formula:

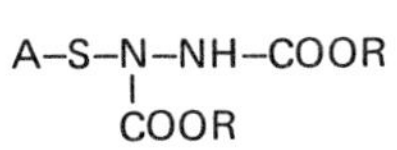

$$\underset{(2)}{\text{A–S–}\underset{\text{COOR}}{\underset{|}{\text{N}}}\text{–NH–COOR}} \qquad \underset{(3)}{\text{A–S–SCN}} \qquad \underset{(4)}{\text{A–S–S–}\overset{\text{S}}{\overset{\|}{\text{C}}}\text{OCH}_3}$$

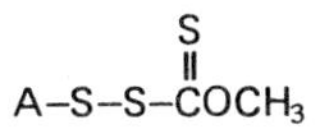

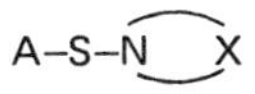

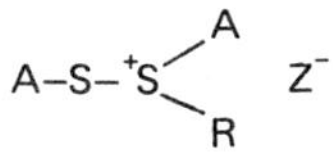

$$\underset{(5)}{\text{A–S–S–A}} \qquad \underset{(6)}{\text{A–S–N}\frown\text{X}} \qquad \underset{(7)}{\text{A–S–}\overset{+}{\text{S}}\begin{smallmatrix}\text{A}\\ \text{R}\end{smallmatrix} \quad \text{Z}^-}$$

with a mercaptan of the general formula (8) A–SH, whereby in the one reaction component A represents a group $R_1(Y)_n$ and in the other reaction component A represents a group R_2, Y and n having the significance given earlier except that n should only stand for 1-4 in formulas (2), (3) and (5), R represents a lower alkyl group, X represents the diacyl residue of an aliphatic or aromatic dicarboxylic acid and Z represents an anion.

As heterocyclic groups containing nitrogen and/or sulfur and, if desired, also oxygen as the hetero atom there come into consideration, in particular, 5-membered or 6-membered heterocyclic groups which contain one or more nitrogen atoms and/or one or more sulfur atoms. Examples of heterocycles forming a basis for these heterocyclic groups include pyrrole, pyridine, pyrazine, imidazole, pyrazole, pyrimidine, thiazole, isothiazole, oxazole, triazole, etc., which heterocycles can be substituted with lower alkyl groups.

The position at which the heterocyclic ring is bound with the rest of the molecule is not critical. The lower alkyl and lower alkenyl groups denoted by R_2 (or R as the case may be) are especially straight-chain or branched-chain groups containing from 1 to 6 carbon atoms (e.g., methyl, ethyl, propyl, isopropyl, butyl, sec.-butyl, isobutyl, amyl, hexyl, etc.).

Y represents an optionally mono- or di-substituted methylene group. Indeed, in addition to the unsubstituted methylene group ($-CH_2-$), Y represents, in particular, a group of the formulas $-CHR^{\circ}-$ or $-CR^{\circ}R^{\infty}-$ wherein R° and R^{∞} represent lower alkyl or lower alkenyl groups.

The disulfides of formula (1) hereinbefore possess useful odorant and/or flavoring properties. In particular, the flavoring spectrum is very broad. They have fruity, spicy (e.g., mustard-like), vegetable-like (e.g., leek, celery, cauliflower, chive, onions, etc.) and mushroom-like notes as well as cheese and meat notes. Of particular interest are the roast and meat notes.

The disulfides of formula (1) can accordingly be used, for example, for the perfuming or flavoring of products such as cosmetics (soaps, salves, powders, etc.), detergents, foodstuffs, luxury goods and drinks, the disulfides preferably not being used alone but rather in the form of compositions containing other odorant or flavoring substances.

The disulfides of formula (1) can be used as flavoring substances, for example, for the production or improvement, intensification, enhancement or modification of fruit, meat, or roast notes in foodstuffs (meat, synthetic meat products, sauces, broths, soups, vegetables, seasoning agents, snack foods, roast products such as coffee or cocoa, milk products such as cheese, curd, yoghurt, etc.), in luxury goods (tobacco, crackers, etc.) and drinks (lemonades, etc.).

The pronounced flavor qualities of the disulfides enable them to be used in low concentrations. A suitable range is 0.001 to 10 ppm, preferably 0.1 to 10 ppm in the finished product (i.e., the aromatized foodstuff, luxury goods or drink).

Example 1: 8.66 g (0.0429 mol) of azidocarboxylic acid diisopropyl ester are dissolved in 40 ml of chloroform. There are then added dropwise 6 g (0.0429 mol) of 2-(2-pyrazinylethyl)-mercaptan and thereafter 4 drops of concentrated sulfuric acid. The mixture is held at reflux temperature for 4 hours. The mixture is then left to cool to room temperature, treated with 20 ml of saturated sodium bicarbonate solution and the resulting mixture extracted four times with 50 ml of chloroform each time. The organic phases are combined, dried over anhydrous sodium sulfate and concentrated on a rotary evaporator at 40°C/12 mm Hg. The yield of crude N-[1'-thia-3'-pyrazin-2-yl-propyl]-N,N'-diisopropoxycarbonylhydrazine amounts to 16.6 g.

The hydrazine derivative is dissolved in 160 ml of isopropanol and treated with 7.5 g (1.2 mol equivalents) of diisopropylethylamine. The mixture is then cooled to 0°C and excess methylmercaptan is passed through. After completion of the addition, the mixture is stirred for a further 2 hours. The mixture is extracted with 50 ml of saturated sodium bicarbonate solution and four times with 100 ml of chloroform each time. The combined organic phases are dried over anhydrous sodium sulfate and concentrated on a rotary evaporator at 40°C/12 mm Hg. The crude product obtained is chromatographed on a column of 150 g of silica gel using hexane/ether (1:4) for the elution. The fractions are combined and subjected to a short-path distillation in a bulb-tube. In this manner, there are obtained 1.2 g of pure 2-(2-pyrazinylethyl)-methyl-disulfide. n_D^{20} = 1.5863; BP 85° to 95°C/0.035 mm Hg. Odor and flavor: sulfur-like, meaty, vegetable-like (leek, chive, garlic, onions, cauliflower).

Example 2: A solution of 8.15 g (0.055 mol) of trimethyloxonium tetrafluoroborate in 20 ml of nitromethane is added to a solution, cooled to 0°C, of 5.17 g (0.055 mol) of freshly distilled dimethyldisulfide in 10 ml of nitromethane. After completion of the addition, the mixture is stirred at this temperature for 90 minutes. There is now added dropwise thereto at 0°C a solution of 7 g (0.05 mol) of 2-(2-pyrazinethyl)-mercaptan as well as 6.45 g (0.05 mol) of diisopropylamine in 10 ml of nitromethane. The mixture is left to come to room temperature, treated with 50 ml of saturated sodium bicarbonate solution and ice and the mixture extracted four times with 100 ml of chloroform each time. The organic phases are combined, dried over anhydrous sodium sulfate and then concentrated on a rotary evaporator at 40°C/12 mm Hg. The crude product obtained is chromatographed on 150 g of silica gel using hexane/ether (1:4) for the elution. The fractions are combined and subjected to a short-path distillation in a bulb-tube at 85° to 88°C/0.03 mm Hg. There are obtained 4.25 g (45.7%) of pure 2-(2-pyrazinethyl)-methyl-disulfide.

According to the same procedure, from dimethyldisulfide and 1-(2-thiazolyl)-2-methylpropyl-mercaptan, there is obtained [1-(2-thiazolyl)-2-methylpropyl]-

methyl-disulfide. Odor and flavor: earthy, fruity (cassis), sulfur-like, vegetable-like (celery).

From dimethyldisulfide and 2-mercapto-pyridine, there is obtained 2-pyridyl-methyl-disulfide. Odor and flavor: earthy, sulfur-like, green, vegetable-like (cauliflower, onion), after broad-leaved garlic, cheese.

From dimethyldisulfide and 2-mercapto-thiazole, there is obtained 2-thiazolyl-methyl-disulfide. Odor and flavor: roast note, sulfur-like, green, cabbage-, chive-like, meat-like (pork).

From dimethyldisulfide and 2-mercapto-pyrimidine, there is obtained 2-pyrimidinyl-methyl-disulfide. Odor and flavor: earthy, spicy, vegetable-like (cauliflower, kohlrabi), of cheese.

From dimethyldisulfide and 4,5-dimethyl-2-thiono-4-oxazoline, there is obtained (4,5-dimethyl-2-oxazolyl)-methyl-disulfide. Odor: spicy Maggi; flavor: spicy, chive-like, reminiscent of lightly fried mushrooms.

Example 3: A solution of 4.18 g (0.022 mol) of triethyloxonium tetrafluoroborate in 12 ml of nitromethane is added to a solution of 2.86 g (0.022 mol) of diethyldisulfide in 8 ml of nitromethane (filtered over basic aluminum oxide) at reflux temperature. After completion of the addition, the mixture is stirred at 30° to 40°C for a further 2 hours. There is then added dropwise at 30°C a mixture (two phases) of 2.22 g (0.02 mol) of 2-mercapto-pyridine and 2.58 g (0.02 mol) of diisopropylethylamine in 15 ml of nitromethane and the resulting mixture is stirred at 30° to 40°C for a further 2 hours.

The working-up is carried out as described in Example 2. There are obtained 2.29 g (67% relative to 2-mercapto-pyridine) of pure 2-pyridyl-ethyl-disulfide. Odor: sulfur-like, after onions, penetrating; flavor: onion-like, sulfur-like, burning.

Example 4: From 3.3 g (0.022 mol) of dipropyldisulfide, 3.26 g (0.022 mol) of trimethyloxonium tetrafluoroborate, 2.22 g (0.02 mol) of 2-mercapto-pyridine and 2.58 g (0.02 mol) of diisopropylethylamine, there is obtained, according to the procedure described in Example 3, 2-pyridyl-propyl-disulfide. Yield: 2.65 g (71.6% relative to 2-mercapto-pyridine). Odor: penetrating, sulfur-like, green, after chive; flavor: sulfur-like, green, of chive.

Example 5: From 3.3 g (0.022 mol) of diisopropyldisulfide, 3.26 g (0.022 mol) of trimethyloxonium tetrafluoroborate, 2.24 g (0.02 mol) of 2-mercapto-pyrimidine and 2.58 g (0.02 mol) of diisopropyl-ethylamine, there is obtained, according to the procedure described in Example 3, 2-pyrimidinyl-isopropyl-disulfide. Yield: 2.5 g (67.2% relative to 2-mercapto-pyrimidine). Odor: penetrating, sharp, green, of onions; flavor: penetrating, burning, of onions.

Example 6: 8 g (0.058 mol) of phenethylmercaptan are added with intensive stirring to a suspension of 12 g (0.058 mol) of N-ethylthiophthalimide in 70 ml of benzene. After a few minutes, a homogeneous solution forms and the phthalimide slowly begins to precipitate out therefrom. After stirring for 48 hours at room temperature, the mixture is filtered through a paper filter, the filtrate is concentrated in a vacuum and the crude 2-phenylethyl-ethyl-disulfide (12.8 g) obtained is distilled through a Vigreux column. There are thus obtained 10.5 g of pure product (91% of theory) which has a purity of 99% (gas-chromatographic estimation). BP 87° to 88°C/0.04 mm Hg. Odor: sulfur-like, spicy, sweetish; flavor: spicy, sweetish, somewhat sulfur-like.

Example 7: Starting from 5.6 g (0.048 mol) of 2-furylethylmercaptan and 9.1 g (0.048 mol) of N-ethylthiophthalimide, there are obtained, according to the procedure described in Example 6 and after chromatography on 100 g of silica gel (elution with toluene), 2.38 g of ethyl-2-(2-furyl)-ethyl-disulfide (purity: 96%). Odor: sulfur-like, penetrating, after onions; flavor: sulfur-like, green, faintly flowery.

The following examples illustrate typical odorant and/or flavoring compositions provided by this process.

Example 8: A broth can be prepared as follows:

	A	B
	. . .(parts by weight). . .	
2-Pyrimidinyl-methyl-disulfide	–	0.5
Diallyl-disulfide	60	60
Allylmercaptan	0.5	0.5
Mustard oil synthetic	2.5	2.5
Dimethyl-disulfide	0.5	–
2,4-Decadienal (10% in ethyl alcohol)	1	1
Capric aldehyde	0.1	0.1
Maltol	2.5	2.5
Butyric acid (10% in ethyl alcohol)	0.2	0.2
Benzyl alcohol	132.7	132.7

20.0 g of a mixture A or B are mixed with 1,000 g of onion oil. At an amount of 20 g per 100 liters of broth, the odor and flavor of the broth, prepared using composition A, are insipid, whereas the broth prepared using composition B possesses an excellent onion flavor.

Example 9: A clear meat soup can be obtained as follows. 20.0 g of a mixture consisting of:

	Parts by Weight
Cooking salt	50.00
Monosodium glutamate	20.04
Caramel powder	0.20
Nutmeg, soluble	0.06
Clove, soluble	0.06
Coriander, soluble	0.08
White pepper, soluble	0.06
HVP, Type 3 H1 (hydrolyzed vegetable proteins)	8.90
HVP, Type RF-B (hydrolyzed vegetable proteins)	2.60
Vegetable fat, melting point 40°C	17.00
Roast onion powder	1.00

are added to 1 liter of hot water. The flavor of this clear meat soup is initially unsatisfactory (faint meat and onion flavor).

By the addition of 3 to 5 ppm of 2-(2-pyrazinylethyl)-methyl-disulfide, the meat and onion flavor are enhanced in an extremely advantageous manner.

By the addition of 3 to 5 ppm of 2-pyridyl-propyl-disulfide, the meat aroma is enhanced in an advantageous manner in that there now appears a spicy-green note which is reminiscent of chive and which harmonizes well with the meat flavor.

Example 10: An odorant substance composition (flowery note) can have the composition A or B:

	A	B
	. . (parts by weight) . .	
Ethyl-2-(2-furyl)-ethyl-disulfide	–	5
Methyl heptenone	5	5
Linalool	5	5
Acetic acid linalyl ester	10	10
Citronellol	30	30
Geraniol	40	40
Ethanol	910	905

A comparison of A with B shows immediately that the traditional composition A falls off strongly. By the addition of ethyl-2-(2-furyl)-ethyl-disulfide, the flowery-green note of the composition is enhanced. It is intensively reminiscent of geranium.

Methylthiobutyryltrimethyl Cyclohexene or Cyclohexadiene

R.A. Wilson, B.D. Mookherjee, A.S. Hruza, M.H. Vock, L.S. Frederick, and J.F. Vinals; U.S. Patents 4,107,209; August 15, 1978; 4,154,693; May 15, 1979; 4,156,662; May 29, 1979; and 4,162,335; July 24, 1979; all assigned to International Flavors & Fragrances Inc. describe the compounds 1-[3-(methylthio)-butyryl] -2,6,6-trimethyl-cyclohexene and the corresponding 1,3-cyclohexadiene analog having the generic structure:

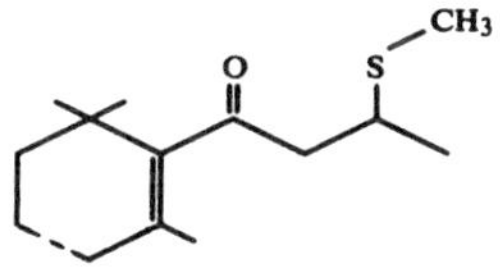

wherein the dashed line could be a single or a double carbon-carbon bond and processes and compositions containing these compounds for use in foodstuffs, chewing gum, toothpastes and medicinal products, tobacco, perfumes, and perfumed articles.

The compounds are produced by reacting methyl mercaptan with β-damascone or β-damascenone.

These compounds have the organoleptic properties given in the table on the following page. In that table, Compound (1) is named 1-[3-(methylthio)butyryl]-2,6,6-trimethyl cyclohexene. Compound (2) is named 1-[3-(methylthio)butyryl]-2,6,6-trimethyl-1,3-cyclohexadiene.

STRUCTURE	FOOD FLAVOR PROPERTIES	TOBACCO FLAVOR PROPERTIES	PERFUMERY PROPERTIES
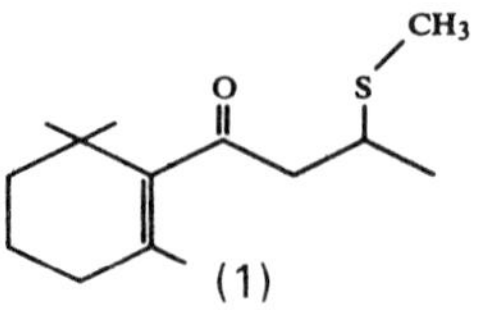 CH_3 O S (1)	Sweet, black-tea-like, cocoa-like and damascenone-like aroma characteristics with sweet, black-tea-like, cocoa-like and damascenone-like flavor characteristics.	Lingering rich, sweet, fruity, dried fruit-like, slightly floral, raspberry-like, blackcurrant-like aroma prior to and on smoking in both the main stream and side stream.	Earthy, potato-like top note, giving way to a minty, tomato vine/ Brussel sprout nuance, with woody undertone.
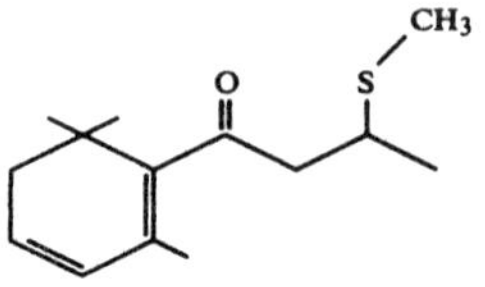 CH_3 O S (2)	Sweet, black-tea-like, tobacco-like, damascenone-like, chocolate-like, dried-fruit-like, rose-petal-like aroma characteristics with sweet, black-tea-like, chocolate-like, damascenone-like, tobacco-like, dried-fruit-like and rose-petal-like flavor characteristics.	Sweet-honey-like, rich, slightly-fruity, hay-tobacco-like, woody-like flavor characteristics prior to and on smoking in the main stream and in the side stream with sweet, floral and fruity nuances.	Floral and natural rose oil-like aroma nuances.

Compounds (1) and (2) are capable of supplying and/or potentiating certain flavor and aroma notes usually lacking in many fruit flavors as well as tobacco flavors heretofore. They are also capable of supplying certain fragrance notes usually lacking in many perfumery materials, for example, floral fragrances.

Food flavoring compositions preferably contain Compounds (1) and/or (2) in concentrations ranging from about 0.1 to 15% by weight, based on the total weight of flavoring composition, and from 0.5 to 20 ppm by weight, based on the total composition.

In another aspect of the process, Compound (1) or Compound (2) is added to tobacco materials as an aroma and flavor additive. Other synthetic or natural flavor and aroma additives may be added to the tobacco together with Compounds (1) and/or (2) as follows:

Synthetic Materials

β-Ethyl-cinnamaldehyde
Eugenol
Dipentene
β-Damascone
β-Damascenone
Maltol
Ethyl maltol
δ-Undecalactone
δ-Decalactone
Benzaldehyde
Amyl acetate
Ethyl butyrate
Ethyl valerate
Ethyl acetate
2-Hexenol-1,2-methyl-5-isopropyl-1,3-nonadiene-8-one
2,6-Dimethyl-2,6-undecadiene-10-one
2-Methyl-5-isopropyl acetophenone
2-Hydroxy-2,5,5,8a-tetramethyl-1-(2-hydroxyethyl)-decahydronaphthalene
Dodecahydro-3a,6,6,9a-tetramethyl naphtho-(2,1-b)-furan
4-Hydroxy hexanoic acid, gamma lactone

Natural Oils

Celery seed oil
Coffee extract
Bergamot oil
Cocoa extract
Nutmeg oil
Origanum oil

An aroma and flavoring concentrate containing Compounds (1) and/or (2) and one or more of the aboveindicated additional flavoring additives may be added to the smoking tobacco material, to the filter or to the leaf or paper wrapper. The smoking tobacco material may be shredded, cured, cased and blended tobacco material or reconstituted tobacco material or tobacco substitutes or mixtures thereof. The proportions of flavoring additives may be varied in accordance with taste but insofar as the augmentation or the enhancement or the imparting of the sweet, honey-like, fruity, woody, cedarwood, dried fruit-like, raspberry-like, black currant-like, Virginia tobacco-like notes are concerned, satisfactory results are obtained if the proportion by weight of the sum total of Compounds (1) and/or (2) to smoking tobacco material is between 250 and 1,500 ppm (0.025 to 1.5%) of the active ingredients to the smoking tobacco material. Satisfactory results are obtained if the proportion by weight of the sum total of Compounds (1) and/or (2) used to flavoring material is between 2,500 and 10,000 ppm (0.25 to 15%).

Any convenient method for incorporating the compounds in the tobacco product may be employed. Thus, they may be dissolved in a suitable solvent as ethanol, n-pentane, diethyl ether and/or other volatile organic solvents and the resulting solution may either be sprayed on the cured, cased and blended tobacco material or the tobacco material may be dipped into such solution. Under certain circumstances, a solution of the compounds, alone or together with other flavoring additives as set forth above, may be applied by means of a suitable applicator such as a brush or roller on the paper or leaf wrapper for the smoking product, or it may be applied to the filter by either spraying or dipping or coating.

Furthermore, it will be apparent that only a portion of the tobacco or substitute therefor need be treated and the thus treated tobacco may be blended with other tobaccos before the ultimate tobacco product is formed.

In accordance with one specific example, an aged, cured and shredded domestic burley tobacco is spread with a 20% ethyl alcohol solution of Compounds (1) and/or (2) in an amount to provide a tobacco composition containing 800 ppm by weight of the experimental compound on a dry basis. Thereafter, the alcohol is removed by evaporation and the tobacco is manufactured into cigarettes by the usual techniques. The cigarette when treated as indicated has a desired and pleasing aroma which is detectable in the main and side streams when the cigarette is smoked. This aroma is described as being sweet, rich, floral, fruity, hay tobacco-like, honey-like, cedarwood-like and Virginia tobacco-like.

Compounds (1) and/or (2) and one or more auxiliary perfume ingredients, including, for example, alcohols, aldehydes, nitriles, esters, other ketones, cyclic esters, synthetic essential oils, and natural essential oils, may be admixed so that the combined odor of the individual components produce a pleasant and desired fragrance, particularly and preferably in floral fragrances. Such perfume compositions usually contain: (a) the main note or the bouquet or foundation stone of the composition; (b) modifiers which round off and accompany the main note; (c) fixatives which include odorous substances which lend a particular note to the perfume throughout all stages of evaporation and substances which retard evaporation; and (d) topnotes which are usually low boiling fresh smelling materials.

In perfume compositions, it is the individual components which contribute particular olfactory characteristics, but the overall effect of the perfume composition will be the sum of the effects of each of the ingredients. Thus, Compounds (1) and/or (2) can be used to alter the aroma characteristics of a perfume composition, for example, by utilizing or moderating the olfactory reaction contributed by another ingredient in the composition.

The amount of the compound which will be effective in perfume compositions depends on many factors, including the other ingredients, their amounts and the effects which are desired. It has been found that perfume compositions containing as little as 0.1% of Compounds (1) or (2) (e.g., 0.05%) can be used to impart floral and natural rose oil-like odor to soaps, cosmetics or other products. The amount employed can range up to 10% of the fragrance components and will depend on considerations of cost, nature of the end product, the effect desired on the finished product and the particular fragrance sought.

These compounds are useful, taken alone or in perfume compositions as an olfactory component in detergents and soaps, space odorants and deodorants, perfumes, colognes, toilet water, bath preparations, such as bath oils, and bath solids; hair preparations, such as lacquers, brilliantines, pomades and shampoos; cosmetics preparations, such as creams, deodorants, hand lotions and sun screens; powders, such as talcs, dusting powders, face powders and the like. When used as an olfactory component, as little as 1% of Compounds (1) and/or (2) will suffice to impart a natural rose oil-like or earthy and woody note(s) to floral formulations. Generally, no more than 3%, based on the ultimate end product, is required.

Example 1: *Preparation of 3-(Methylthio)-1-(2,6,6-Trimethyl-1,3-Cyclohexadien-1-yl)-1-Butanone* – In a 50-ml, 3-necked, round-bottom flask equipped with a magnetic stirring assembly, a tube submerged for the introduction of gas, an immersion thermometer and a dry ice isopropanol condenser, the outlet of which leads through a 50% alkali trap to the hood sink, is placed 8.3 g of β-damascenone and approximately 1 g of triethylamine. Gaseous methyl mercaptan is introduced into the stirred reaction mixture which is maintained at 10°C. Addition is continued for approximately 1 hour and then stopped.

The stirred reaction mix is allowed to attain room temperature, and the reaction mix is stirred overnight at room temperature. The next morning, the light yellow reaction mix is transferred to 50 ml of ether in a separatory funnel and washed twice with dilute sulfuric acid, twice with saturated sodium bicarbonate solution, and then with saturated saline solution until the wash is neutral. The dried ether layer is stripped of solvent on a rotovap and GLC of the residue of an ⅛" x 10' stainless steel, 5% Carbowax 20 M column isothermally at 190°C shows one major late eluting peak which is trapped for infra-red and nuclear magnetic resonance-mass spectral analyses.

Mass spectral analysis shows a satisfactory match with the cyclohexadiene compound [Compound (2)] which had previously been prepared from black tobacco lamina. Since the product still smells strongly of methyl mercaptan, it is subjected to GLC on a 10' x ¼" Carbowax 20 M column.

Example 2: *Preparation of 1-[3-(Methylthio)Butyryl]-2,2,6-Trimethylcyclohexene-1* – In the same fashion, this compound [Compound (1)] is prepared by the treatment of β-damascone with methyl mercaptan. Mass spectral analysis affirmed its production by matching it with the mass spectrum of Compound (1) as prepared from black tobacco lamina.

Example 3: The following basic raspberry formulation is prepared:

Ingredients	Parts by Weight
Vanillin	2
Maltol	4
p-Hydroxybenzylacetone	5
α-Ionone (10% in propylene glycol)	2
Ethyl butyrate	6
Ethyl acetate	16
Dimethylsulfide	1
Isobutylacetate	14
Acetic acid	10
Acetaldehyde	10
Propylene glycol	930

The basic raspberry formulation is divided into two parts. To the first part 0.1% by weight of Compound (2), prepared according to Example 1, is added. Both flavors with and without the additional material are compared at the rate of 100 ppm in water by a bench panel consisting of four people. The raspberry flavor with the addition of Compound (2) is considered not only stronger but as having a more raspberry juice-like character and having more of the desired raspberry seed or kernel note. The taste is closer to the ripe raspberry with pleasant tea and raspberry kernel notes. Therefore, all panel members prefer the flavor with the addition of this flavor chemical.

Example 4: The following floral perfume formulation is prepared:

Ingredients	Parts by Weight
Phenylethyl alcohol	25
Rhodinol	22
Hydroxycitronellal	6
Linalool	5
Cinnamic alcohol	3
α-Ionone	15
Amyl acetate (1% in diethyl phthalate)	6
Vetiverol	2
Ylang-ylang oil	3
Nerol	2
Musk ketone	3
Vanillin (10% in diethyl phthalate)	2
Styrax essence	1
2,2,6-Trimethyl-1-[3-(methylthio)-butyryl]-1,3-cyclohexadiene*	5

*Produced according to Example 1

The product of Example 1 itself has a sweet, rosey, floral, cured tobacco note. It gives lift and a more natural floral note to the fragrance in which it is incorporated. Although it was incorporated above at a level of 5% by weight, it may be effectively used at from 0.1 up to 10% by weight in such floral formulations.

For special effects, it may be used as high as 50% by weight. Addition of the product of Example 1 to the fragrance greatly increases its esthetic qualities and gives the olfactory illusion of the presence of natural rose oil.

Example 5: Concentrated liquid detergents with floral and rose oil-like aromas are prepared containing 0.10, 0.15 and 0.20% of Compound (2) prepared according to Example 1. They are prepared by adding and homogeneously mixing the appropriate quantity of Compound (2) in the liquid detergents. The detergents all possess a floral, and rose oil-like fragrance, the intensity increasing with greater concentrations of Compound (2)

Example 6: A tobacco mixture is produced by admixing the following ingredients:

Ingredients	Parts by Weight
Bright	40.1
Burley	24.9
Maryland	1.1
Turkish	11.6
Stem (flue-cured)	14.2
Glycerin	2.8
Water	5.3

Cigarettes are prepared from this tobacco.

The following flavor formulation is prepared:

Ingredients	Parts by Weight
Ethyl butyrate	0.05
Ethyl valerate	0.05
Maltol	2.00
Cocoa extract	26.00
Coffee extract	10.00
Ethyl alcohol	20.00
Water	41.90

The abovestated tobacco flavor formulation is applied at the rate of 0.1% to all of the cigarettes produced using the above tobacco formulation. Half of the cigarettes are then treated with 500 or 1,000 ppm of Compound (1), produced according to the process of Example 2. The control cigarettes and the experimental cigarettes containing Compound (1) are evaluated by paired comparison and the results are as follows.

The experimental cigarettes, prior to smoking, are found to have a rich sweet flavor effect with lingering sweetness with fruity, black currant-like, raspberry-like, dried fruit-like, and damascenone-like notes, and to have, on smoking, more tobacco-like, sweet-fruity and damascenone-like notes and to have more body than the control cigarettes.

The tobacco of the experimental cigarettes, prior to smoking, has sweet, floral, fruity, earthy and green notes. All cigarettes are evaluated for smoke flavor with a 20 mm cellulose acetate filter. Compound (1) enhances the sweet tobacco-like, fruity aromatic taste and aroma of the blended cigarette.

OTHER COMPOUNDS

Certain Substituted Carboxaldehydes

J.B. Hall, M.A. Sprecker, M.H. Vock, E.J. Shuster, J. Vinals and R.M. Novak; U.S. Patents 4,068,012; January 10, 1978; 4,081,483; March 28, 1978; 4,122,122; October 24, 1978; and 4,097,416; June 27, 1978; all assigned to International Flavors and Fragrances Inc. describe a process for making the following compounds and mixtures of compounds: 1-(2-propenyl)-3-(4-methyl-3-pentenyl)-Δ^3-cyclohexene-1-carboxaldehydes [A], 1-(2-propenyl)-4-(4-methyl-3-pentenyl)-Δ^3-cyclohexene-1-carboxaldehydes [B], and 1-(2-methyl-2-propenyl)-4-(4-methyl-3-pentenyl)-Δ^3-cyclohexene-1-carboxaldehydes [C] by reacting myrac aldehyde having the structure:

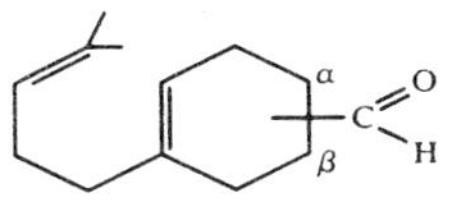

with an allylic or methallylic halide in the presence of an inert solvent and an alkali metal hydroxide and in the presence of a phase transfer agent. The reaction is carried out in a two phase system and may be illustrated as follows:

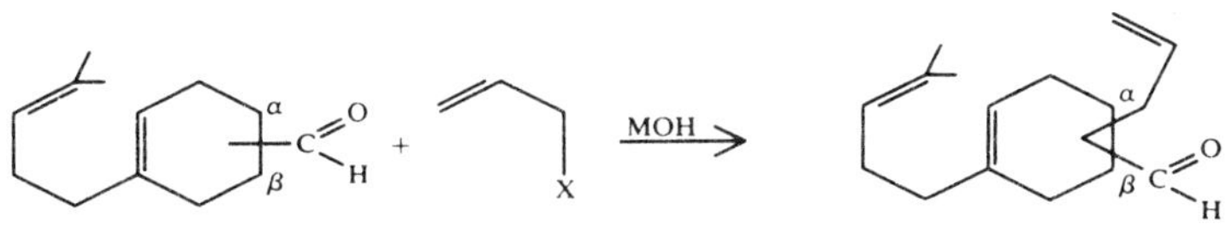

wherein the carboxaldehyde moiety is bonded to the α-carbon atom or the β-carbon atom, X is chloro or bromo and M is alkali metal.

One aspect of the process comprises the step of placing the reactants for the process and the base, respectively, in two immiscible phases; an organic phase and either (1) an aqueous base phase or (2) a solid base phase with the reactants being located substantially entirely in the first mentioned organic phase and the base being located substantially entirely in the second mentioned phase; and adding to this two phase system a phase transfer agent which may be one or more of several organic quaternary ammonium salts.

Examples of phase transfer agents which may be used are such quaternary ammonium salts as cetyltrimethylammonium bromide or tricaprylmethylammonium chloride; tertiary amines such as trimethylamine; or crown ethers such as 18-crown-6 or dibenzo-18-crown-6 having the respective structures:

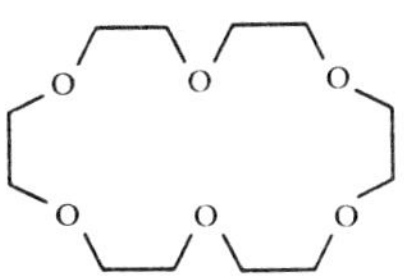

and

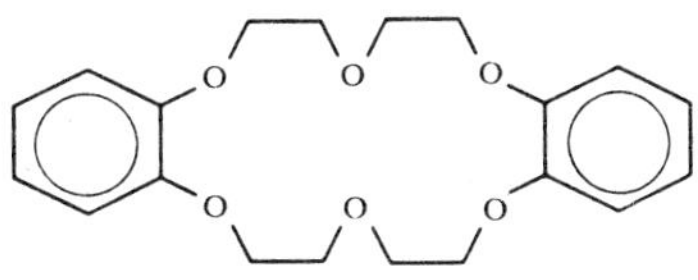

The process is carried out in an inexpensive solvent which is inert to the reaction system such as toluene, benzene, o-xylene, m-xylene, p-xylene, ethylbenzene, n-hexane, cyclohexane, methylene chloride and o-dichlorobenzene, at a preferred temperature in the range of 50° to 120°C. The reaction time is inversely proportional to the reaction temperature, with lower reaction temperatures giving rise to greater reaction times; and, accordingly, the reaction time ranges from 30 minutes up to 10 hours.

The preferred mol ratio of myrac aldehyde to the allyl or methallyl halide reactant is from 1:1 up to 1:1.2. The preferred mol ratio of base to allylic or methallylic halide in the reaction mass is from 1:1 up to 1:1.2.

The quantity of phase transfer agent in the reaction mass, based on the amount of myrac aldehyde in the reaction mass, may vary from 0.5 g/mol of the myrac aldehyde up to 25 g/mol, with a preferred concentration of phase transfer agent being in the range of from 2.5 up to 7.5 g of phase transfer agent per mol of myrac aldehyde.

The reaction is preferably carried out at atmospheric pressure. The particular base used in the reaction is critical and preferred are sodium hydroxide and potassium hydroxide.

When the compounds 1-(2-propenyl)-(4-methyl-3-pentenyl)-Δ^3-cyclohexene-1-carboxaldehydes and 1-(2-methyl-2-propenyl)-(4-methyl-3-pentenyl)-Δ^3-cyclohexene-1-carboxaldehydes are used as a food flavor adjuvant or fragrance enhancers, the nature of the coingredients in formulating the product composition will also serve to alter, modify, augment or enhance the organoleptic characteristics of the ultimate products treated therewith.

These compounds may be used for their organoleptic properties in perfumes, perfumed articles, foods, tobaccos and medicinal products. They are particularly useful because they do not discolor with age.

Such aldehydes as [A] and [B] when added to foodstuffs and medicinal products augment and enhance sweet, fatty, cocoa butterlike, green/grassy, dairy, apricotlike and carrotlike aroma characteristics and fatty, cocoa butterlike, green/grassy, dairylike, peach and coconut flavor characteristics.

Aldehydes, such as [C], when added to foodstuffs and medicinal products, augment and enhance floral, coriander, fatty, green and creamy aromas and flavor characteristics.

When added to perfumes and perfumed articles, aldehydes [A] and [B] impart, augment and/or enhance sweet, fruity and lavender notes with citrus floral and balsam nuances, while aldehydes [C] impart citrus, floral and fatty nuances.

In tobaccos and tobacco flavor and aroma imparting or enhancing compositions, cocoa liquorlike, sweet-rich and creamy notes are imparted prior to and during smoking by compositions containing the aldehydes [A] and [B], and sweet, spicy, corianderlike, citrus, and fruity notes are imparted by aldehydes [C].

Example 1: *Preparation of Compounds [A] and [B] –*

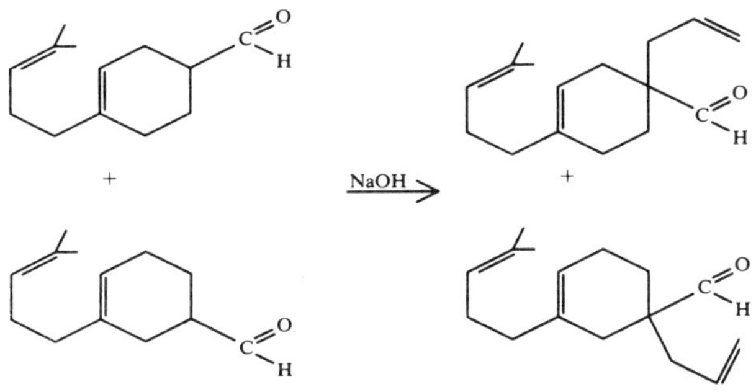

A slurry containing 289 g of allyl chloride, 180 g of granular sodium hydroxide, 400 ml of toluene, 25 g of Aliquat 336, and 576 g of a mixture of 4-(4-methyl-3-pentenyl)-3-cyclohexene-1-carboxaldehyde and 3-(4-methyl-3-pentenyl)-3-cyclohexene-1-carboxaldehyde is heated at reflux for 7 hours. Water is added to the cooled reaction mass and the resulting organic phase is separated and washed twice with water. Distillation through a short column affords an oil which contained 381 g (55%) of product (boiling point 140° to 144°C, 1.5 mm). The product is purified by steam vacuum fractional distillation through a 1" x 12" Goodloe packed column (boiling point 113° to 120°C, 1.5 mm at a steam rate of 1.6 ml liquid H_2O/min).

Example 2: *Preparation of a Cologne and Handkerchief Perfume* – The 1-(2-propenyl)-(4-methyl-3-pentenyl)-Δ^3-cyclohexene-1-carboxaldehyde(s), prepared according to the process of Example 1 is incorporated in a cologne at a concentration of 2.5% in 85% aqueous ethanol; and into a handkerchief perfume at a concentration of 20% (in 95% aqueous ethanol). A distinct and definite fruity, lavenderlike note with citrus and balsam nuances is imparted to the cologne and to the handkerchief perfume.

Example 3: *Basic Chocolate Flavor Formulation* – The following basic chocolate flavor formulation is produced:

	Parts by Weight
Dimethyl sulfide	1.0
Isobutyl acetate	1.0
Isoamyl acetate	1.0
Phenylethyl acetate	0.5
Diacetyl (10% in ethanol)	0.5
Furfural (50% in propylene)	0.5
Isoamyl alcohol	1.0
γ-Butyrolactone	5.0
Acetophenone	0.5
Benzaldehyde	1.0
Phenyl acetic acid	2.0
Maltol	3.0
Acetaldehyde	2.0
Isobutyraldehyde	8.0
Isovaleraldehyde	15.0
Phenylethyl alcohol	8.0
Vanillin	15.0
Propylene glycol	40.0
Total	100.0

To the above chocolate formulation 15% by weight of the mixture of Compounds [A] and [B] is added. A portion of the formulation contains no such additive. The formulation containing the carboxaldehydes is compared to the formulation without the carboxaldehydes in water at the rate of 5 ppm and submitted for evaluation to a bench panel. The opinion of the panel is unanimous; that the flavor with the Compounds [A] and [B], at the level of 15%, has characteristic cocoa butterlike aroma and taste characteristics.

When the Compounds [A] and [B] are added at the rate of 5% to the same basic chocolate formulation, fuller, more cocoa powderlike notes in aroma and taste are created as compared to the same formulation without the compounds at the level of 5 ppm in water.

Example 4: 10 parts by weight of 50 Bloom pigskin gelatin is added to 90 parts by weight of water at a temperature of 150°F. The mixture is agitated until the gelatin is completely dissolved and the solution is cooled to 120°F. 20 parts by weight of the liquid flavor composition of Example 3 is added to the solution which is then homogenized to form an emulsion having a particle size typically in the range of 2 to 5 μ. This material is kept at 120°F, under which conditions the gelatin will not jell.

Example 5: *Chewing Gum* – 100 parts by weight of chicle are mixed with 18 parts by weight of the flavor prepared in Example 4. 300 parts of sucrose and 100 parts of corn syrup are then added. Mixing is effected in a ribbon blender with jacketed side walls of the type manufactured by the Baker Perkins Co.

The resultant chewing gum blend is then manufactured into strips 1" in width and 0.1" in thickness. The strips are cut into lengths of 3" each. On chewing, the chewing gum has a pleasant long lasting chocolate flavor.

Example 6: *Tobacco Formulation* – A tobacco mixture is produced by admixing the following ingredients:

	Parts by Weight
Bright	40.1
Burley	24.9
Maryland	1.1
Turkish	11.6
Stem (flue-cured)	14.2
Glycerine	2.8
Water	5.3

Cigarettes are prepared from this tobacco. The following flavor formulation is prepared:

	Parts by Weight
Ethyl butyrate	0.05
Ethyl valerate	0.05
Maltol	2.00
Cocoa extract	26.00
Coffee extract	10.00
Ethyl alcohol	20.00
Water	41.90

The abovestated tobacco flavor formulation is applied at the rate of 1.0% to all of the cigarettes produced using the above tobacco formulation. Half of the cigarettes are then treated with 500 or 1,000 ppm of the mixture produced according to Example 1. The control cigarettes not containing the Compounds [A] and [B] and the experimental cigarettes are evaluated by paired comparison and the results are as follows: The experimental cigarettes are found to have more smoke body and a fuller smoke body sensation in the mouth on smoking. The tobaccolike notes are enhanced and the flavor of the tobacco on smoking is more aromatic with cocoa liquorlike, sweet-rich and creamy oily notes.

The tobacco of the experimental cigarettes, prior to smoking, has cocoa liquorlike, sweet-rich and creamy oily notes. All cigarettes are evaluated for smoke flavor with a 20 mm cellulose acetate filter.

Alkyl-2-Methyl-3,4-Pentadienoates

J.B. Hall, D.E. Hruza, M.H. Vock, J. Vinals and E.J. Shuster; U.S. Patent 4,094,823; June 13, 1978; assigned to International Flavors & Fragrances Inc. have found that C_1 to C_6 alkyl-2-methyl-3,4-pentadienoates, more particularly isobutyl-2-methyl-3,4-pentadienoate and n-hexyl-2-methyl-3,4-pentadienoate alter the flavor and/or aroma of consumable materials, e.g., tobacco, foodstuffs, perfume compositions and perfumed articles.

The C_1 to C_6 alkyl-2-methyl-3,4-pentadienoates useful in this process are formed by reaction of propargyl alcohol and a tri-C_1 to C_6 alkyl o-propionate in the presence of a lower alkanoic acid according to the reaction:

OH + OR OR OR → O OR

(wherein R is one of C_1 to C_6 alkyl, e.g., methyl, ethyl, isobutyl or n-hexyl) at pressures of from 2 up to 10 atmospheres, and temperatures of 140° to 180°C. The mol ratios of triethyl o-propionate:propargyl alcohol may vary from 1:2 up to 2:1 with a slight molar excess of propargyl alcohol being preferred. The lower alkanoic acid present during the reaction may be any acetic acid, propionic acid, butyric acid, valeric acid or isovaleric acid.

Example 1: *Preparation of Ethyl-2-Methyl-3,4-Pentadienoate* – Into a 2 liter autoclave, the following materials are placed:

	Amount [g (mol)]
Triethyl o-propionate	880 (5)
2-Propyn-1-ol	336 (6)
Propionic acid	25 -

The autoclave is closed and the reaction mass is heated to 160°C over a period of 50 minutes. The reaction mass is maintained at a temperature of between 150° and 160°C at a pressure of 40 up to 90 psig for a period of 3 hours. At the end of this 3 hour period, the autoclave is opened and the reaction mass is cooled to room temperature. 25.0 g of sodium bicarbonate is then added to the reaction mass in order to neutralize the propionic acid. 30 g of Primol and 0.1 g of Ionol are added and the resulting reaction product is distilled in a rush-over column yielding seven fractions at 100 mm Hg pressure.

Fraction	Vapor Temperature (°C)	Liquid Temperature (°C)	Weight of Fraction (g)
1	37-39	41-43	171.0
2	42	47	165.3
3	47	58	137.6
4	82	98	179.3
5	114	113	203.4
6	120	150	217.3
7	75	200	13.7

Fractions 2 through 4 are combined, and 20 g Primol and 1 g Ionol are added thereto. The resulting mixture is distilled in a 1' x 1" Goodloe distillation column yielding twenty fractions at 120 mm Hg pressure.

Fraction	Vapor Temperature (°C)	Liquid Temperature (°C)	Reflux Ratio	Weight of Fraction (g)
1	35-104	102-110	19:1	24.1
2	104	110	9:1	6.6
3	104	110	9:1	6.1
4	104	110	9:1	21.1
5	105	111	9:1	18.6
6	105	111	9:1	18.7
7	105	111	9:1	16.7
8	105	112	9:1	19.3
9	105	112	9:1	21.4
10	105	112	9:1	15.8
11	104-105	112	9:1	25.1
12	105	113	9:1	25.0
13	105	114	9:1	21.4
14	105	115	9:1	23.0
15	105	117	9:1	23.8
16	105	121	9:1	27.6
17	105	130	9:1	28.7
18	105	172	9:1	28.5
19	107	181	9:1	11.2
20	108	210	9:1	8.9

Fractions 9 through 18 are combined and confirmed by IR, NMR and mass spectral analyses to be ethyl-2-methyl-3,4-pentadienoate. At a concentration of 0.2 ppm, the resulting product, ethyl-2-methyl-3,4-pentadienoate has a fruity, strawberry, creamy aroma with berry, apple and pineapple notes and a sweet, fruity, strawberry flavor with woody and creamy nuances.

Example 2: To a portion of a basic strawberry flavor formulation, 4% by weight of ethyl-2-methyl-3,4-pentadienoate is added. The formulation with the ethyl-2-methyl-3,4-pentadienoate is compared to the same formulation without it.

Both flavors are evaluated in water at the rate of 50 ppm. Both beverages are tasted by an expert panel. The beverage containing the strawberry formulation with the addition of ethyl-2-methyl-3,4-pentadienoate is unanimously preferred as having fresh strawberry and a sweet-strawberry aroma and taste missing in the basic strawberry formulation.

Example 3: When the compound prepared in Example 1 is applied at the rate of 0.1% by weight to cigarettes of a standard formulation, the cigarettes containing the ethyl-2-methyl-3,4-pentadienoate in the tobacco or in the filter were found to have a sweeter and fruitier aroma.

In smoke flavor, the cigarettes containing the ethyl-2-methyl-3,4-pentadienoate were more aromatic, sweeter, fruitier and slightly less harsh in the mouth and throat. In addition, those cigarettes containing the ethyl-2-methyl-3,4-pentadienoate in the tobacco give rise to a woody nuance in the taste and aroma in smoking.

Example 4: Concentrated liquid detergents with a fruity, strawberry odor are prepared containing 0.10%, 0.15% and 0.20% of ethyl-2-methyl-3,4-pentadienoate, prepared according to Example 1. They are prepared by adding and homogeneously mixing the appropriate quantity of ethyl-2-methyl-3,4-pentadienoate in the liquid detergent. The detergents all possess a fruity, strawberry fragrance, the intensity increasing with greater concentrations of ethyl-2-methyl-3,4-pentadienoate.

Acetals of Conjugated Alkenals

W.L. Schreiber and A.O. Pittet; U.S. Patent 4,107,217; August 15, 1978; assigned to International Flavors & Fragrances Inc. have developed a process for the use of acetal intermediates to produce 2-alkylidene-3-alkynals and 2-alkylidene-3-alkenals. These aldehydes are useful for augmenting or enhancing the organoleptic properties (aroma or taste) of consumable materials such as foodstuffs, tobacco products, and perfumes.

The processes using these acetals provide a relatively straightforward and convenient synthesis for obtaining 2-alkylidene-3-alkynals, acetals thereof and 2-alkylidene-3-alkenals and acetals thereof in good yields. Briefly, the processes comprise reacting an alkyl metallo acetylide with a dialkoxy acetonitrile or dialkoxy dialkyl acetamide and hydrolyzing the imine salt so obtained to form the corresponding 1,1-dialkoxy-3-alkyne-2-one.

Then, one alternative procedure involves treating the dialkoxy alkynone with an alkylidene triaryl substituted phosphorane, or an alkylidene phosphorus triamide or an alkyl phosphoramide anionic compound to form a 1,1-dialkoxy-2-alkylidene-3-alkyne and either (1) hydrolyzing the 1,1-dialkoxy-2-alkylidene-3-alkyne with aqueous acid to provide a 2-alkylidene-3-alkynal, or (2) reducing the triple bond of the 1,1-dialkoxy-2-alkylidene-3-alkyne thus forming a 1,1-dialkoxy-2-alkylidene-3-alkene.

The second alternative procedure involves first reducing (as by hydrogenation) the dialkoxy alkynone to form a 1,1-dialkoxy-3-alken-2-one and then treating the 1,1-dialkoxy-3-alken-2-one with an alkylidene triaryl substituted phosphorane, or an alkylidene phosphorus triamide or an alkyl phosphoramide anionic compound to form a 1,1-dialkoxy-2-alkylidene-3-alkene. The alkynals so obtained have the formula:

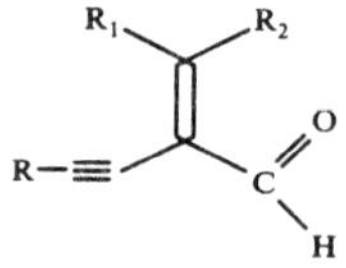

wherein one of R_1 and R_2 is hydrogen and the other is lower alkyl and R is lower alkyl. These alkynals have a number of uses, one of which is hydrogenation of the triple bond to a double bond for production of the corresponding 2-alkylidene-3-alkenal compounds.

The 1,1-dialkoxy-2-alkylidene-3-alkenes so obtained are compounds having the formula:

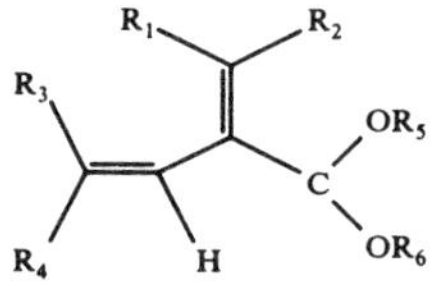

wherein one of R_1 and R_2 is hydrogen and the other is lower alkyl; wherein one of R_3 and R_4 is hydrogen and the other is lower alkyl; and wherein R_5 and R_6 are each the same or different lower alkyl (e.g., methyl, ethyl, n-propyl, isopropyl, isobutyl and n-butyl). These 2-alkylidene alkenal acetals can either be hydrolyzed to the corresponding 2-alkylidene-3-alkenal compounds or they can be used as precursors for 2-alkylidene-3-alkenal compounds.

Example 1: *Preparation of 1,1-Dimethoxy-3-Hexyne-2-One* – An ether solution of ethylmagnesium bromide is prepared from 7.3 g magnesium turnings and 32.7 g ethyl bromide. About 20 g of ethyl acetylene is admitted as a gas under dry ice condenser and the mixture is refluxed for 2 hours until gas evolution ceases. The mixture is then cooled below 0°C, and 30.3 g of dimethoxyacetonitrile is added in ether solution.

The mixture is allowed to come to room temperature and stirred for 2 hours, during which time the lower layer of the two-phase mixture becomes almost solid. The mixture is again cooled and treated with 16 ml sulfuric acid diluted with 300 ml water. The layers are separated, and the organic layer is washed successively with saturated aqueous sodium chloride solution and saturated aqueous sodium bicarbonate solution and then dried over 4 A molecular sieves. Thorough removal of solvent gives 35.3 g of yellow oil, 99% pure by GLC (gas-liquid phase chromatography).

Example 2: *Preparation of 1,1-Dimethoxy-cis-3-Hexene-2-One* – 6 g of the 1,1-dimethoxy-3-hexyne-2-one (of Example 1) is stirred under hydrogen gas at one atmosphere in 40 ml hexane containing 0.6 g Lindlar catalyst (palladium-on-calcium carbonate poisoned with lead acetate) and 4.0 g quinoline. The reaction is terminated when 1% of the starting material (1,1-dimethoxy-3-hexyne-2-one) remains after about 1½ hours.

The mixture is filtered and the quinoline washed out with dilute aqueous hydrochloric acid. The organic layer is washed with saturated aqueous sodium bicarbonate and then brine; and the solvent is evaporated. GLC and NMR of the crude material show the product is substantially 1,1-dimethoxy-cis-3-hexene-2-one.

Example 3: *Preparation of 1,1-Dimethoxy-trans-3-Hexene-2-One* – The crude product produced in Example 2 is dissolved in 6 ml of acetic acid with 0.1 g of sodium iodide. By GLC analysis on Carbowax (polyethylene glycol) it is clear that the 1,1-dimethoxy-cis-3-hexene-2-one is converted to a new material of later retention time. After ½ hour, less than 5% of cis- material remains.

The material is isolated by partitioning between water and ether, washing the ether layer successively with aqueous sodium bicarbonate and brine and then drying over 4 A molecular sieves. Evaporation of the solvent provides 5.0 g of yellow oil. NMR and GLC indicate essentially all trans material having the structure:

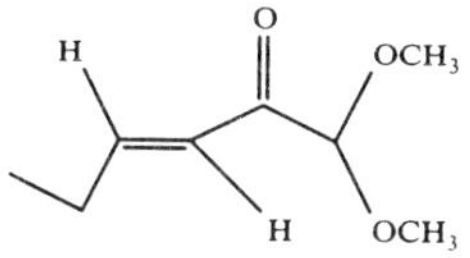

Example 4: *Preparation of trans-2-Ethylidene-trans-3-Hexenal* – Ethyltriphenylphosphonium bromide (3.71 g) and 6.3 ml of 1.6 N n-butyl lithium are mixed in ether solution and 1.58 g of the 1,1-dimethoxy-trans-3-hexene-2-one of Example 3 is added, while keeping the internal temperature below 30°C. After a few minutes the mixture is filtered and the solvent evaporated. A small amount of solid is present so the residue is dissolved in isopentane, filtered, and again evaporated to give 1.10 g of a yellow-orange oil.

GLC and NMR indicate the presence of two acetals of 2-ethylidene-trans-3-hexenal: cis- and trans-isomers at the ethylidene group, namely, cis-2-ethylidene-trans-3-hexenal dimethyl acetal and trans-2-ethylidene-trans-3-hexenal dimethyl acetal. The acetal material is dissolved in 2 ml water and 3 ml acetic acid with a small amount of sodium iodide. After a few minutes GLC obtained on a 10' x ⅛" DC-710 (20%) shows complete hydrolysis. (In the absence of sodium iodide a mixture of cis- and trans-ethylidene isomers of 2-ethylidene-trans-3-hexenal is obtained.)

The product is isolated by partitioning between water and ether. The organic layer is washed successively in water, aqueous sodium bicarbonate, and aqueous sodium chloride and finally evaporated to give 0.70 g of an orange oil. The major peak (80%), isolated by preparative GLC is trans-2-ethylidene-trans-3-hexenal having the structure on the following page.

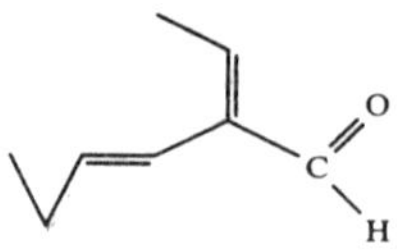

This material has a musty, harsh nuance when added to a tobacco flavor formulation, which in turn is added to tobacco; the compound enhances the sweet, maple, nutlike character and the natural smell of the tobacco.

Certain Tricyclic Alcohols

K.K. Light, E.J. Shuster, J.F. Vinals and M.H. Vock; U.S. Patent 4,139,650; February 13, 1979; assigned to International Flavors & Fragrances Inc. have determined that certain tricyclic alcohols are capable of imparting a variety of flavors and fragrances to various consumable materials. These are certain tricyclic alcohols having the structure:

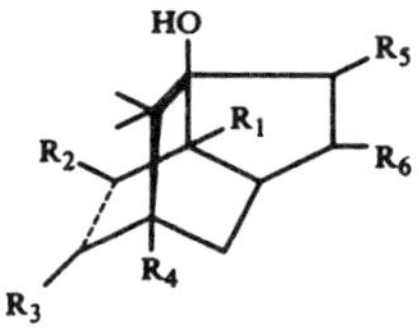

wherein each of R_1, R_2, R_3, R_4, R_5 and R_6 is selected from the group consisting of hydrogen and methyl; wherein the dashed line is a carbon-carbon single bond or a carbon-carbon double bond; and wherein when the dashed line is a carbon-carbon single bond at least one of R_2 or R_3 is hydrogen.

These tricyclic alcohols are actually racemic mixtures rather than individual stereoisomers.

A special example of a product which has been found to be useful for these purposes is the compound having the structure:

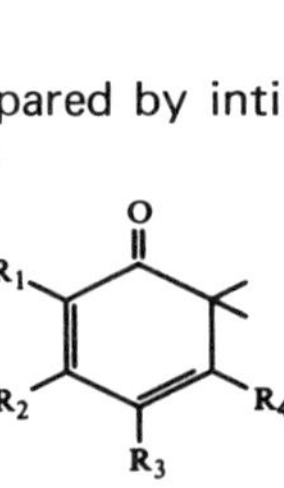

This compound has a strong patchoulilike, slightly minty, sweet, warm, earthy, camphoraceous note and an earthy, patchouli, nutty and woodylike taste and aroma in food flavors.

The tricyclic alcohols may be prepared by intimately admixing a methylcyclohexadienone having the structure:

with an acetylenic compound having the structure:

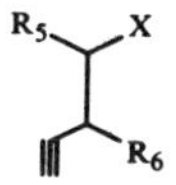

wherein X may be either hydroxyl, bromo and chloro, thereby forming a diene compound having the structure:

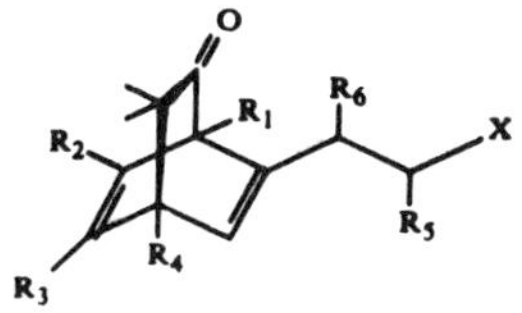

In this reaction, the most preferred temperature range is 200° to 210°C. The reaction may be carried out in the presence of an inert solvent such as benzene, hexane or cyclohexane (or any other inert solvent) or in the absence of solvent. It is preferred to use equimolar quantities of each reactant.

The diene compound is then hydrogenated with hydrogen in the presence of a catalyst such as palladium, platinum, nickel or other suitable hydrogenation catalyst.

A reaction temperature of 100° to 200°C is preferred. The reaction is preferably carried out at superatmospheric pressures and preferred pressures range from 5 to 15 atmospheres.

The hydrogenation reaction gives rise to a ketone product having the structure:

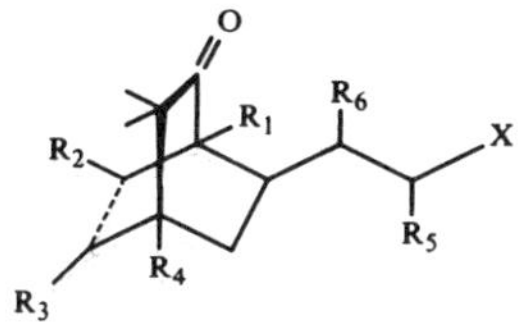

but it is noteworthy that the compound produced is one where the dashed line is a carbon-carbon single bond if one of R_2 or R_3 is hydrogen and the compound is primarily one where the dashed line is a carbon-carbon double bond if R_2 and R_3 are both methyl.

When X is halogen, the ketone thus produced may then be immediately cyclized by treating same with an alkali metal selected from the group consisting of sodium, potassium or lithium. The cyclization may be carried out in diethyl ether, tetrahydrofuran or benzene.

Prior to cyclization, in the event that X is OH, the ketone must be halogenated with thionyl chloride or any other suitable halogenating agent. The halogenation reaction may be carried out in the presence or in the absence of an inert solvent such as benzene, toluene, cyclohexane or pyridine. A reaction temperature of 80°C is preferred. The mol ratio of halogenating agent:ketone of 3:1 is preferred when using thionyl chloride and a ratio of 10:1 is preferred when using aqueous HCl and HBr.

8,8-Dimethyloctahydro-1,5-methano-1H-inden-1-ol (Compound A) may be made, for example, by starting with 6,6-dimethylcyclohexadienone and 3-butyn-1-ol. Compound A imparts a warm patchoulilike character to a woody cologne composition, which may be used to perfume soaps, detergents, etc.

Example 1: *Flavor Composition* – The following basic walnut flavor formulation is prepared:

	Parts by Weight
Ethyl-2-methyl butyrate	10
Vanillin	40
Butyl valerate	40
2,3-Diethylpyrazine	5
Methylcyclopentenolone	80
Benzaldehyde	60
Valerian oil Indian (1% in 95% aqueous ethanol alcohol)	0.5
Propylene glycol	764.5

Compound A is added to the above formulation at the rate of 1.5%. This formulation is compared to a formulation which does not have Compound A added to it, at the rate of 20 ppm in water. The formulation containing Compound A has a woody-balsamic, fresh walnut kernel and walnut skinlike taste and, in addition, has a fuller mouth-feel and longer lasting taste. The flavor that has Compound A added to it is preferred by a group of flavor panelists who consider it to be a substantially improved walnut flavor.

Example 2: To 50% of cigarettes made and flavored by the standard International Flavors & Fragrances formulations, 10 and 20 ppm of Compound A are added. These cigarettes are referred to as experimental cigarettes, and the cigarettes without Compound A are referred to as control cigarettes. The control and experimental cigarettes are then evaluated by paired comparison and the results are as follows:

(a) In aroma, the experimental cigarettes are found to be more aromatic.

(b) In smoke flavor, the experimental cigarettes are found to be more aromatic, sweeter, more bitter, greener, richer and slightly less harsh in the mouth and more cigarette tobaccolike than the control cigarettes.

The experimental cigarettes containing 20 ppm of Compound A are found to be woody, slightly chemical and mouth-coating in the smoke flavor.

All cigarettes, both control and experimental, are evaluated for a smoke flavor with 20 mm cellulose acetate filter. Compound A enhances the tobaccolike taste of the blended cigarette.

Terpenyl Ethers

C.J. Mussinan, B.D. Mookherjee, M.H. Vock, F.L. Schmitt, E.J. Shuster; J.M. Sanders; B.M. Light and E.J. Granda; U.S. Patents 4,163,068; July 31, 1979; and 4,131,687; December 26, 1978; both assigned to International Flavors & Fragrances Inc. have found that the flavors and/or fragrances of such materials as perfumes, perfumed articles, colognes, foodstuffs, chewing gums, toothpastes and medicinal products can be augmented or enhanced by adding thereto a small but effective amount of at least one C_{10}-terpenyl ether having the structure: $(C_{10}\text{-Terpenyl})\text{—}(O-R_1)_n$ wherein n is 1 or 2 and wherein the terpene moiety has one of the structures:

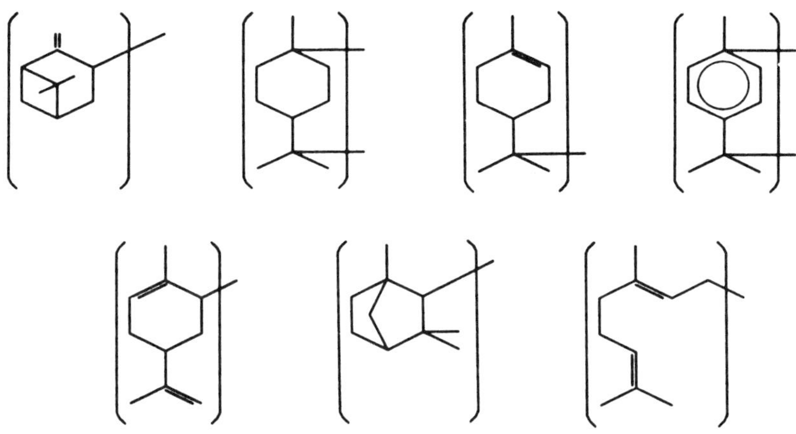

and R_1 is C_1 to C_4 alkyl, C_3 or C_4 alkenyl, C_3 haloalkenyl, or C_3 alkynyl.

Examples of the C_{10}-terpenyl ethers and their organoleptic characteristics are as follows:

Structure of Compound	Name of Compound	Flavor Characteristics	Fragrance Characteristics
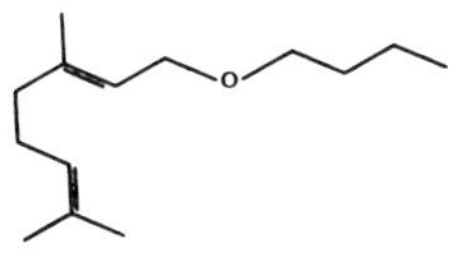	Geranyl butyl ether	Citrus, floral, rose aroma characteristics with citrus, nutty, floral, rose and bitter flavor characteristics	A green citrus, lemony, floral aroma
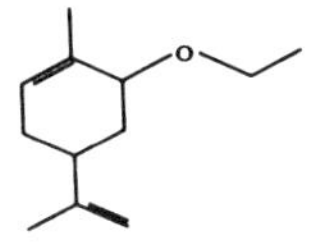	Carvyl ethyl ether	A woody, green, parsley-like, celery-like and grapefruit-like aroma characteristic with woody, green, green pea-like, celery-like, citrus flavor characteristics and a sour and astringent effect	An earthy, carrot topnote giving way to a petitgrain character with basil/carvone nuances

(continued)

Structure of Compound	Name of Compound	Flavor Characteristics	Fragrance Characteristics
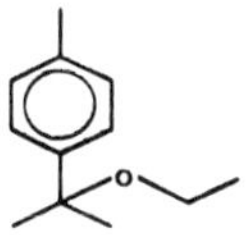	8-p-cymenyl-ethyl ether	An herbal, cumarin-like and green aroma characteristic with herbal, cumin-like, lemon juice-like, green, and oily flavor characteristics	A sweet phenolic, cymene-like, acetophenone-like aroma with a musty, vanilla, anthranilate note
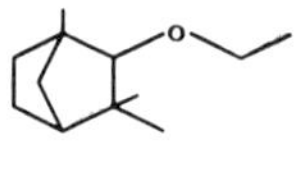	fenchyl ethyl ether	A woody, cedarwood-like, oakwood-like, lemon juice-like, camphoraceous, herbaceous, citrus-like, blueberry-like, sassafras and rooty aroma characteristic with woody, cedarwood-like, steamed oak, camphoraceous, herbaceous, citrusy, woody, blueberry, sassafras-like and rooty flavor characteristics and a sour mouthfeel and astringent effect.	A sweet, camphoraceous, minty, herbaceous, woody, green aroma with a eucalyptus-like undertone.

These C_{10}-terpenyl ethers can be produced using any one of several known techniques. Thus, the terpenyl alcohol can be reacted with a lower alkanol in the presence of an acid catalyst such as p-toluenesulfonic acid at reflux conditions.

As a further example, 8-p-cymenol may be reacted with an excess of ethanol (e.g., from 2 up to 10-fold excess) in the presence of p-toluenesulfonic acid, and the reaction mass may then be refluxed for a period of from 2 to 10 hours. By the same token, carvyl ethyl ether may be produced by reacting ethanol with carveol, the ratio of ethanol:carveol being about 1 mol ethanol:0.1 mol carveol, and the reaction may take place in the presence of p-toluenesulfonic acid.

The ethers may also be formed by reacting the corresponding terpenic alcohol with the appropriate sulfate in the presence of sodium hydride in anhydrous media and in the presence of an inert solvent such as dimethylformamide. Thus, for example, fenchyl alcohol is reacted with diethyl sulfate, the fenchyl alcohol and diethyl sulfate being in molar proportion of about 1:1 in the presence of sodium hydride and dimethylformamide.

Compounds having the generic structure:

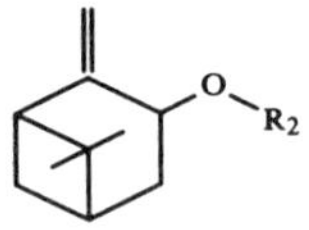

wherein R_2 represents C_3 or C_4 alkenyl, C_3 haloalkenyl or C_3 alkynyl, may be prepared by reacting pinocarveol with the appropriate organic halide.

This reaction is carried out under the influence of a base comprising the step of placing the reactants for the process and the base, respectively, in two immiscible phases; an organic phase and either (1) an aqueous base phase or (2) a solid base phase with the reactants being located substantially entirely in the first mentioned organic phase and the base being located substantially entirely in the second mentioned phase; and adding to the two phase system a phase transfer agent which may be one or more of several organic quaternary ammonium salts, such as tricaprylmethylammonium chloride, cetyltrimethylammonium bromide, and benzyltrimethylammonium hydroxide.

Example 1: *Synthesis of Fenchyl Ethyl Ether* – The following chemicals were used: fenchyl alcohol, 154 g (1 mol); sodium hydride, 48 g, 50% in mineral oil (1 mol); diethyl sulfate, 169.4 g (1 mol); and dimethylformamide (DMF), 500 ml.

The NaH and DMF (400 ml) are added to a 2 liter, round-bottom flask equipped with an N_2 bubbler, vented Friedrich's condenser, mechanical stirrer and addition funnel. $CaCl_2$ drying tubes are placed on all open ends. N_2 is bubbled into the flask to evacuate the air. Fenchyl alcohol is then dissolved in 100 ml of DMF and placed in the addition funnel and added dropwise to the reaction mass. When the addition is complete, the mixture is heated until no more H_2 is given off. The reaction mass is then cooled to room temperature, and the diethyl sulfate is added dropwise. The reaction is slightly exothermic. After all of the diethyl sulfate is added, GC analysis indicates that less than 0.3% of the diethyl sulfate is present.

The mixture is cooled to room temperature, and 1 liter of water is slowly added thereto over a period of one hour. Two layers are present, the top layer being drawn off and saved. The aqueous layer is extracted with two 250 ml portions of diethyl ether. The ether layer is combined with the top layer and dried over anhydrous Na_2SO_4; then stripped by rotary evaporator. Yield of crude material: 170.80 g; GC yield = 43.4%. The 170.8 g of crude material is added to 74.2 g of triethyl borate and distilled at atmospheric pressure.

The material left in the pot is then transferred to a fractional distillation apparatus and distilled at 66° to 67°C at 5 mm Hg pressure. Fraction 5 is subjected to MS, NMR and IR analyses. The percent yield based on Fractions 4 through 6 combined is 23.3%.

Example 2: *Herbal Fragrance Produced Using Fenchyl Ethyl Ether* – The following mixture is prepared:

	Percent by Weight
Amyl cinnamic aldehyde	20
Phenylacetaldehyde dimethylacetal	4
Thyme oil, white	8
Sauge sclaree French	8
Galbanum oil	4
Geranyl acetate	10
Juniper berry oil	4
Methyl octin carbonate	2
Linalyl acetate	10
Dihydro methyl jasmonate	20
Fenchyl ethyl ether (produced according to Example 1)	10

The fenchyl ethyl ether produced according to Example 1 adds a strong herbaceous character to this herbal fragrance formulation.

Example 3: *Preparation of a Detergent Composition* – A granular detergent composition is prepared according to Example 9 of Canadian Patent 1,004,566 containing the following ingredients:

	Percent by Weight
Anhydrous sodium carbonate	30.0
Hydrated sodium silicate (81.5% solids, SiO_2:Na_2O ratio 2.1:1 by weight)	20.0
Coconut alcohol condensed with 6 molar proportions of ethylene oxide	10.0
Sodium citrate dihydrate	10.0
Sodium dichlorocyanurate dihydrate	3.8
Polyethylene glycol (Carbowax 4000)	2.0
Dimethyl silicone	0.8
Anhydrous sodium sulfate	15.5
Herbal fragrance produced according to Example 2	5.9

This composition has an excellent herbaceous aroma.

Example 4: *Flavor Utility of Fenchyl Ethyl Ether* – The following lemon flavor formulation is prepared:

	Parts by Weight
Natural lemon oil, terpeneless	10
Acetaldehyde	0.6
α-Terpineol	2.1
Citral	1.8
Carvone	0.24
Terpinolene	1.2
α-Terpinene	0.25
Diphenyl	0.25
α-Fenchyl alcohol	0.25
Limonene	0.35
Linalool	0.25
Geranyl acetate	0.25
Nootkatone	0.25
Neryl acetate	0.25

The flavor formulation is divided into two portions. 4 ppm of fenchyl ethyl ether is added to 200 ppm of the first portion of the lemon flavor prepared above; to the second portion of the lemon flavor nothing is added.

A definite aroma improvement, a more natural lemon juice aroma and taste as well as a pleasant sour effect and generally improved taste is created as a result of the addition of the fenchyl ethyl ether to the lemon flavor. In general, the fenchyl ethyl ether supplies a natural lemon juice note to this lemon flavor.

FOR TOBACCO PRODUCTS AND PERFUMES

Bicyclo[2.2.2]-Octene Derivatives

J.M. Sanders, J.F. Vinals and F.L. Schmitt; U.S. Patents 4,128,729; December 5, 1978; 4,139,484; February 13, 1979; and 4,163,737; August 7, 1979; all assigned to International Flavors & Fragrances Inc. describe the process for preparation of various unsaturated bicyclooctenemethanols and their use for augmenting and enhancing the organoleptic properties of perfumes, perfumed products and tobaccos.

The bicyclooctenemethanol compounds useful for these purposes have one of the following formulas:

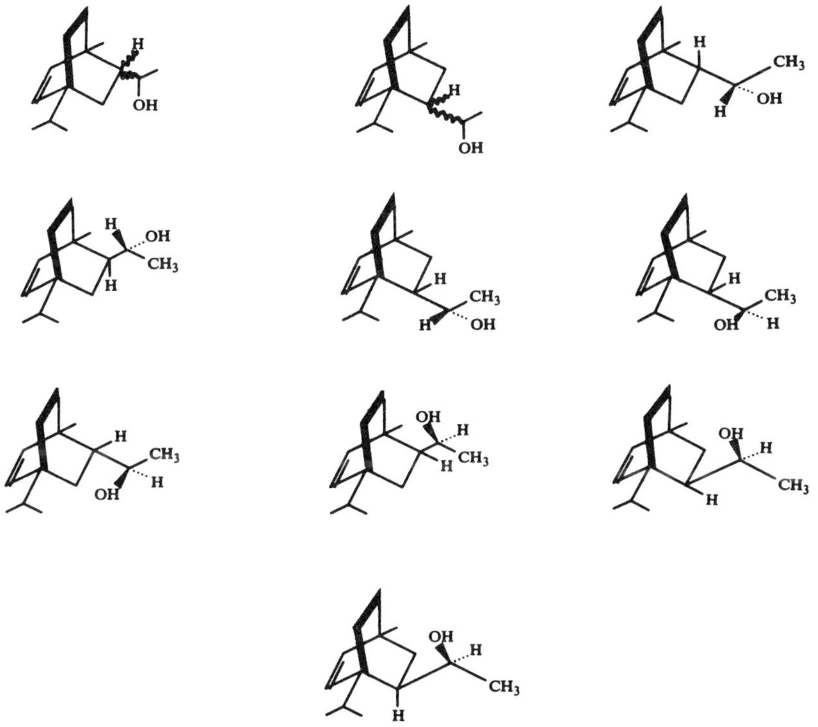

in which the wavy line represents exo or endo configurations of the ethanol moiety with respect to the carbon-carbon double bond of the bicyclooctene moiety of the molecule and wherein, in those representations where wavy lines appear, mixtures are intended to be represented; and wherein in the ethanol moiety a heavy line indicates that the substituent hydroxyl or hydrogen group lies above a plane defined by the two carbon atoms of the ethanol side chain and the site of attachment of the side chain to the bicyclooctene moiety and a dotted line indicates that the substituent hydroxyl or hydrogen groups lies below this plane.

These bicyclooctenemethanol derivatives are prepared according to a process whereby the compound 1-methyl-4-(2-propenyl)cyclohexene-1 (also referred to herein as limonene) is isomerized to form 1-methyl-4-isopropyl-1,3-cyclohexadiene (also referred to herein as α-terpinene) and other double bond isomers.

The 1-methyl-4-isopropyl-1,3-cyclohexadiene thus formed is separated from the reaction mixture and is then reacted (in admixture without being isolated, or if desired, after isolation) via a Diels-Alder reaction with acrolein thereby forming a mixture of carboxaldehydes. This mixture of carboxaldehydes is then reacted with a methylmagnesium halide thereby forming a mixture of oxymagnesium halide compounds. This mixture of oxymagnesium halide compounds is then hydrolyzed to form the desired products.

This process provides an organoleptically improved smoking tobacco product and additives therefor as well as methods of making them which overcome specific problems heretofore encountered in which specific desired sweet, oriental, spicy, woody, tobaccolike, and hay tobaccolike flavor and aroma characteristics of natural tobacco (prior to smoking and on smoking; in the mainstream and in the sidestream) are created or enhanced or modified or augmented and may readily be controlled and maintained at the desired uniform level regardless of variations in the tobacco components of the blend.

An aroma and flavoring concentrate containing one or more of the bicyclo[2.2.2] octenemethanol derivatives of this process and, if desired, one or more additional flavoring additives may be added to the smoking tobacco material, to the filter or to the leaf or paper wrapper. The smoking tobacco material may be shredded, cured, cased and blended tobacco material or reconstituted tobacco material or tobacco substitutes or mixtures thereof.

The proportions of flavoring additives may be varied in accordance with taste but insofar as enhancement or the imparting of natural and/or sweet, oriental, spicy, woody and tobacco/haylike notes, satisfactory results are obtained if the proportion by weight of the sum total of the bicyclo[2.2.2] octenemethanol derivatives to smoking tobacco material is between 250 ppm and 1,500 ppm (0.025 to 0.15%) of the active ingredients to the smoking tobacco material. It has further been found that satisfactory results are obtained if the proportion by weight of the sum total of bicyclo[2.2.2] octenemethanol derivatives used as flavoring material is between 2,500 and 15,000 ppm (0.25 to 1.5%).

While this process is particularly useful in the manufacture of smoking tobacco, such as cigarette tobacco, cigar tobacco and pipe tobacco, other tobacco products formed from sheeted tobacco dust or fines may also be used. Likewise, the bicyclo[2.2.2] octenemethanol derivatives can be incorporated with materials such as filter tip materials, seam paste, packaging materials and the like which are used along with tobacco to form a product adapted for smoking.

The bicyclo[2.2.2] octenemethanol derivatives and one or more auxiliary perfume ingredients including, for example, other alcohols, aldehydes, ketones, nitriles, esters, cyclic esters (lactones), natural essential oils and other synthetic essential oils may be admixed so that the combined odors of the individual components produce a pleasant and desired fragrance, particularly and preferably in rose fragrances.

Examples 1 through 3 following, illustrate a method for preparing the bicyclo-[2.2.2] octene derivatives of this process. Examples 4 through 6 illustrate the uses of the bicyclo [2.2.2] octene derivatives for their organoleptic properties. All parts and percentages given herein are by weight unless otherwise specified.

Example 1: *Isomerization of Limonene –*

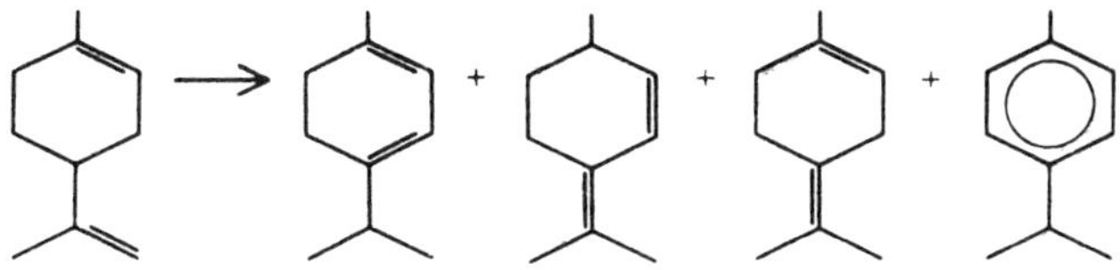

A mixture of 6,000 g of limonene and 3,000 g of 50% wt/wt sulfuric acid is stirred at reflux (110° to 116°C) for approximately 4½ hours. The mass is cooled to room temperature, the organic layer is separated, and the aqueous layer is extracted with 1,000 g of chloroform. The combined organic layer and chloroform extract are washed successively with 10% salt solution, 10% sodium carbonate solution and 10% salt solution.

The chloroform is recovered by distillation at atmospheric pressure through a short column and the stripped crude product is distilled rapidly at 50 mm Hg. In this way, 4,713 g of material is obtained, boiling point 94° to 103°C/mm Hg, which contains approximately 40% α-terpinene and is suitable for use without further purification.

Example 2: *Preparation of 1(and 4)-Isopropyl-4(and 1)-Methylbicyclo [2.2.2]-Oct-5-Ene-2-Carboxaldehyde –*

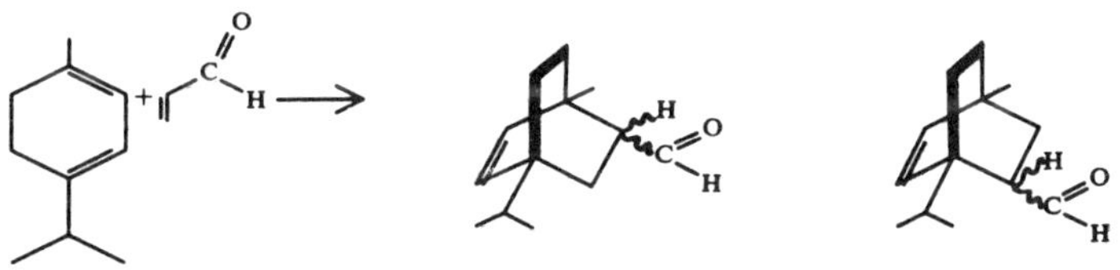

A mixture of 470 g of acrolein, 400 g of toluene, and 2,380 g of α-terpinene (approximately 40%, prepared according to Example 1) is added to a mixture of 33 g of anhydrous stannic chloride and 200 g of toluene over a period of two hours with external cooling to maintain 5° to 10°C. The reaction mass is stirred for an additional two hours at 5° to 10°C and then is washed with three 500 g portions of 10% salt solution.

After addition of 50 g of triethanolamine and 3 g of Ionol to the washed organic solution, the toluene is removed and the crude product is distilled rapidly through a short column under reduced pressure to give 1,567 g of oil, boiling point 72° to 156°C/3 to 4 mm Hg. Redistillation of this material using a 12" x 1½" Goodloe packed column gives 793 g of product, boiling point 105° to 120°C/3 to 4 mm Hg, which is shown to be a mixture of isomers in the reaction equation, wherein the wavy lines represent exo or endo configurations of the carboxaldehyde moiety with respect to the carbon-carbon double bond of the bicyclo[2.2.2]-octene moiety.

Example 3: *Preparation of α,1(and 4)-Dimethyl-4(and 1)-Isopropylbicyclo[2.2.2]-Oct-5-Ene-2-Methanol* –

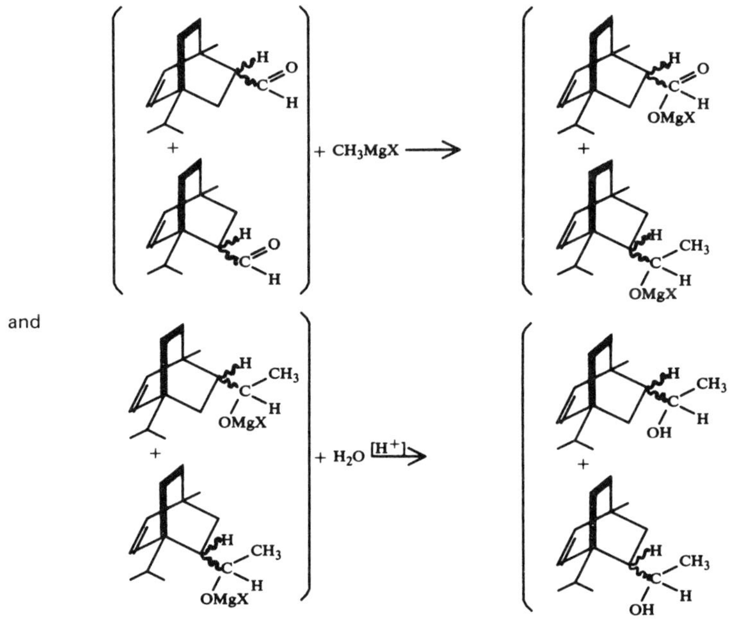

and

(wherein X is chloro)

A solution of 256 g of 1(or 4)-isopropyl-4(or 1)-methylbicyclo[2.2.2]oct-5-ene-2-carboxaldehyde (prepared according to Example 2) in 167 g of tetrahydrofuran is added in 65 minutes to 436 g of 3 M methylmagnesium chloride in tetrahydrofuran at reflux (69° to 71°C). The reaction mass is stirred an additional two hours at reflux and then is cooled to room temperature.

The Grignard complex is decomposed by slow addition of 158 g of saturated ammonium chloride solution, and the resulting magnesium salts are removed by filtration. The filtrate is concentrated under reduced pressure (20 to 50 mm Hg) to remove the tetrahydrofuran, and the stripped crude product is then distilled quickly through a short column to give 228 g of oil, boiling point 100° to 119°C/3 mm Hg. Redistillation through a 12" x 1" Goodloe packed column gives 132 g of oil, boiling point 106° to 113°C/3 mm Hg, which is shown to be a mixture of isomers having the structure shown, as the product in the above reaction equation.

Example 4: *Rose Formulation* – To demonstrate the use of the bicyclo[2.2.2]-octenemethanol derivatives produced according to Example 3 in a rose formulation, the following formula is provided:

Ingredients	Parts by weight
Phenylethyl alcohol	175
Geraniol	400
Trichloromethylphenyl carbinyl acetate	20
Phenylethyl acetate	60
Undecylenic aldehyde (10% in diethyl phthalate)	5
n-Nonyl aldehyde (10% in diethyl phthalate)	2
Musk ketone	10
Musk ambrette	10
Eugenol phenyl acetate	20
Citronellol	100
Vanillin (10% in diethyl phthalate)	6
Eugenol	30
Citronellyl formate	30
Geranyl acetate	10
Linalool	40
Geranyl phenyl acetate	50
cis,β,γ-Hexenyl acetate	2
Bicyclo[2,2,2] octenemethanol derivative mixture prepared according to Example 3	25
Total	1,000

The addition of 2.5% of bicyclo[2.2.2] octenemethanol derivatives produced according to Example 3 lends a great deal of strength and character to the rose fragrance. It contributes great floralcy and woodiness with clary-sage and violet leaf nuances to this rose aroma. At lower concentrations (0.5%) its contribution is more subtle; however, it still gives an interesting natural effect.

This product may normally be used from approximately 0.01 to 10% in perfume compositions. For special effects, however, higher concentrations (50% plus) can be used.

Example 5: *Tobacco Formulation* – A tobacco mixture is produced by admixing the following ingredients:

Ingredients	Parts by Weight
Bright	40.1
Burley	24.9
Maryland	1.1
Turkish	11.6
Stem (flue-cured)	14.2
Glycerin	2.8
Water	5.3

Cigarettes are prepared from this tobacco. The following flavor formulation is prepared:

Ingredients	Parts by Weight
Ethyl butyrate	0.05
Ethyl valerate	0.05
Maltol	2.00

(continued)

Ingredients	Parts by Weight
Cocoa extract	26.00
Coffee extract	10.00
Ethyl alcohol	20.00
Water	41.90

The abovestated tobacco flavor formulation is applied at the rate of 0.1% to all of the cigarettes produced using the above tobacco formulation. Half of the cigarettes are then treated with 500 or 1,000 ppm of bicyclo[2.2.2] octenemethanol derivative produced according to the process of Example 3.

The control cigarettes not containing the bicyclo[2.2.2] octenemethanol derivatives and the experimental cigarettes which contain the bicyclo[2.2.2] octenemethanol derivatives produced according to the process of Example 3 are evaluated by paired comparison and the results are as follows: The experimental cigarettes are found, on smoking, to have a sweeter, spicy, woody-oriental, Virginia hay tobaccolike taste with much more body and much more natural tobaccolike aroma prior to smoking and on smoking in the mainstream and in the sidestream.

Furthermore, the tobacco of the experimental cigarettes, prior to smoking, has strong, sweet, spicy and woody-hay tobaccolike notes. All cigarettes are evaluated for smoke flavor with a 20 mm cellulose acetate filter.

Example 6: In a vessel associated with a source of heat, a quantity of diced Versamide 930 amounting to 88 pbw is heated to about 130°C at which temperature the resin is a pourable and stirrable body of liquid. A quantity of the perfume formulation of Example 4 having a rose bouquet amounting to 12 pbw is stirred into the liquid resin until a uniform blend is achieved at which time the mixture is poured into standing cold water to facilitate rapid cooling and solidification and minimize loss of perfume formulation.

The product is a clear amber solid having a highly polished surface with a pronounced odor faithfully reproducing the fragrance of the rose perfume formulation used in making it. The material has a rose character with excellent and very strong, sweet, woody and floral notes and clary-sage, violet leaf undertones. The resulting product is in the form of a solid solution which blends itself to molding under heat and pressure into objects of jewelry such as pendant earrings, to casting in molds to form decorative art objects, and to spreading as a film on substrates such as Christmas tree ornaments, glass light bulbs, and the like.

Substituted Dimethyl Dihydroxy Benzenes and Cyclohexadienes

J.B. Hall, M.A. Sprecker, E.J. Shuster, F.L. Schmitt and J.F. Vinals; U.S. Patent 4,115,431; September 19, 1978; assigned to International Flavors & Fragrances Inc. describe certain substituted dimethyl dihydroxy benzene and cyclohexadiene compounds and mixtures thereof which may be used for augmenting or enhancing the taste and/or aroma of perfumes, perfumed articles, tobaccos and/or tobacco flavoring compositions.

These substituted dimethyl dihydroxy benzene and cyclohexadiene compounds are represented by the generic structures:

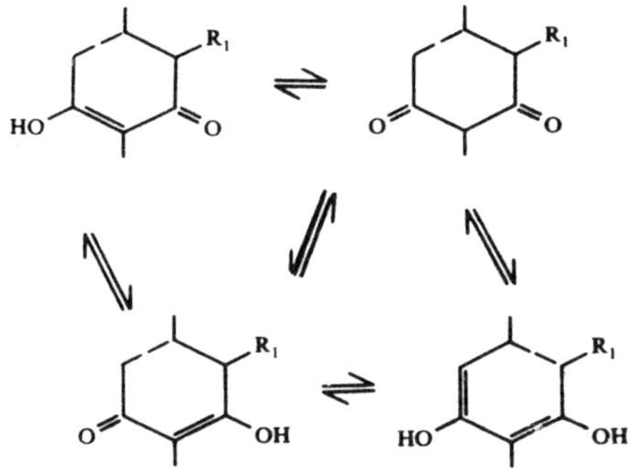

(and other tautomers) wherein the dashed line represents a carbon-carbon single bond or a carbon-carbon double bond and R_1 is either acetyl having the structure –CO– or nitrile having the structure –C≡N with the proviso that when the dashed line is a carbon-carbon single bond, R_1 is nitrile.

The compounds 3,6-dimethyl-2,4-dioxocyclohexanecarbonitrile; 2,4-dihydroxy-3,6-dimethylbenzonitrile; and 2,4-dihydroxy-3,6-dimethyl-1-acetophenone, when combined with other perfume compositions, serve as substitutes for oakmoss in the combination of components which determine the odor note of various perfume compositions.

In addition, each of these chemicals, taken alone or together, not only simulates with great fidelity the characteristic odor note of oakmoss, but are less expensive than natural oakmoss or its prior art synthetic simulations. Each of the compounds is prepared from readily available materials in commercial syntheses.

Example 1: *Preparation of 2,4-Dihydroxy-3,6-Dimethylacetophenone* – Into a 12 liter reaction flask equipped with stirrer, thermometer, and reflux condenser are added 300 g of potassium hydroxide dissolved in 3 liters of methanol. To the methanol/potassium hydroxide mixture is added 3 mols (588 g) of 2,4-dihydroxy-3,6-dimethyl-carbomethoxybenzene (Veramoss). The reaction mixture is heated to 68°C and refluxing is commenced. At this point, an additional amount of KOH solution is added over a period of 45 minutes while maintaining the reaction temperature at 68° to 75°C.

The reaction mass continues to be heated and refluxed for a period of 4 hours, at the erd of which time 3,600 ml of 12% hydrochloric acid solution is added over a period of 30 minutes. At the end of the addition of the hydrochloric acid the methanol is removed and the reaction mass is allowed to cool down, yielding crystals of β-orcinol (20 g/0.143 mol).

This compound is then placed in a 250 ml reaction flask equipped with gas inlet tube and gas outlet tube and cooling bath. Acetonitrile (12 g/0.28 mol), zinc chloride (4 g) and anhydrous diethyl ether (100 ml) are added to the reaction mass and hydrogen chloride gas is passed through the reaction mass for a period of 2 hours at a temperature of 0°C. At the end of the 2 hour period, the hydrogen chloride gas addition is complete and the reaction mass is permitted to warm up to room temperature with stirring. 22 g of crystals are filtered and dissolved in 100 ml water in a 250 ml flask equipped with stirrer and reflux condenser. The reaction mass is heated to mild reflux for a period of 1.2 hours and then cooled to room temperature. The resulting crystals are filtered. NMR, IR and Mass Spectral analyses yield the information that the structure of the resulting material is 2,4-dihydroxy-3,6-dimethylacetophenone, having the formula:

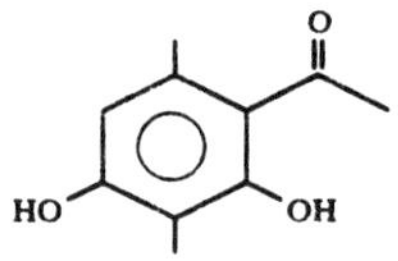

HO OH O

Example 2: A tobacco mixture is prepared by the usual International Flavors & Fragrances testing formulation, and cigarettes are made up. The usual flavor formulation was prepared and applied at the rate of 0.1% to all of the cigarettes produced. Half of the cigarettes are then treated with 500 or 1,000 ppm of 2,4-dihydroxy-3,6-dimethylacetophenone.

The control cigarettes not containing the 2,4-dihydroxy-3,6-dimethylacetophenone and the experimental cigarettes which contain it are evaluated by paired comparison and the results are as follows: The experimental cigarettes are found to have a sweet, slightly smokey, slightly woody, oakmosslike aroma prior to smoking and a woody, oakmosslike aroma on smoking in both the mainstream and in the sidestream. The tobacco of the experimental cigarettes, compared with the control cigarettes, is more aromatic and more tobaccolike. In addition, the control cigarettes have none of the interesting and pleasant oakmoss aroma in the mainstream or in the sidestream on smoking.

Examples 3 and 4: 2,4-dihydroxy-3,6-dimethylacetophenone, as produced in Example 1, was used to replace, partially or completely, the oakmoss extract used in a number of Chypre and Fougere perfume compositions. It was found that by adding any quantity of the 2,4-dihydroxy-3,6-dimethylacetophenone, the odor quality of natural oakmoss may be achieved with great fidelity but with great reduction in cost.

2-Butyl-3,5,5-Trimethyl-2-Cyclohexen-1,4-Dione

K.K. Light, B.M. Spencer, J.F. Vinals, J. Kiwala and E.J. Shuster; U.S. Patent 4,126,641; November 21, 1978; assigned to International Flavors & Fragrances Inc. have found that perfume compositions, colognes, and perfumed articles having deep, slightly woody, herbaceous aromas, and tobacco and tobacco flavor compositions having pungent, sweet, hay-, tea-, tobaccolike, floral and fruity notes, both prior to and on smoking in both the mainstream and sidestream, may be provided by the utilization in such compositions of 2-butyl-3,5,5-trimethyl-2-cyclohexen-1,4-dione (Compound Y) having the structure:

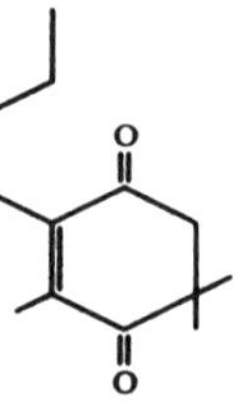

O O

The following examples will illustrate the preparation and uses of Compound Y. All parts and percentages are by weight unless otherwise specified.

Example 1: *Preparation of Compound Y* – To a 2 liter flask fitted with a reflux condenser, a water separator, and a stirrer is charged 265 g of 2,6,6-trimethylcyclohex-2-en-1,4-dione [prepared by the method described in *Helv. Chim. Acta* 39, 2041 (1956)], 2,669 g of propylene glycol, 500 ml of benzene, and 1 g of p-toluene-sulfonic acid. The reactants are stirred and refluxed until no additional water is collected in the water separator. The product is washed with water to a pH of 6, and the solvent is stripped off. The residue is vacuum distilled through a 9" Goodloe column to yield 257 g of product, 2,7,9,9-tetramethyl-1,4-dioxaspiro[4.5]-dec-6-en-8-one, boiling point 92°C at 1 mm Hg (80% yield).

A 1 liter flask fitted with stirrer, thermometer, nitrogen inlet tube, reflux condenser and dropping funnel is charged with 120 g (0.3 mol) of n-butyltriphenylphosphonium bromide and 500 ml benzene. A solution of 2.3 M n-butyllithium in hexane (126 ml, 0.29 mol) is added dropwise over 30 minutes. The temperature rises from 20° to 42°C, and the solution turns dark red. The mixture is stirred for 1 hour at ambient temperature, followed by the addition of 42 g (0.2 mol) of the just prepared 2,7,9,9-tetramethyl-1,4-dioxaspiro[4.5] dec-6-en-8-one over a 20 minute period. A slight exotherm is noted, and the color of the solution changes from red to brown.

After stirring for an additional 6 hours at ambient temperature, the inorganic salts are removed by washing with water. The product is then stirred at room temperature with 200 ml of 1% hydrochloric acid to deketalize the product. The progress of the deketalization is monitored by GLC (10' x ⅛" SE 30 column, programmed from 80° to 220°C at 8°/min). When no further change is noted (36 hours at room temperature), the organic layer is separated, and the solvent is stripped off. The residue is distilled, and 29.4 g of material (boiling point 80° to 108°C at 0.6 mm Hg) is obtained. This material is redistilled on a spinning band column yielding 18 g of 2-butyl-3,5,5-trimethyl-2-cyclohexen-1,4-dione (Compound Y) (boiling point 88°C at 0.5 mm Hg) with a purity of greater than 90% (39% yield).

Example 2: Compound Y, prepared according to the process of Example 1, is incorporated in a cologne at a concentration of 2.5% in 85% aqueous ethanol; and into a handkerchief perfume at a concentration of 20% (in 95% aqueous ethanol). A distinct and definite deep, slightly woody, herbaceous aroma is imparted to the cologne and to the handkerchief perfume.

Example 3: A tobacco blend is made up and flavored according to the formulations given in Example 7 of U.S. Patent 4,076,854 and made up into cigarettes. One-third of these model cigarettes are treated in the tobacco section with Compound Y produced according to Example 1, at 100 ppm per cigarette. Another one-third of these model cigarettes are treated in the filter with Compound Y at the rate of 2×10^{-5} and 3×10^{-5} g.

When evaluated by paired comparison, the cigarettes treated both in the tobacco and in the filter with Compound Y are found, in smoke flavor, to be more tobaccolike with enhanced Virginia tobaccolike notes.

1-(2,6,6-Trimethyl-1,3-Cyclohexadien-1-yl)-1,3-Butanedione

In two closely related patents, one by *R.A. Wilson, B.D. Mookherjee and W.I. Taylor; U.S. Patent 4,157,350; and the other by B.D. Mookherjee, R.A. Wilson, F.L. Schmitt, J.F. Vinals and J. Kiwala; U.S. Patent 4,157,351; both granted June 5, 1979 and both assigned to International Flavors & Fragrances Inc.,* there are described two compounds–1-(2,6,6-trimethyl-1,3-cyclohexadien-1-yl)-1,3-butanedione (Compound X) and its enolate–and the process for making these compounds and compositions using them for augmenting, altering, modifying and enhancing the flavors and/or aromas of tobacco and perfumed articles.

Compound X is prepared by reacting 1-acetyl-2,6,6-trimethyl-1,3-cyclohexadiene with a methylmagnesium halide; reacting the resulting organometallic compound with an acetyl halide, and hydrolyzing the product with a dilute acid.

The reaction of the methylmagnesium halide with the 1-acetyl-2,6,6-trimethyl-1,3-cyclohexadiene takes place at a temperature preferably from 0° to 5°C at atmospheric pressure in the presence of an inert anhydrous solvent such as tetrahydrofuran or diethyl ether. The reaction of the first organometallic compound with the acetyl halide takes place at a temperature in the range of from 0° to 30°C, preferably at about 20°C. It is preferred that the acetyl halide be in slight molar excess with respect to the first organometallic compound.

The hydrolysis of the second organometallic compound preferably takes place using aqueous acetic acid (in excess) preferably at about 10°C at atmospheric pressure.

Following the hydrolysis, which produces Compound X in equilibrium with its enolate, the reaction mass is worked up by standard extraction and distillation procedures.

The reaction sequence may be illustrated as follows:

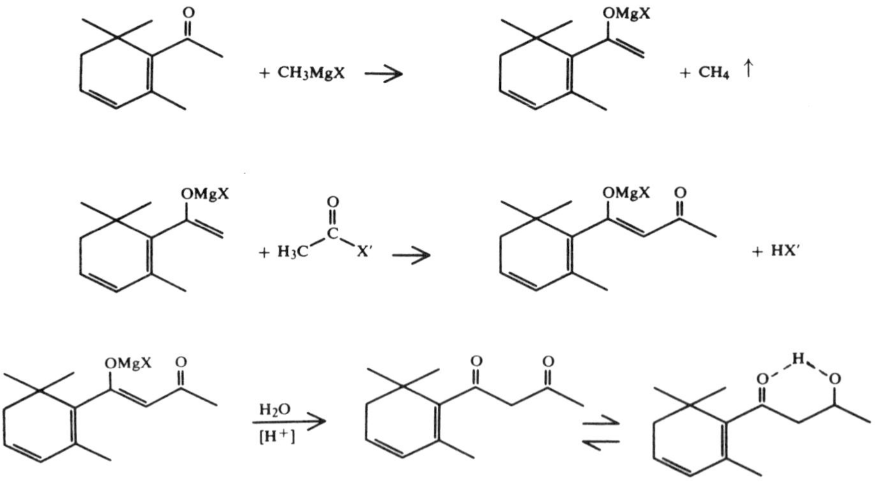

O + CH_3MgX → OMgX + CH_4 ↑

OMgX + H_3C–C(=O)–X′ → OMgX O + HX′

OMgX O $\xrightarrow[\text{[H}^+]]{H_2O}$ O O ⇌ O H O

wherein X is halogen selected from the group consisting of chloro, bromo, and iodo, and X' is halogen selected from the group consisting of chloro and bromo.

The preparation of Compound X and its inclusion in various tobacco and perfumed compositions are illustrated by the examples, in which all parts and percentages are by weight, unless otherwise specified.

Example 1: *Preparation of Compound X* – Into a 250 ml three-necked, round-bottom flask equipped with a mechanical stirrer, an immersion thermometer, a nitrogen purge, a dry ice/isopropanol cooling bath, a 125 ml addition funnel and a water-cooled condenser with gas bubbler, is placed 42.7 ml of a tetrahydrofuran solution of methylmagnesium chloride (3 M). The methylmagnesium chloride/THF solution is chilled to 0° to 5°C with stirring, and 20 g (0.122 mol) of 1-acetyl-2,6,6-trimethyl-1,3-cyclohexadiene is added dropwise over a one minute period.

Following completion of the addition, the reaction mass is stirred at about 20°C until no further methane is observed to evolve (approximately 30 minutes). The reaction mass is then chilled to 0°C and 10.14 g (0.13 mol) of freshly distilled acetyl chloride is added dropwise over a 20 minute period. Following the addition of the acetyl chloride, the reaction mass is stirred at a temperature of 20°C for 15 minutes and then once more chilled to 10°C, and a solution of 8.1 g of acetic acid in 10 ml water is added dropwise during a 5 minute period.

Following the addition of the acid, water is added to dissolve salts, and the reaction mass is transferred to a separatory funnel. Diethyl ether is added, and the ether layer is separated and washed twice with saturated sodium bicarbonate solution and then twice with saturated sodium chloride solution; dried and concentrated in a vacuum. The concentrate is then analyzed using GLC (conditions: 20' x ¼" 5% SE-30 column). NMR, IR and mass spectral data confirm that the last eluting major peak is 1-(2,6,6-trimethyl-1,3-cyclohexadien-1-yl)-1,3-butanedione (Compound X). The residue is then submitted to fractional distillation after adding thereto 50 g Primol, and fraction 3 is shown to contain 95.4% Compound X.

Example 2: *Rose Formulation* – To demonstrate the use of Compound X in a rose formulation, the following formula is provided:

Ingredients	Parts by Weight
Phenylethyl alcohol	200
Geraniol	400
Trichloromethylphenyl carbinyl acetate	20
Phenylethyl acetate	60
Undecylenic aldehyde (10% in diethyl phthalate)	5
n-Nonyl aldehyde (10% in diethyl phthalate)	2
Musk ketone	10
Musk ambrette	10
Eugenol phenyl acetate	20
Citronellol	100
Vanillin (10% in diethyl phthalate)	6
Eugenol	30
Citronellyl formate	30
Geranyl acetate	10

(continued)

Ingredients	Parts by Weight
Linalool	40
Geranyl phenyl acetate	50
cis-β,γ-hexenyl acetate	2
Compound X	5

The addition of 0.5% of Compound X lends a great deal of strength and character to the rose fragrance. It contributes great floralcy and the heady natural sweetness of the red rose flower together with honeylike, fruity, grapelike, and haylike nuances and sea-amber undertones with sweet-floral topnotes. At lower concentrations (0.01%) its contribution is more subtle; however, it still gives an interesting natural effect with the ambergrislike undertone.

This product may normally be used from approximately 0.01% to 10% in perfume compositions. For special effects, however, higher concentrations (50% plus) can be used.

Example 3: *Tobacco Formulation* – A tobacco mixture is produced by admixing the following ingredients:

Ingredients	Parts by Weight
Bright	40.1
Burley	24.9
Maryland	1.1
Turkish	11.6
Stem (flue-cured)	14.2
Glycerin	2.8
Water	5.3

Cigarettes are prepared from this tobacco.

The following flavor formulation is prepared:

Ingredients	Parts by Weight
Ethyl butyrate	0.05
Ethyl valerate	0.05
Maltol	2.00
Cocoa extract	26.00
Coffee extract	10.00
Ethyl alcohol	20.00
Water	41.90

The abovestated tobacco flavor formulation is applied at a rate of 0.1% to all of the cigarettes produced using the above tobacco formulation. Half of the cigarettes are then treated with 50 or 150 ppm of Compound X produced according to the process of Example 1. The control cigarettes not containing Compound X and the experimental cigarettes which contain Compound X are evaluated by paired comparison, and the results are as follows.

The experimental cigarettes are found, on smoking, to have improved body, enhanced tobacco character, to be sweeter, and to have more pronounced hay/tea character in the mainstream. The experimental cigarettes are also found, on smoking, to be sweeter, more tobaccolike, having a fresh cigarette pack aroma in the sidestream.

The tobacco of the experimental cigarettes, prior to smoking, has sweet, fruity, berrylike, fresh squeezed fruit (apple and grape) juicelike, haylike and tealike aroma characteristics.

All cigarettes are evaluated for smoke flavor with a 20 mm cellulose acetate filter.

Example 4: *Preparation of a Detergent Composition* – A total of 100 g of a detergent powder is intimately admixed with 0.15 g of Compound X until a substantially homogeneous composition is obtained. This composition has an excellent sweet, honeylike, fruity, grapelike, roselike, iononelike, haylike and floral aroma with sea-amber/ambergrislike undertones and sweet-floral topnotes.

FRAGRANCES—WOODY

WOODY–GENERAL

Substituted 1-Alkenynyl-Cyclohexanols

M. Baumann and W. Hoffmann; U.S. Patent 4,088,681; May 9, 1978; assigned to BASF Aktiengesellschaft, Germany have obtained the 1-alkenynyl-cyclohexanols of the general formula:

(1)

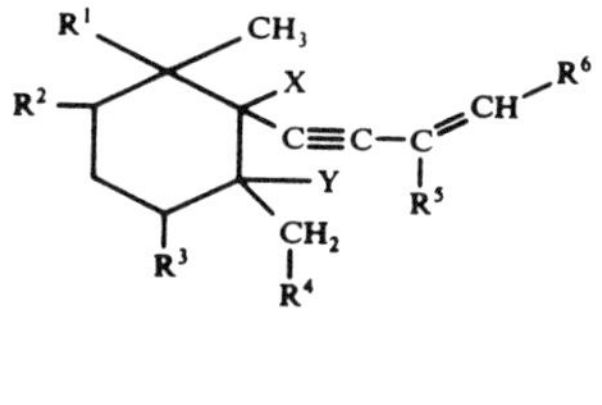

where R^1 to R^4 are $-H$ or $-CH_3$; R^5 and R^6 are $-H$ or a saturated or olefinically unsaturated aliphatic hydrocarbon radical of 1 to 6 carbon atoms, which is not substituted by hetero-atoms, and are preferably $-H$, $-CH_3$ or $-C_2H_5$, or alternatively R^5 and R^6 together are optionally alkyl-substituted alkylene of 3 to 7 carbon atoms and X is $-OH$ if Y is $-H$ or X and Y together are an additional bond between the carbon atoms on which they are present.

These compounds of Formula (1) have a fresh, woody and tart coniferous odor and can be used as constituents of scent compositions, or to improve the odor of industrial products. Furthermore, they offer a new and economical method of obtaining the β-damascones, which are popular scents, since they can be converted simply, by heating in the presence of acids, to the corresponding β-damascones.

In the process for the manufacture of these substituted 1-alkenynyl-cyclohexanols of Formula (1), cyclohexanones of the general Formula (2):

(2)

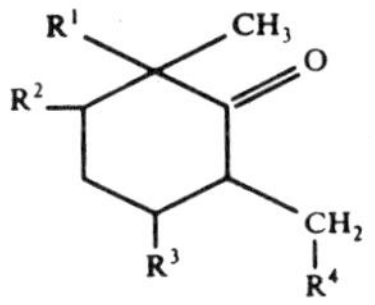

$$\text{cyclohexanone with } C(R^1)(CH_3) \text{ adjacent to } C=O,\ R^2,\ R^3,\ CH_2R^4$$

where R^1 to R^4 have the previously described meanings, are reacted with an enyne of the general Formula (3):

(3) $$H{-}C{\equiv}C{-}\underset{\displaystyle R^5}{\underset{|}{C}}{=}CH{-}R^6$$

where R^5 and R^6 have the above meanings, in the presence of a strongly basic condensing agent, in an inert solvent, or are reacted with an alkali metal salt of an enyne of the Formula (3) in an inert solvent.

The cyclohexanones of Formula (2) required as starting materials for the process are conventional compounds which may be manufactured by alkylating cyclohexanones, or hydrogenating cyclohexanones which in turn may be manufactured from aliphatic ketones and α,β-unsaturated carbonyl compounds.

Enynes of Formula (3) which are preferable for use in the process are vinylacetylene and methylbutenyne.

The enynes of Formula (3) need not be employed as such. Instead of using the enynes in the presence of a strongly basic condensing agent, the active salts of the enyne may be used directly. This is very advisable, for example, when using the unsubstituted enyne of the Formula (3), since vinylacetylene itself is difficult to handle. In that case sodium vinylacetylide, for example, is used; it can be obtained simply, by the action of sodium on 1,4-dichloro-2-butene in liquid ammonia.

Example 1: A solution of CH_3MgCl in tetrahydrofuran (THF) is prepared by passing methyl chloride into a suspension of 19 g of Mg filings in 500 ml of (THF). 50 g of methylbutenyne are added dropwise thereto in the course of 50 minutes at 0°C. The reaction mixture is then stirred until the evolution of gas has ceased. 105 g of 2,3,6-trimethylcyclohexanone are then added dropwise at room temperature and while cooling, and the reaction mixture is left overnight, while being stirred, in order to complete the reaction.

120 ml of water are then added dropwise and the organic phase is decanted, concentrated and subjected to fractional distillation. 93.5 g (61% of theory) of 2,3,6-trimethyl-1-(3'-methyl-but-1'-yn-3'-en-1'-yl)-cyclohexanol of boiling point 72°C/0.1 mm Hg are obtained. The spectroscopic data confirm the structure.

Fragrance note: fresh, herbaceous.

Example 2: 28 g of 2,2,6-trimethyl-cyclohexanone are added dropwise to a suspension of sodium vinylacetylide (prepared from 25 g of 1,4-dichlorobut-2-ene and 15 g of Na in liquid ammonia) in 200 ml of THF, while cooling, and the reaction mixture is stirred overnight at room temperature. 50 ml of water are then added, the aqueous phase is extracted with ether and the resulting organic phase is dried and concentrated. Subsequent fractional distillation gives 5.5 g of unconverted trimethylcylohexanone and 20.7 g (corresponding to 68% of theory, based on ketone converted) of 2,2,6-trimethyl-1-(but-1'-yn-3'-en-1'-yl)-cyclohexanol of boiling point 62° to 63°C/0.3 mm Hg and refractive index n_D^{25} 1.4982. Fragrance note: fresh, tart-woody, coniferous.

Polycyclic Alcohols

J.J. Bloomfield and D.C. Owsley; U.S. Patent 4,119,575; October 10, 1978; assigned to Monsanto Company have found that a specific class of compounds having characteristic aromas which are useful in the preparation of fragrances are represented by the structural formulas:

(1)

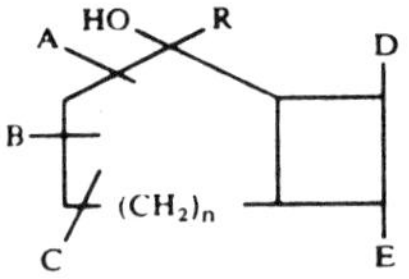

(2)

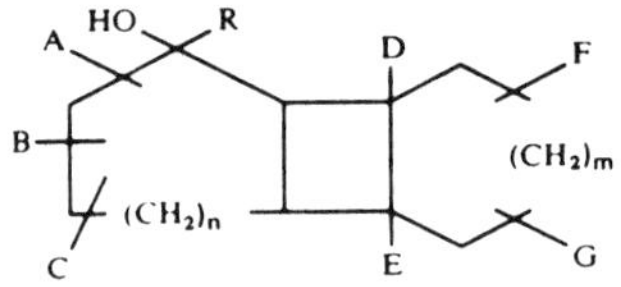

(3)

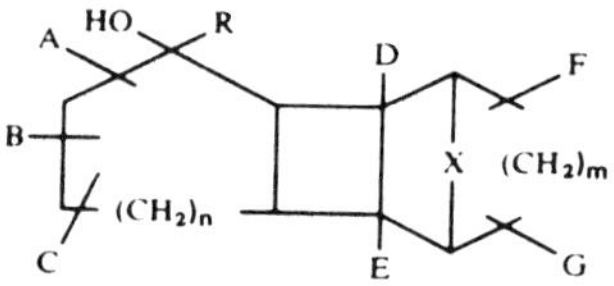

wherein n is an integer 0 or 1; A, B and C each independently represent hydrogen or alkyl having from 1 to 3 carbon atoms, provided that when n is 0 at least one of A, B or C cannot be hydrogen; R represents hydrogen or alkyl having from 1 to 6 carbon atoms; D and E each independently represent hydrogen or alkyl having from 1 to 6 carbon atoms, provided that the sum of the carbon atoms in D and E does not exceed 6, provided that, in the bicyclo compounds, at least one of A, B, C, D or E must be an alkyl; m is an integer 1 through 8; F and G represent hydrogen or alkyl having from 1 to 3 carbon atoms; X represents $I-(CH_2)_p-J$ wherein p is an integer 0 through 2 and I and J each independently represent hydrogen or methyl, provided that if p is 0 then m must be greater than 2; provided that the sum of the carbon and oxygen atoms in the compound is no greater than 23.

Preferred embodiments are those compounds represented by the following structural formulas:

(4)

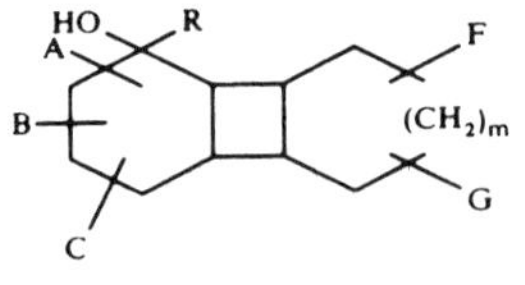

(5)

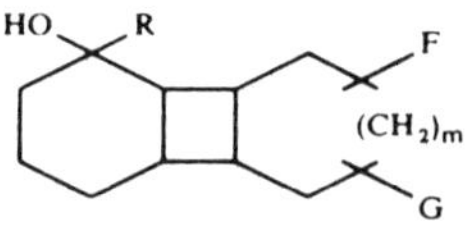

(6)

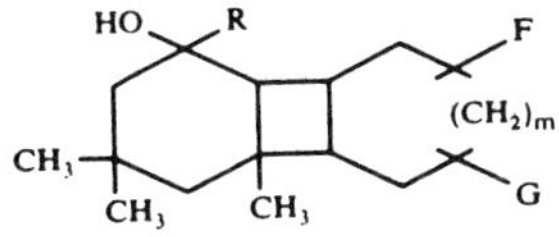

wherein A, B, C, F, G, m and R have the same meanings as set forth above. The compounds of Formula (6) are particularly preferred compounds.

The following examples will illustrate the preparation of two of these compounds.

Example 1: *4,4,6-Trimethylbicyclo[4.2.0] Octane-2-ol* – To a 2-liter 3-necked flask fitted with a mechanical stirrer, addition funnel, reflux condenser and drying tube was added 15 g (0.39 mol) of sodium borohydride and 500 ml 2-propanol. To this solution was slowly added 218.6 g (1.317 mols) of 4,4,6-trimethylbicyclo[4.2.0] octane-2-one in 50 ml of 2-propanol. After 2 hours, about 200 ml of water was added. The mixture was stirred an additional 1 hour, then about 35 ml concentrated hydrochloric acid was added. After another 1 hour of stirring, 400 ml of toluene and 500 ml of water were added.

The layers were separated and the organic layer was washed three times with 100 ml of water. The combined water layers were extracted with toluene. The toluene layers were combined and dried over potassium carbonate. The solution was concentrated in vacuo and the residue distilled through a 25 cm Vigreux column to yield 215.8 g (1.284 mols, 97.5% yield) of 4,4,6-trimethylbicyclo-[4.2.0] octane-2-ol. IR: 3360 cm^{-1}. The aroma of the compound had camphor, minty, patchouly and woody notes.

Example 2: *5,5,7-Trimethyltricyclo[6.4.0.$0^{2,7}$] Dodecane-3-ol* – To 7.0 g (0.19 mol) of sodium borohydride in 400 ml of isopropyl alcohol in a 1 liter round bottomed flask equipped with magnetic stirrer, reflux condenser and drying tube is added 140 g (0.636 mol) of 5,5,7-trimethyltricyclo[6.4.0.$0^{2,7}$] dodecane-3-one in 100 ml of isopropyl alcohol. The mixture is stirred for about 15 hours and then 300 ml of a saturated aqueous sodium chloride solution containing 70 ml of concentrated hydrochloric acid is cautiously added. The entire mixture is transferred to a 1-liter separatory funnel and the layers are separated.

The aqueous layer is extracted three times with 100 ml of petroleum ether. The organic layers are combined, extracted three times with 100 ml of water, dried over magnesium sulfate and concentrated in vacuo to yield the product as a viscous oil. 127.5 g of 5,5,7-trimethyltricyclo[6.4.0.$0^{2,7}$] dodecane-3-ol was recovered (0.572 mol, 90% yield). IR: 3380 cm^{-1}, BP 104° to 110°C/0.2 mm. This compound had a sandalwood aroma.

Other compounds of interest are tricyclo[8.4.0.$0^{2,9}$] tetradecane-11-ol; 2,4,4,6-tetramethylbicyclo[4.2.0] octane-2-ol; and 3,5,5,7-tetramethyltricyclo[6.4.0.$0^{2,7}$]-dodecane-3-ol.

Cyclohexene-3-Nitriles

R.S. DeSimone; U.S. Patent 4,146,507; March 27, 1979; assigned to Polak's

Frutal Works, Inc. has found that the following compounds are useful in perfume compositions: 3,5-dimethyl-3-cyclohexene nitrile, 2,4-dimethyl-3-cyclohexene nitrile, mixtures of the 2,4- and 3,5-dimethyl nitriles, 2,4,6-trimethyl-3-cyclohexene nitrile, 3,5,6-trimethyl-3-cyclohexene nitrile and mixtures of the 2,4,6-trimethyl and 3,5,6-trimethyl nitriles.

The compounds are readily synthesized via the well-known Diels-Alder reaction using 2-methyl-1,3-pentadiene and either acrylonitrile or crotonitrile via the following general reaction:

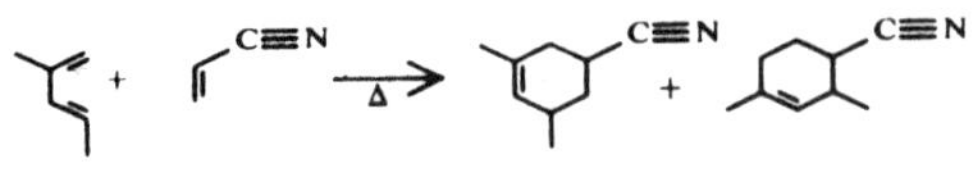

Use of crotonitrile in place of acrylonitrile produces the mixture of 2,4,6- and 3,5,6-trimethyl-3-cyclohexene nitriles.

These nitriles can be employed as mixtures in the isomer proportions obtained from the reactions noted above or the isomer mixtures can be fractionated to recover individual isomers which can be used as such or blended with each other in other proportions. The isomers or blends thereof can be employed as perfumes per se in a suitable carrier or they can be used in combination with other ingredients in perfumes having woody, cinnamic notes. The amount present in a perfume can be about 0.01 to about 6%, preferably about 0.1 to 4% of either an individual nitrile or of an isomer mixture based on the weight of the perfume composition.

Compared to many compounds, both natural and synthetic, these nitriles possess a relatively high degree of stability to acid or basic conditions, as well as to oxidative and thermal effects. This stability lends special utility in fragranced products where the aroma of a perfume not containing stable components would not maintain its integrity in such bases as detergents, cleaners, soaps, and personal care products. In these product bases, many fragrance compositions designed specifically for colognes or fine perfumes would not exhibit satisfactory odor integrity either prior to or during use.

Example 1: *Preparation of Mixed 3,5-Dimethyl- and 2,4-Dimethyl-3-Cyclohexene Nitriles* – A one-gallon stainless steel autoclave equipped with a steam jacket and magnetically driven turbine-type stirrer was charged with 492 g of 50% purity 2-methyl-1,3-pentadiene containing about 25% of 2-methyl-2,4-pentadiene and approximately 25% of mixed C_6 monoolefinic alcohols and C_8 cyclic ethers. [See S.A. Ballard et al., *J. Amer. Chem. Soc.*, Vol. 72, 5734 (1950) in which the synthesis of 2-methyl-1,3-pentadiene with its by-products is described.]

At the same time was charged 212 g of acrylonitrile. The autoclave was sealed and the contents maintained between 90° and 123°C with stirring for 3 hours and then cooled. The mixture was withdrawn from the autoclave (693 g) and charged to a one-liter still flask. Rectification was performed on a 23-mm by 4-foot spinning band distillation column to yield 19 fractions.

The odor of combined fractions 8 to 18 has a strong, green, cuminic note with a herbal, cinnamic, woody background. The dry-out odor on a blotter after 24 hours is strong, warm, woody.

The nitriles of the example can be used in woody perfume compositions, such as sandalwood, patchouli, vetivert, oakmoss, cedarwood, etc., and its primary effect is as a base modifier in such perfume oils. They can also be effective when blended with florals, such as ylang, jasmin, tuberose, muguet and rose. They can also be used to modify top notes, particularly in citrus or herbal citrus compositions.

Example 2: *Green, Woody, Spicy Perfume* – The following perfume composition incorporates the mixed 3,5-dimethyl- and 2,4-dimethyl-3-cyclohexene nitriles:

Component	Parts by Weight
Base Notes	
Cedarwood oil	200
1,1,2,3,3,6-Hexamethyl-5-acetyl indan	35
1,1,2,4,4,7-Hexamethyl-6-acetyl Tetralin	10
3,5-Dimethyl and 2,4-dimethyl-3-cyclohexene nitriles*	30
Isobornyl acetate	160
Hexylcinnamic aldehyde	60
Styralyl acetate	70
Alpha-terpineol	70
Top Notes	
2-Ethylpyridine	1
CP formate (IFF register)	5
1,4-Cineole	25
1,8-Cineole	25
Eugenol	20
Linalool	40
Linalyl acetate	40
Methyl hexyl ketone	4
Lavandin oil	40
Spearmint oil	10
Modifiers	
Lemon oil messina	55
Phenylethyl alcohol	50
Hydroxycitronellal	50

*Blend of fractions 6 to 19 of Example 1

Example 3: *Preparation of 2,4,6-Trimethyl and 3,5,6-Trimethyl-3-Cyclohexene Nitriles* – An autoclave as described in Example 1 was charged with 368 g of 50% 2-methyl-1,3-pentadiene, as in Example 1, and 201 g of crotonitrile. The mixture was heated with stirring at about 115°C for a total of 25 hours. After cooling, 522 g of crude reaction mixture were recovered from the autoclave. Rectification was performed on a one-inch diameter by one-foot, 7-plate Goodloe column, and 11 fractions were obtained.

Fractions 6 through 9, when combined, had a warm, spicy, fresh character with a soft, green background. The dry-out after 24 hours is mild, woody, cinnamic. The product of this example can be used in spicy fragrance compositions, such as cinnamon, ylang, lilac, carnation and jasmin. It can also be effectively blended with the balsamic resin group, as well as the more woody class of materials, such as sandalwood, vetivert and patchouli oils.

Example 4: *Woody, Balsamic, Citrus Fragrance –*

Component	Parts by Weight
Base Notes	
Coumarin	150
2,4,6-Trimethyl and 3,5,6-trimethyl-3-cyclohexene nitriles*	20
Heliotropin	30
Musk ambrette	10
Myrrh resin	10
Benzoin, 50% diethyl phthalate	40
Styrax resin	20
Ethyl vanillin	25
Amyris oil	40
Top Notes	
Citral dimethyl acetal	70
Tetrahydrolinalool	90
Cuminyl alcohol	10
Dibenzyl ether	200
Linalool	70
Modifiers	
Benzyl acetate	7.5
Hexylcinnamic aldehyde	7.5
Orange oil	85
Lemon oil	85
Beta-methyl naphthyl ketone	30

*Fractions 6 through 9 of Example 3

VETIVER

Tricyclo Ketones

G.H. Büchi and A. Hauser; U.S. Patent 4,124,642; November 7, 1978; assigned to Firmenich, SA, Switzerland describe procedures for the synthesis of two compounds useful in perfumery:

(1) 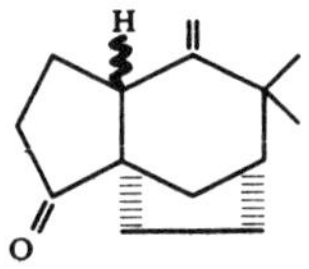 (2)

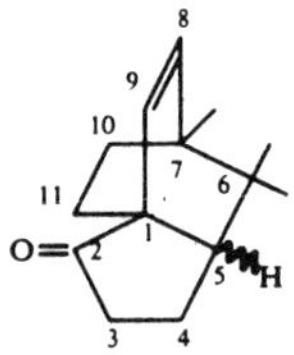

Compound (1), 7,7-dimethyl-6-methylene-tricyclo[6.2.1.$0^{1,5}$]undecan-2-one, is khusimone which is known to be mainly responsible for the characteristic scent of vetiver oil. Its synthesis from zizanoic acid was previously reported, but the process is not practical because zizanoic acid is a rare, expensive product of nature.

Compound (2), 6,6,7-trimethyl-tricyclo[5.2.2.$0^{1,5}$]undec-8-en-2-one, is the starting material for the described synthesis of khusimone and also has useful odoriferous properties of its own. The scent is typically woody and is reminiscent of oriental sandalwood with certain nuances of vetiver oil.

Both of these compounds have two epimeric forms but are useful as racemic mixtures of the epimers.

Compound (2) is synthesized through promotion of an internal Diels-Alder reaction on a ketal of Formula (3) by subjecting it to a thermal treatment, then hydrolyzing the obtained reaction mixture.

(3)

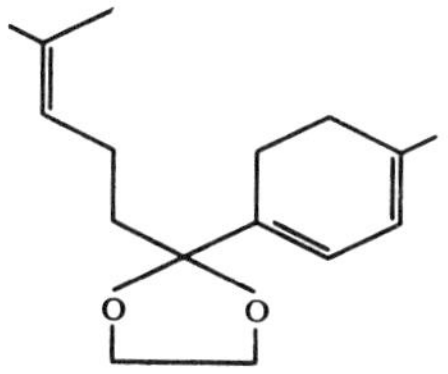

A process is also described for making the ketal (3) using available starting materials, i.e., 2-chloroacrylonitrile and isoprene.

Compound (1) is synthesized by (a) isomerizing 6,6,7-trimethyl-tricyclo[5.2.2.$0^{1,5}$] undec-8-en-2-one by means of an acidic agent to give 6,7,7-trimethyl-tricyclo-[6.2.1.$0^{1,5}$] undec-5-en-2-one; (b) reacting the compound thus obtained with singlet oxygen followed by workup with a reducing agent to yield a tricyclo carbinol of formula:

(4)

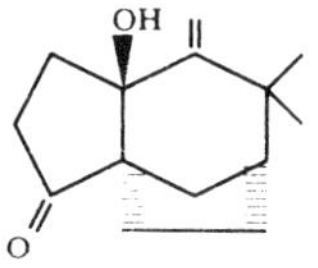

and (c) reducing the tricyclo carbinol (4) by means of a metal in an acidic aqueous medium.

The following example shows the use of Compound (2). A perfume composition for a men's eau-de-toilette was prepared by admixing the following ingredients:

Ingredients	Parts by Weight
Coumarin	80
Musk xylene	70
Synthetic bergamot	100
Lavandin oil	150
Linalyl acetate	100
Coriander	30
Benzyl acetate	50
Amyl salicylate	50
Canaga oil	20
Benzyl salicylate	50
Patchouli	30
Synthetic geranium	100
Clove oil	20
Benzoin resinoid	50
Compound 2	100

The addition of Compound (2) to the composition brings about a character of originality and distinct elegance.

Substituted Acetonaphthones

W.L. Schreiber, J.N. Siano and E.J. Shuster; U.S. Patents 4,076,749; Feb. 28, 1978; 4,107,066; August 15, 1978; 4,108,799; August 22, 1978 and 4,156,695; May 29, 1979; all assigned to International Flavors and Fragrances, Inc. describe a genus of substituted hexahydro acetonaphthones having the generic structure:

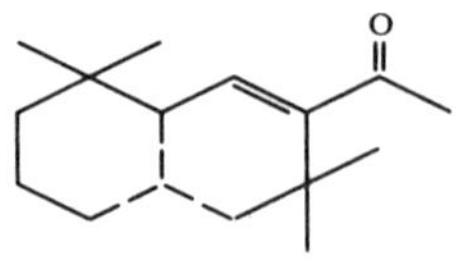

wherein one of the dashed lines is a carbon-carbon double bond and each of the other of the dashed lines is a carbon-carbon single bond. Such compounds are useful as fragrance ingredients for perfumes and perfumed articles including soaps, detergents, cosmetic powders and colognes. More particularly, these compounds have aromas with sweet woody, citrusy, vetiver-like, musky, woody/peppery, woody/leathery, hay and green nuances and/or woody fragrance notes.

The mixtures and isomers of this process, as well as the known individual isomeric component of these mixtures, are produced by first forming 3-chloromesityl oxide from chlorine and mesityl oxide and then reacting the 3-chloromesityl oxide with myrcene via a Diels-Alder reaction thereby forming a chlorine containing Diels-Alder adduct. The chlorine containing Diels-Alder adduct may then either: (1) be dehydrochlorinated to form an acetyl conjugated cyclohexadiene which may be cyclized to form a mixture of compounds represented by the structure:

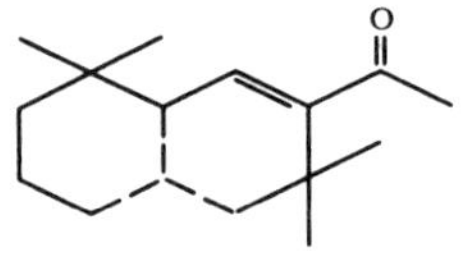

wherein one of the dashed lines is a carbon-carbon double bond and each of the other of the dashed lines is a carbon-carbon single bond; or (2) it may first be cyclized to form an acetyl chlorinated octahydronaphthalene which may then be dehydrochlorinated to form a composition containing a substantial quantity of the known isomer having the structure:

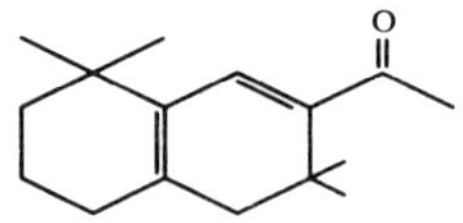

Isomer A

When the mixture of isomers is formed as in method (1), these isomers may, if desired, be separated by means of preparative vapor phase chromatography. As a result of such a separation, the isomer having the structure shown on the following page is produced.

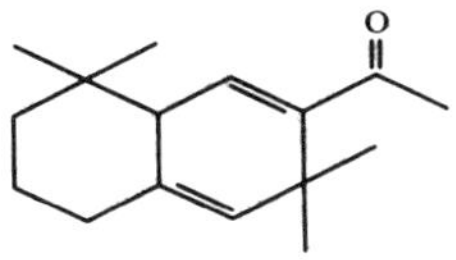

Isomer B

One or more of the hexahydrotetramethyl acetonaphthone derivatives of the process (including Isomer A) and one or more auxiliary perfume ingredients, including, for example, alcohols, aldehydes, ketones, terpenic hydrocarbons, nitriles, esters, lactones, natural essential oils and synthetic essential oils, may be admixed so that the combined odors of the individual components produce a pleasant and desired fragrance, particularly and preferably in vetiver fragrances.

Such perfume compositions usually contain (a) the main note or the "bouquet" or foundation stone of the composition; (b) the modifiers which round off and accompany the main note; (c) fixatives which include odorous substances which lead a particular note to the perfume throughout all stages of evaporation and substances which retard evaporation; and (d) topnotes which are usually low boiling fresh smelling materials.

It has been found that perfume compositions containing as little as 0.01% of hexahydrotetramethyl acetonaphthone derivative(s) or even less (e.g., 0.005%) can be used to impart a vetiver aroma with sweet woody, citrusy, musky, woody/peppery, woody/leathery, hay and green nuances to soaps, cosmetics, detergents, powders and colognes. The amount employed can range up to 70% of the fragrance components and will depend on considerations of cost, nature of the end product, the effect desired on the finished product and the particular fragrance sought.

Example 1: *(A) Preparation of 3-Chloromesityl Oxide –*

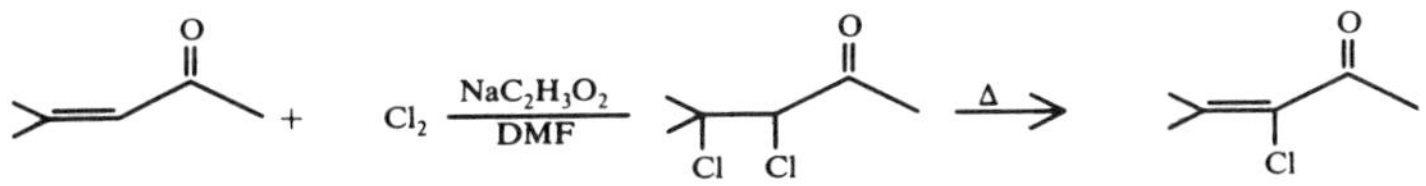

$$+ \; Cl_2 \xrightarrow[DMF]{NaC_2H_3O_2} \quad \xrightarrow{\Delta}$$

Into a 12 liter reaction flask equipped with stirrer, condenser, thermometer and gas addition tube, 1,963 g (20.0 mols) mesityl oxide, 2,050 g (25.0 mols) of sodium acetate and 2,000 ml dimethyl formamide are placed. The contents of the flask are cooled to 0°C. At a rapid rate over a 3½ hour period, while maintaining the reaction mass at 0°C, 888 g (25.0 mols) chlorine is added to the reaction mass. The mixture is then heated to 100°C for a period of 2 hours. Sufficient quantities of water are added to the reaction mass to cause it to separate into two phases. The organic phase is washed with two 2,000 cc portions of water, and then dried over anhydrous magnesium sulfate, filtered and distilled through a 12" x 1½" stone packed column yielding 1,740 g of product, BP 68° to 69°C (45 mm Hg).

(B) Preparation of 3-Chloromesityl Oxide – A 22 liter three necked reaction flask equipped with stirrer, thermometer, cooling bath and gas inlet tube is charged with 4,900 g of mesityl oxide and 2,500 ml dimethylformamide. The mixture is stirred at 0° to 5°C, as 4,438 g of chlorine is added over a 5 hour period. Nitrogen is then passed through the mixture as it is heated to 100°C and held at that temperature for 3 hours. The cooled reaction mass is washed with three volumes of water (methylene chloride is added to aid in separation of the layers). The organic phase is then distilled through a 2' x 1½" column packed with porcelain saddles to give 3,231 g product, BP 70° to 80°C (50 mm Hg).

Example 2: *(A) Diels-Alder Reaction of Myrcene with 3-Chloromesityl Oxide* –

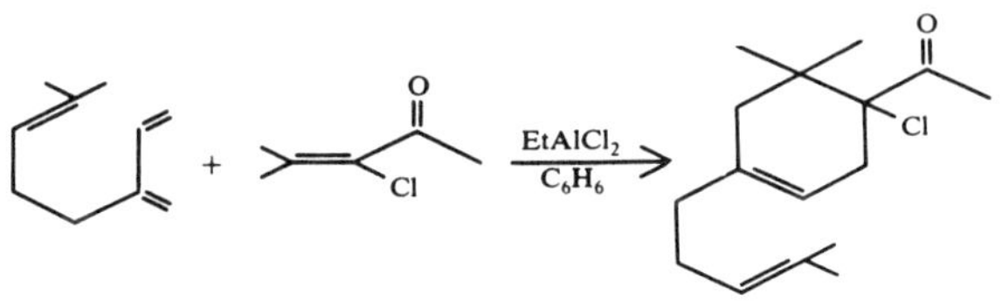

$\xrightarrow[C_6H_6]{EtAlCl_2}$

Into a 500 cc reaction flask equipped with stirrer, thermometer, addition funnel, heating mantle and nitrogen blanket is placed 13 g of 25% ethyl aluminum dichloride in heptane (0.025 mol) and a solution of 35 g (0.25 mol) of 3-chloromesityl oxide in 50 ml anhydrous benzene. While maintaining the resulting mixture at a temperature of 4°C, over a period of 30 minutes, a solution of 53 g of 85% myrcene (0.30 mol) in 50 ml benzene is added. The reaction mass is then allowed to warm to room temperature and then stirred at a temperature of between 25° and 45°C for a period of 5 hours. 250 ml water and 300 ml diethyl ether is then added to the reaction mass.

The organic phase is separated, washed with water, dried and stripped of solvent. The resulting oil is distilled through a micro Vigreux column to give 47.0 g product, BP 112° to 123°C (0.8 mm Hg).

Fraction 7, BP 120° to 123°C (0.8 mm Hg), is analyzed using NMR and infrared analyses, showing the structure to be that of the Diels-Alder adduct shown above.

(B) Dehydrohalogenation of Diels-Alder Adduct –

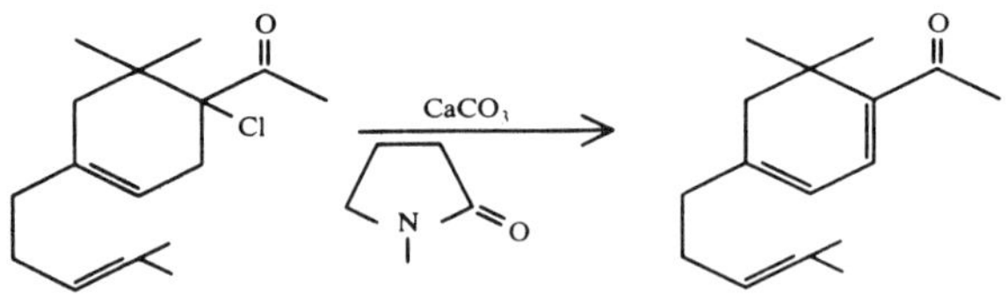

$\xrightarrow{CaCO_3}$

Into a 250 cc reaction flask equipped with stirrer, thermometer, reflux condenser and heating mantle is placed 48 g (0.22 mol) of the Diels-Alder adduct produced according to Example 2(A), 125 ml N-methylpyrrolidinone and 35 g calcium carbonate. The reaction mass is heated and maintained at 150° to 180°C for a period of 1 hour. The reaction mass is then cooled to room temperature, after which time it is filtered and the solids washed with diethyl ether. The solution is washed with water and the aqueous phase is extracted with diethyl ether. The ether extracts are bulked, dried over anhydrous magnesium sulfate, stripped of solvent and distilled through a micro Vigreux column to give 28.0 g product, BP 122° to 128°C (1.0 mm Hg).

NMR and Infrared analyses confirm that the resulting product has the structure shown in the reaction equation.

(C) Cyclization of Acetyl Conjugated Cyclohexadiene Derivative –

H_3PO_4 / Toluene

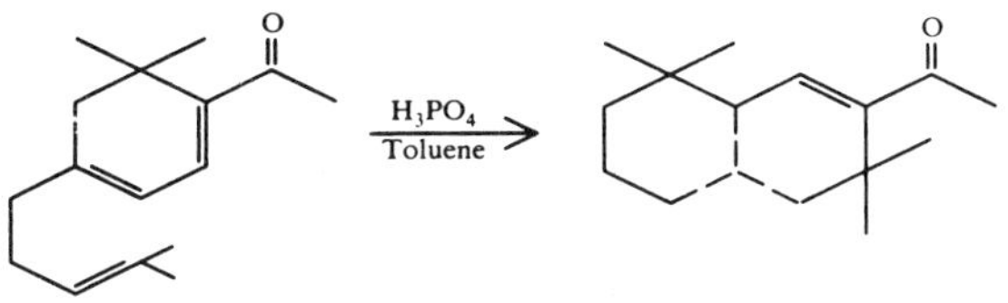

Into a 100 ml reaction flask equipped with stirrer, reflux condenser, thermometer and heating mantle are placed 25 g of the acetyl conjugated cyclohexadiene derivative produced according to Example 2(B), 25 g 85% phosphoric acid and 10 g of toluene.

The reaction mass is stirred at a temperature of 25° to 35°C for a period of 2 hours, after which time it is heated at 50° to 65°C over a period of 1½ hours. At the end of this period of time, the reaction mass is cooled to room temperature and two volumes of water are added. The aqueous phase is separated from the organic phase and then extracted with an equal volume of toluene.

The combined organic layers are washed with water, aqueous sodium bicarbonate and then washed neutral with water. The organic phase is then stripped of solvent (crude weight: 30 g) and distilled on a 4 inch micro Vigreux column yielding 19.0 g material, BP 126° to 145°C (0.8 to 0.9 mm Hg).

Fraction 4 is confirmed by NMR and infrared analyses to be a mixture of compounds having the structure of the product shown in the reaction equation, wherein one of the dashed lines is a carbon-carbon double bond and each of the other of the dashed lines is a carbon-carbon single bond.

The components of fraction 4 are separated using preparative GLC (6' x ¼" Carbowax 20M column). Peak 1 is confirmed by NMR and infrared analyses to have the structure of Isomer A, and peak 2 to have the structure of Isomer B. Peak 1 has a woody, musky note, peak 2, a woody, peppery note, and peak 3, a woody, leathery, "cymene" note. The entire mixture in fraction 4 has a sweet, woody, herbal, vetiver aroma with a citrus, vetivone undertone.

Example 3: *Vetiver Fragrance* – The following mixture is prepared:

Ingredients	Parts by Weight
Hexahydro acetonaphthone derivative*	50

(continued)

Ingredients	Parts by Weight
Cedrol	25
Cedryl acetate	5
Isobutyl quinoline	1
Beta-ionone	2
Caryophyllene	15
Eugenol	2

*Prepared according to Example 2(C)

The hexahydro acetonaphthone derivative prepared according to Example 2(C) imparts the rich, deep, green, woody note of vetiver to this fragrance.

Example 4: *Preparation of a Cologne and Handkerchief Perfume* – The composition of Example 3 is incorporated in a cologne at a concentration of 2.5% in 85% aqueous ethanol; and into a handkerchief perfume at a concentration of 20% (in 95% aqueous ethanol). The use of the hexahydro acetonaphthone derivative in the composition affords a distinct and definite deep, green, musky, grapefruit and woody aroma to the handkerchief perfume and cologne.

Example 5: *Preparation of a Detergent Composition* – A total of 100 g of a detergent powder (lysine salt of n-dodecyl benzene sulfonic acid, as more specifically described in U.S. Patent 3,948,818; April 6, 1976) is mixed with 0.70 g of the composition of Example 3 until a substantially homogeneous composition is obtained. This composition has an excellent vetiver aroma with rich, deep, green and woody notes.

SANDALWOOD

Catechol-Camphene Reaction Products

In perfumery, perhaps the most desired of the popular wood notes, as well as the most expensive and least available, is the sandalwood odor. Oil of sandalwood, therefore, while being highly prized in the perfume world because of its desirable odor, has enjoyed only limited use, on account of its limited availability and high cost.

J.B. Hall and W.J. Wiegers; U.S. Patents 4,104,203; August 1, 1978; 4,131,555 and 4,131,557; December 26, 1978; all assigned to International Flavors & Fragrances Inc. have been able to synthesize mixtures of compounds having an intense sandalwood aroma by reacting catechol with camphene in the presence of a Friedel-Crafts catalyst and hydrogenating the mixture, producing intermediate cyclohexane diols, and continuing the hydrogenation to produce a mixture of compounds having an intense sandalwood aroma. The total reaction can be summarized by the equation:

OH OH + Friedel-Crafts Catalyst → OH OH + OH OH

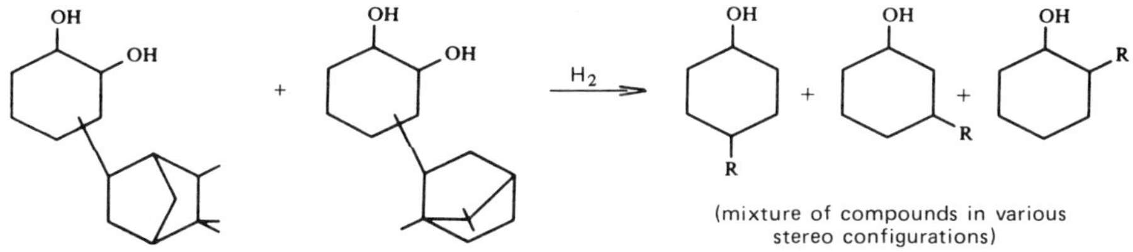

(mixture of compounds in various stereo configurations)

As Friedel-Crafts catalyst, various substances may be used, including boron trifluoride etherate; sulfuric acid; acid clay (e.g. Filtrol); molecular sieve catalysts (SK-500, Union Carbide Co.); aluminum trichloride; boron trifluoride; ferric chloride; and zinc chloride.

The preferred reaction temperature range for the reaction of catechol with camphene is 120° to 175°C. The preferred mol ratio range of catechol to camphene is from 0.8:1 up to about 1.2:1.

The ratio of Friedel-Crafts catalyst:catechol is dependent upon the particular catalyst used. Thus, when using a boron trifluoride-etherate catalyst, the preferred range is from about 8 to about 12 g catalyst per mol of catechol. When using sulfuric acid, the preferred range is from about 0.1 to about 0.5 g of sulfuric acid per mol of catechol (based upon 100% sulfuric acid); etc.

It is preferred to carry out the reaction at atmospheric pressure. The time of reaction of catechol with camphene is inversely proportional to the temperature of reaction. Thus, at the higher range of temperatures of reaction, shorter times of reaction are required and at the lower temperatures of reaction, longer times of reaction are required. Accordingly, the time of reaction varies from about 2 up to about 25 hours with the usual time of reaction being between 5 and 12 hours.

Insofar as the hydrogenation reaction is concerned, it may be conducted as a one-stage reaction or a two-stage reaction. When carrying out the two-stage reaction, the first-stage reaction temperature is preferably 125° to 165°C and the second-stage reaction temperature is preferably from 225° to 250°C. The odoriferous sandalwood composition is produced at the higher temperature range during the carrying out of the second stage.

Both stages of the hydrogenation reaction may be carried out at pressures between 200 and 2,500 pounds per square inch of hydrogen. The hydrogenation is carried out in the presence of a hydrogenation catalyst.

The preferred catalysts are Raney nickel, 5% palladium-on-carbon and 5% rhodium-on-carbon. With regard to the use of Raney nickel catalysts, the percent of catalyst in the camphene-catechol reaction product may vary from about 3% up to about 20% (based on weight of camphene-catechol reaction product) with a preferred percent of hydrogenation catalyst in the camphene-catechol reaction product being from about 5% up to about 10% by weight. With regard to the palladium-on-carbon and rhodium-on-carbon catalysts, the percent of catalyst in the reaction product may vary from about 0.1% up to about 5% (based on the total weight of camphene-catechol reaction product).

Example 1: Into a 1 liter reaction flask equipped with stirrer, thermometer, reflux condenser, dropping funnel, and heating mantle are placed: 188 g catechol; 110 g toluene; 16 g boron trifluoride etherate. The resulting mixture is heated to 45°C. 136 g camphene and 150 g toluene are placed in a dropping funnel. The camphene-toluene solution is added to the catechol-toluene-boron trifluoride etherate mixture over a period of 2 hours while maintaining the reaction mass at 45°C. The reaction mass is then heated to 95°C and maintained at that temperature with stirring for an additional 12 hour period.

100 g of 5% sodium hydroxide are then added to the reaction mass followed by the stirring of the reaction mass for a period of 15 minutes. The reaction mass is then separated and the organic layer is washed neutral. Solvent is then stripped off yielding 343 g of an oil. The resulting oil is rushed-over yielding 10 fractions.

Into a 1 liter autoclave are placed 250 g of fractions 2 through 7 boiling at 214° to 225°C at 2.3 to 2.7 mm Hg (bulked), 100 ml of isopropyl alcohol, and 15 g of Raney nickel.

After sealing, the autoclave is then purged with nitrogen followed by hydrogen, and heated to 220°C under 450 psi hydrogen pressure. While maintaining the autoclave pressure between 450 and 500 psig and stirring at a rate of 1,200 rpm, the autoclave is periodically repressurized with hydrogen over a period of 16 hours. At the end of the 16 hour period, the reaction mass is removed from the autoclave and filtered. The autoclave is rinsed with two 200 g portions of isopropyl alcohol. The isopropyl alcohol is then stripped from the reaction mass and the resulting material (240 g) is distilled on an 8 inch Vigreaux column yielding ten fractions. Fractions 2 through 9, boiling at 186° to 207°C at 2.5 mm Hg, are bulked and the resulting product has a strong sandalwood aroma. The resulting product is a mixture of several chemical compounds.

Example 2: *Perfumed Liquid Detergent* – Concentrated liquid detergents with intense sandalwood aromas are prepared containing 0.10%, 0.15% and 0.20% of the composition produced by means of the hydrogenation of the reaction product of camphene and catechol prepared according to Example 1. The detergents are prepared by adding and homogeneously mixing the appropriate quantity of hydrogenation product (of reaction product of catechol and camphene) in the liquid detergent. The detergents all possess intense sandalwood aromas, with the intensity increasing with greater concentrations of hydrogenation product.

Example 3: *Preparation of a Cologne and Handkerchief Perfume* –The composition produced by means of hydrogenation of the reaction product of camphene and catechol prepared according to the process of Example 1 is incorporated into a cologne at a concentration of 2.5% in 85% aqueous ethanol; and into a handkerchief perfume at a concentration of 20% (in 95% aqueous ethanol). A distinct and definite sandalwood fragrance with musky and woody nuances is imparted to the cologne and to the handkerchief perfume.

Example 4: *Preparation of a Soap Composition* – 100 g of soap chips are mixed with 1 g of the composition produced by means of hydrogenation of the reaction product of catechol and camphene produced according to Example 1 until a substantially homogeneous composition is obtained. The perfumed soap composition manifests an excellent sandalwood, woody, musky aroma.

Example 5: *Reaction of Camphene and Catechol (Using H_2SO_4 Catalyst)* – Into a 1 liter reaction flask equipped with stirrer, thermometer, reflux condenser, dropping funnel and heating mantle are placed 330 g catechol and 0.4 g concentrated sulfuric acid (95%).

The resulting mixture is heated to 120°C. Placed in a dropping funnel fitted with a heating tape is 375 g of molten camphene. The molten camphene is added to the catechol-sulfuric acid mixture over a period of 6 hours while maintaining the reaction mass at 120°C. At the end of the 6 hour period, the reaction mass is stirred at 120°C for another 2 hours. 0.5 g of sodium hydroxide dissolved in 1 g water is then added, followed by the stirring of the reaction mass for a period of 15 minutes. The resulting crude product is then rushed-over yielding 10 fractions. Fractions 4 through 9 boiling at 215° to 254°C at 3 mm Hg pressure are bulked.

Into a 1 liter autoclave are placed 350 g of reaction product of catechol and camphene produced above; 100 ml isopropyl alcohol and 30 g Raney nickel. After sealing, the autoclave is purged with nitrogen followed by hydrogen, and then heated to a temperature in the range of 225° to 230°C at a pressure of 500 psig. The autoclave is operated at this set of conditions for 3 hours. After the 3 hour period the autoclave pressure rises to about 850 psig and the autoclave is periodically vented back to 500 psig for a period of 12 hours.

At the end of the 12 hour period, the reaction mass is removed from the autoclave and filtered. The autoclave is then rinsed with two 200 g portions of isopropyl alcohol. The isopropyl alcohol is then stripped from the reaction mass and the resulting material is distilled on a 10 inch Vigreaux column yielding 15 fractions. Fractions 6 through 10, boiling at 164° to 189°C at 2.5 mm Hg pressure are bulked. The resulting product has a strong sandalwood aroma, with woody nuances. The resulting product is a mixture of several chemical compounds.

Example 6: *Floral-Woody Formulation* – The following mixture is prepared:

Ingredients	Parts by Weight
Rhodinol	40
Phenylethyl alcohol	60
Linalol	20
Bergamot oil	30
Hydroxycitronellal	110
Alpha-amylcinnamic aldehyde	30
p-isopropyl-alpha-methyl phenyl-propyl aldehyde (10% in ethyl alcohol)	45
Undecyl aldehyde (10% in ethyl alcohol)	45
Benzyl acetate	75
Indole (10% in ethyl alcohol)	15

(continued)

Ingredients	Parts by Weight
Undecalactone (1% in ethyl alcohol)	30
Ylang	20
Eugenol	10
Methylionone	90
Vetiver acetate	30
Product produced according to example 5	6
Musk ketone	60
Heliotropin	40
Coumarin	20
Natural civet (10% in ethyl alcohol)	25
Orange absolute	30
Jasmin absolute	35
Rose absolute	30

The product produced according to Example 5 adds the intense sandalwood note to this floral woody formulation.

Cyclopentene Derivatives

W. Hoffmann and K. von Fraunberg; U.S. Patent 4,069,258; January 17, 1978; assigned to BASF AG, Germany describe certain cyclopentene derivatives of the general formula 1:

(1)

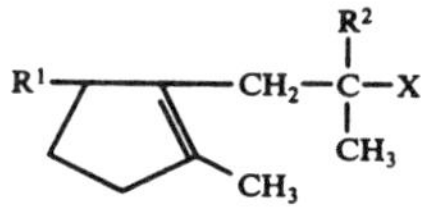

where R^1 is isopropyl or isopropenyl, R^2 is alkyl of 1 to 3 carbon atoms, X is CHO, $CH(OR^4)_2$,

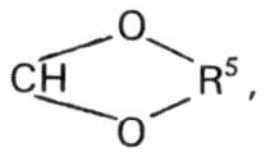

$CH{<}^{O}_{O}{>}R^5$,

CH_2OH or CH_2OCOR^3, R^3 is hydrogen or alkyl or alkenyl of up to 3 carbon atoms, R^4 is alkyl of 1 to 3 carbon atoms and R^5 is 1,2-alkylene or 1,3-alkylene of 2 to 6 carbon atoms. The derivatives are valuable fragrance materials and aromatics, the acetals and aldehydes in particular being stable.

Aldehydes of the general formula 1 are obtained when compounds of the general formula 2:

(2)

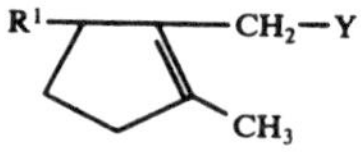

where Y is halogen, preferably chlorine or bromine, are reacted with aldehydes of the general formula 3:

(3) $H{-}C(R^2)(CH_3){-}CHO$

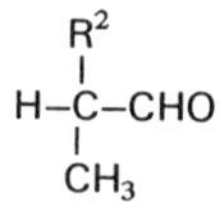

in the presence of bases and quaternary ammonium salts and these may be acetalized, e.g., by conventional methods to form the dialkyl acetals and 1,2- and 1,3-alkylene acetals of general formula 1.

Further, the free aldehydes (1) may be reduced by conventional methods, to the corresponding alcohols of formula 1 and that these may, if desired, be converted to the R^3–COOH esters.

The free aldehydes (1), which, themselves, have valuable fragrance characteristics, may be used for the manufacture of further fragrance materials by converting them, in the conventional manner, with reducing agents such as sodium boranate, to the alcohols (1a):

(1a) R^1–(cyclopentene ring, CH_3)–CH_2–C(R^2)(CH_3)–CH_2OH

and further converting these to their esters with carboxylic acids R^3–COOH. Another very suitable method for the preparation of the alcohols (1a) is reduction with aluminum triisopropylate by the Meerwein-Ponndorf method.

In general, compounds 1 which are preferred both in respect for their properties as end products, and as intermediates, are those where R^1 is isopropenyl, since the compounds derived therefrom have a fresher note than the isopropyl derivatives. The isopropyl derivatives can be prepared from the corresponding isopropenyl compounds by conventional partial hydrogenation.

With increasing size of the alkyl groups R^2 and R^3, the fragrance becomes suggestive of musk. Parallel thereto, its tenacity increases.

The perfumes according to these formulations, the inexpensive availability of which is an important advantage, are generally suitable as ingredients in cosmetics of all kinds and in soaps and detergents, the concentrations used being from 0.1 to 5% by weight. They may also be used for flavoring foodstuffs and beverages. In view of their chemical stability, they are of particular importance in compositions which are exposed to light or, as in the case of detergents, are exposed to alkaline reagents.

Example 1: *1-Methyl-2-(2,2-Dimethyl-2-Formylethyl)-3-Isopropenylcyclopent-1-ene* – A mixture of 857 g (5 mols) of 1-methyl-2-chloromethyl-3-isopropenylcyclopentene and 360 g (5 mols) of isobutyraldehyde was added in the course of 5 hours, while stirring, to a mixture of 480 g of 50% strength sodium hydroxide solution, 24 g of tetrabutylammonium iodide and 500 ml of toluene, at 80°C. The organic phase was then separated off and washed repeatedly with 200 ml portions of water, the toluene was stripped off and the residue was distilled in a rotary evaporator; boiling point 75° to 77°C/0.3 mm Hg; n_D^{25}: 1.4788; the structure was confirmed by the nuclear resonance spectrum; yield: 76%.

The compound has a floral fragrance suggestive of sandalwood. The starting compound was manufactured by chlorinating 760 g (5 mols) of 1-methyl-2-methylene-3-isopropenylcyclopentanol with phosgene in a mixture of 440 g of dimethylformamide and 1.5 liters of toluene in the course of 7 hours at from 0° to

5°C. After washing the organic phase with sodium carbonate solution and water, the chloromethylcyclopentene was obtained in 89% yield by distillation. BP: 45°C/4 mm Hg; n_D^{25}: 1.4928; the structure was confirmed by the nuclear resonance spectrum.

Example 2: *1-Methyl-2-(2,2-Dimethyl-3-Hydroxypropyl)-3-Isopropenylcyclopent-1-ene* – A solution of 1 liter of ethanol and 27 g (0.7 mol) of sodium boranate was added in the course of 2 hours to 412 g (2 mols) of the product from Example 1, at from 25° to 30°C. The ethanol was then stripped off and the residue was acidified with dilute sulfuric acid, washed and then worked up by distillation in the conventional manner. Boiling point: 77°C/0.1 mm Hg; n_D^{25}: 1.4894; yield 82%; the structure was confirmed by the nuclear resonance spectrum. The compound has a typical sandalwood odor.

Example 3: *1-Methyl-2-(2,2-Dimethyl-3-Acetoxypropyl)-3-Isopropenylcyclopent-1-ene* – This compound was prepared by esterifying 32 g (0.25 mol) of the product from Example 2 with 24 g (0.3 mol) of acetyl chloride and 24 g (0.3 mol) of pyridine at from 15° to 20°C. On working up in the conventional manner, the ester was obtained in 95% yield; BP: 88° to 90°C/25 mm Hg; n_D^{25}: 1.4743. The structure was confirmed by the nuclear resonance spectrum. Fragrance was musk-like, floral; good tenacity.

2,3-Dimethyl-5-(Substituted Cyclopenten-1-yl)-2-Pentanol

V. Kamath, B.D. Mookherjee and F.L. Schmitt; U.S. Patents 4,149,020; April 10, 1979 and 4,170,577; October 9, 1979; both assigned to International Flavors & Fragrances Inc. have provided a compound, 2,3-dimethyl-5-(2,2,3-trimethyl-3-cyclopenten-1-yl)-2-pentanol (Compound A), having the structure:

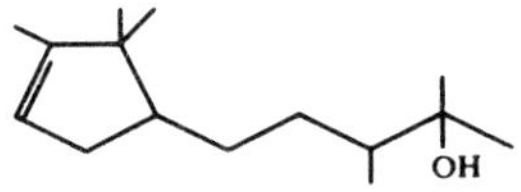

OH

which is characterized by a long lasting woody, oily sandalwood aroma reminiscent of sandalwood oil and the naturally occurring santalols.

Example 1: *Preparation of Compound (A)* – 2.1 g of 3-methyl-5-(2,2,3-trimethylcyclopent-3-en-1-yl)pentan-2-ol (0.01 mol) and 25 ml toluene are admixed in a 50 ml round bottom flask equipped with thermometer, addition funnel and a magnetic stirrer. The resulting mixture is cooled to 5°C and a solution of 3.3 g sodium dichromate (0.011 mol) in 5 ml water and 1 ml concentrated sulfuric acid is slowly added thereto.

The resulting reaction mass is stirred for a period of 0.5 hour at 5°C using an ice bath to cool the reaction mass. The ice bath is then removed and the reaction mass is stirred at room temperature for an additional half hour. The reaction mass is then poured into a beaker containing 25 g of ice. The resulting slurry is stirred and transferred to a 125 ml separatory funnel. The organic phase is washed with 20 ml of a 5% sodium carbonate solution followed by 3 25 ml portions of saturated sodium chloride solution. The resulting material is dried over anhydrous magnesium sulfate and filtered. The solvent is evaporated therefrom, thereby obtaining 1.8 g of crude intermediate ketone having the structure:

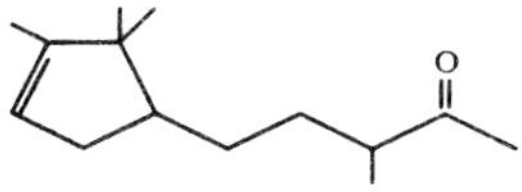

When the reaction is scaled up fivefold (using 10.5 g of 3-methyl-5-(2,2,3-trimethylcyclopent-3-en-1-yl)pentan-2-ol), 9.5 g of the ketone is obtained.

2.5 ml of methyl magnesium bromide (3 molar solution in diethylether) is placed in a 25 ml reaction flask fitted with a stirrer, thermometer, condenser and cooling bath. The methyl magnesium bromide solution is cooled to 5°C and 1.5 g of the intermediate ketone is added to the methyl magnesium bromide solution.

The resulting reaction mass is stirred for a period of 30 minutes at 5°C. The cooling bath is then removed and the reaction mass is stirred for 1 hour at room temperature. The reaction mass is then poured into a beaker containing 20 g of ice. The resulting organic layer is separated from the aqueous layer, and the organic layer is washed with a 5 molar aqueous solution of hydrochloric acid followed by washes of sodium chloride solution. The resulting material is dried and the solvent is removed yielding 1.5 g of crude alcohol having the structure:

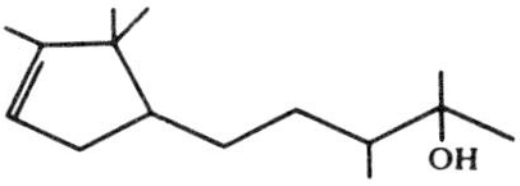

OH

The crude material is injected onto a 10% Carbowax column (10 ft = ⅛) and the major peak is collected. The major peak has a good sandalwood note which is described as woody, oily and santalol isomer-like. The structure is confirmed using NMR and IR analyses.

Example 2: The following synthetic sandalwood oil formulation is produced:

Ingredient	Parts by Weight
Amyris oil	100
Amyris acetate	220
Cedarwood oil	150
Trans decahydro beta-naphthol formate	100
Guaiophene (1% in diethyl phthalate)	50
Eugenol (10% in diethyl phthalate)	50
Galaxolide (2.5% in diethyl phthalate)	30
Geranyl phenyl acetate	50
Compound A	250

The addition of 25% of Compound (A) to the sandalwood formulation contributes the main sandalwood note to the fragrance. The odor of the fragrance without Compound (A) is far distant from the desired odor of sandalwood.

Example 3: *Liquid Detergent Containing Tricyclene-9-Butenone* – Concentrated liquid detergents with a sandalwood-like odor containing 0.2%, 0.5% and 1.2% of the product produced in accordance with the process of Example 1 are prepared by adding the appropriate quantity of Compound (A) to the liquid detergent known as P-87. The sandalwood aroma of the liquid detergent increases with increasing concentration of Compound (A).

Example 4: *Cologne and Handkerchief Perfume* – Compound (A) produced by the process of Example 1 is incorporated into a cologne having a concentration of 2.5% in 85% ethanol; and into a handkerchief perfume in a concentration of 10% (in 95% ethanol). Compound (A) affords a distinct and definite sandalwood aroma with woody, oily and santalol-like notes to the handkerchief perfume and to the cologne.

CEDARWOOD

Certain Isolongifolene Compounds

H.R. Ansari, N. Unwin and H.R. Wagner; U.S. Patent 4,100,110; July 11, 1978; assigned to Bush Boake Allen Limited, England have synthesized a number of compounds useful in perfumery from the terpene isolongifolene (1).

(1)

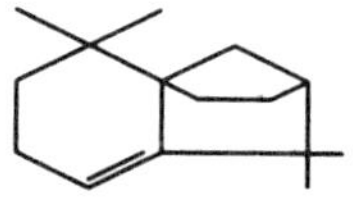

This is accomplished by means of a Prins reaction followed optionally by a reduction or saponification. The Prins reaction involves the reaction of an aldehyde with an olefin, preferably in the presence as catalyst of a Lewis acid or mineral acid. The reaction may be carried out: (a) using formaldehyde in a carboxylic acid as solvent to convert an olefin to an ester $R''CH_2OR$ (where R is an acyl residue of the carboxylic acid solvent; (b) using a formaldehyde condensate such as paraformaldehyde in an autoclave, to provide a primary alcohol $R''CH_2OH$ or (c) using a higher aldehyde to provide a secondary alcohol R"(CHOH)R. Under certain conditions, this reaction may alternatively or additionally provide the ketone R"(CO)R.

Example 1: Isolongifolene (204 g, 1 mol) was treated with paraformaldehyde (36 g, 1.2 mols) in formic acid (138 g, 3 mols) at 90°C over a period of 16 hours. After this duration, about 75% of isolongifolene had been converted into a mixture of C_{16} formates. The product was washed twice with water, dried and distilled mainly in two fractions, namely, the hydrocarbon fraction and the formate fraction as indicated by the boiling points given below:

	Grams	Temperature (°C)	Millimeters
Fraction 1*	75	70-100	0.5
Fraction 2**	110	110-120	0.5
Residue	35		

*mainly isolongifolene **mainly C_{16} formates

This represents w/w yield of about 84.5% based on consumed isolongifolene. The formate fraction comprised a mixture of

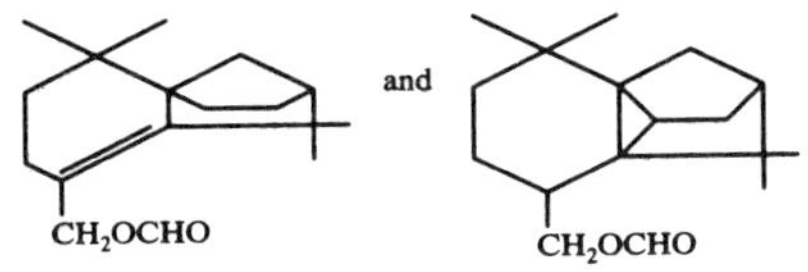

CH_2OCHO and CH_2OCHO

determined by NMR and IR analysis, and had a pronounced cedarwood-vetiver odor of great tenacity, with amber, clary sage dry-out.

Example 2: Example 1 was repeated using 180 g acetic acid and 2 g sulfuric acid instead of the formic acid. The mixture was heated at 60°C. A similar yield of the corresponding acetate esters was obtained having a strong cedar-vetiver odor reminiscent of cedryl and vetiveryl acetates.

Example 3: The mixtures of formates (265 g) and acetates (276 g), prepared according to Examples 1 and 2, respectively, were each saponified by boiling with 30% aqueous sodium hydroxide (48 g) over a period of 6 hours to give a mixture of corresponding alcohols. The product (230 g) was distilled through a short column and in each case had a soft, woody odor, reminiscent of cedrol.

Example 4: A mixture of alcohols (234 g) prepared according to Example 3 was dissolved in 250 cc of glacial acetic acid, and cooled in a water bath while sodium dichromate (170 g) in 750 cc of acetic acid was added over a period of 2 hours. After this, the reaction mixture was allowed to stand at ambient temperature for 20 hours. The mixture of aldehydes thus formed was isolated by work-up with water and distilled to give the finished product (175 g) with a strong cedar-vetiver odor and a pronounced amber note on dry-out.

Example 5: The product of Example 4 was used in a chypre formulation, as follows:

Ingredient	Parts by Weight
Musk ambrette	40
Musk ketone	60
Coumarin	50
Oil of bergamot	150
Oil of lemon	100
Methyl ionone	50
Hexyl cinnamic aldehyde	100
Hydroxycitronellal	100
Oil of lavender	50
Oil of sandalwood	40
Osyrol preparation (Bush Boake Allen)	60
Isoeugenol	20
Eugenol	10
Benzyl acetate	30
Phenyl ethyl alcohol	40
Oakmoss absolute	30
Oil of vetiver	20
Product of Example 4	50

Example 6: The products of Examples 1 and 3 were used in a Fougere formulation, as follows:

Ingredient	Parts by Weight
Musk ambrette	100
Musk ketone	50
Coumarin	50
Oil of lavender	100
Oil of patchouli	30
Oil of geranium	40
Oil of sandalwood	30
Oil of bergamot	100
Linalool	50
Linalyl acetate	50
β-methyl ionone	60
Anisaldehyde	30
Methyl anthranilate	5
Vanillin	55
Rose absolute synthetic	20
Jasmin absolute synthetic	20
Oil of ylang	10
Hydroxycitronellal	100
Product of Example 3	50
Product of Example 1	100

FRAGRANCES—FLORAL AND FRUITY

FLORAL

Octadienes

P.A. Ochsner; U.S. Patent 4,066,710; January 3, 1978; assigned to Givaudan Corporation describes octadiene, olfactory derivatives having the general formula

$$\text{(I)} \qquad H_3C-\underset{OR}{\overset{CH_3}{\overset{|}{\underset{|}{C}}}}-CH_2-CH_2-CH{=}CH-\overset{CH_3}{\overset{|}{C}}{=}CH_2$$

wherein R represents a hydrogen atom or a $C_{1\text{-}5}$-alkyl, $C_{2\text{-}5}$-alkenyl or $C_{1\text{-}5}$-alkanoyl group.

Examples of $C_{1\text{-}5}$-alkyl groups are methyl, ethyl, etc. Examples of $C_{2\text{-}5}$-alkenyl groups are allyl, crotyl, etc. Examples of $C_{1\text{-}5}$-alkanoyl groups are acetyl, propionyl, etc.

The octadiene derivatives of formula I are manufactured by hydrating 2,7-dimethyl-1,3,7-octatriene of the formula

$$\text{(II)} \qquad H_2C{=}\overset{CH_3}{\overset{|}{C}}-CH_2-CH_2-CH{=}CH-\overset{CH_3}{\overset{|}{C}}{=}CH_2$$

with intermediate protection of the conjugated double-bonds to give 2,7-dimethyl-5,7-octadien-2-ol and, if desired, subjecting the octadien-2-ol to esterification to give an octadiene derivative of formula I in which R represents a $C_{1\text{-}5}$-alkanoyl group or to etherification to give an octadiene derivative of formula I in which R represents a $C_{1\text{-}5}$-alkyl or $C_{2\text{-}5}$-alkenyl group.

As shown in the following reaction scheme, intermediate protection of the conjugated double-bonds in 2,7-dimethyl-1,3,7-octatriene of formula II can be expediently achieved by converting the octatriene into the sulfone of formula III. Water can then be added according to known methods at the terminal double-bond of this sulfone and the resulting hydrated sulfone of formula IV can be finally converted into 2,7-dimethyl-5,7-octadien-2-ol (formula I; R = H) by removal of the protecting group.

(II) $H_2C{=}C(CH_3){-}CH_2{-}CH_2{-}CH{=}CH{-}C(CH_3){=}CH_2$

↓

(III) $H_2C{=}C(CH_3){-}CH_2{-}CH_2{-}CH$ — ring: $CH{-}SO_2{-}CH_2{-}C(CH_3){=}CH{-}$

↓

(IV) $H_3C{-}C(CH_3)(OH){-}CH_2{-}CH_2{-}CH$ — ring: $CH{-}SO_2{-}CH_2{-}C(CH_3){=}CH{-}$

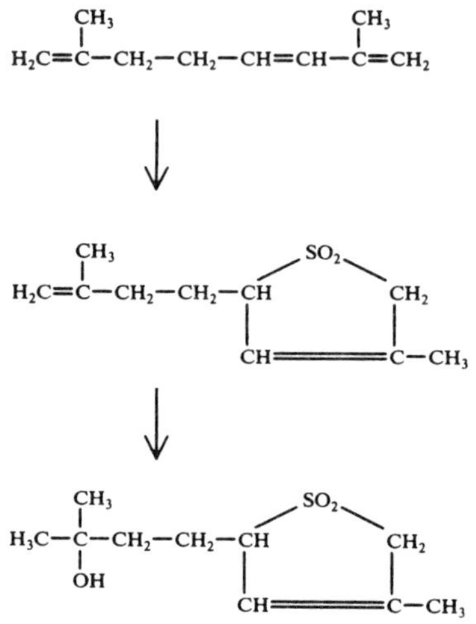

In the foregoing reaction scheme, 2,7-dimethyl-1,3,7-octatriene of formula II is converted into the sulfone of formula III using an excess of sulfur dioxide in the presence of about 1% of a polymerization inhibitor, for example, hydroquinone, 3-tert-butyl-4-methoxyphenol, 2,6-di(tert-butyl)-4-methylphenol (BHT) etc. After removal of the excess sulfur dioxide, this sulfone of formula III is hydrated at the terminal double-bond; for example, by the action of 40 to 60% aqueous sulfuric acid, expediently at temperatures of 10° to 25°C. The resulting product of formula IV is neutralized (e.g., with sodium hydroxide) and separated with a solvent (e.g., benzene).

The removal of the SO_2-protecting group and, concommitantly, the reintroduction of the conjugated double-bond can be carried out by heating the compound of formula IV, expediently in a vacuum and at temperatures 120° to 130°C.

It has proved advantageous to carry out the heating of the compound of formula IV in the presence of 1 to 2% of a high-boiling organic base; for example, triethanolamine or a tertiary amine such as a trialkylamine (e.g., trimethylamine). Inorganic compounds which have a weak basic reaction (e.g., calcium carbonate) can also be used.

The esterification of the resulting alcohol (2,7-dimethyl-5,7-octadien-2-ol) can be carried out according to known methods, expediently by reacting the alcohol with a compound yielding the desired $C_{1\text{-}5}$-alkanoyl group, especially using an appropriate acid anhydride such as acetic anhydride, in the presence of a base (e.g., pyridine, sodium acetate, etc). A corresponding acid halide can, however, also be used for this esterification.

The octadiene derivatives of formula I possess particular odorant properties. They can accordingly be used in the perfume industry for the manufacture of perfumes and perfume products; for example, for the perfuming of soaps, solid and liquid detergents, aerosols and cosmetic products of all kinds such as toilet waters, ointments, face milks, make-ups, lipsticks, bath salts and bath oils. In the finished perfumes or perfumed products, the content of the octadiene derivatives can lie within wide limits; for example, between 1% (detergents) and 20% (alcoholic solutions). In perfume bases or concentrates, the octadiene derivatives can, of course, also be present in amounts greater than 20%.

These octadiene derivatives provide, in general, a flowery, especially lavender-like, odor without a fatty note. The free alcohol, namely the octadiene derivative of formula I in which R represents a hydrogen atom, possesses outstanding fragrance qualities, the odor thereof being pleasantly flowery (reminiscent of lavender), linalool-like, piquant, earthy, slightly metallic and long-lasting. The compositions which contain this alcohol have a powerful, fresh action; the alcohol being especially suitable for flowery, woody or hesperidine notes.

The octadiene derivative of formula I in which R represents an acetyl group possesses a natural flowery, fruity, green, slightly woody odor which is somewhat reminiscent of grapefruit and neroli and which is vetiverlike in the background. Compositions containing this octadiene derivative accordingly have a very natural action.

The ethers of formula I, especially the methyl ether, are readily volatile compounds and they can accordingly be used in perfume compositions, especially for head-notes.

The octadiene derivatives of formula I can be advantageously incorporated into odorant compositions of the flowery type. Such compositions thereby acquire strength and cohesion and are thus modified in an advantageous manner.

The following examples illustrate typical odorant compositions containing the octadiene derivatives provided by this process.

Example 1: *Rose Composition –*

	Parts by weight
2,7-Dimethyl-5,7-octadien-2-ol	100
C_{10}-Aldehyde 10% in diethylphthalate	2
C_8-Aldehyde 10% in diethylphthalate	3
C_9-Aldehyde 10% in diethylphthalate	5
Guayl acetate	10
Phenylethyl acetate	15
Benzyl acetate	20
Methylionone (Isoraldeine 70)	20
Phenylacetaldehyde (10% in diethyleneglycol monoethyl ether)	20
Dimethylbenzyl carbinyl acetate	20

(continued)

Example 1: (continued)

	Parts by weight
Rosacetol	25
Eugenol	30
Nerol	50
Geraniol	80
Citronellol	100
Rhodinol pure	120
Phenylethyl alcohol extra	180
Total	800

Example 2: *Gardenia Composition –*

	Parts by weight
2,7-Dimethyl-5,7-octadien-2-ol-acetate	130
Isoeugenol	5
C_{14}-Aldehyde 10% in diethylphthalate	10
Musk ambrette	10
Gardenol	10
Methyl benzoate	20
Benzyl acetate	20
Musk ketone	20
Nonalactone 10% in diethylphthalate	30
Ylang Ylang extra	30
Hydroxycitronellal	35
Dimethylbenzyl carbinyl acetate	50
Phenylethyl alcohol extra	80
α-Ionone	100
Hexylcinnamic aldehyde	100
Total	650

Hexyloxyacetonitrile

K. Kulka, T. Zazula and J.M. Yurecko; U.S. Patent 4,077,916; March 7, 1978; assigned to Fritzsche Dodge & Olcott, Inc. have found that desirable perfume compositions can be made which contain from 0.1 to 1% by weight hexyloxyacetronitrile and at least 1% by weight of a perfume component which modifies the olfactory properties of the hexyloxyacetonitrile. The hexyloxyacetonitrile may be incorporated in the perfume composition in the form of a solution of dipropylene glycol. Desirably, the compositions contain at least 0.1% by weight of hexyloxyacetaldehyde dimethyl acetal.

The hexyloxyacetonitrile utilized in these perfume compositions is unique in its olfactory properties. This nitrile has a flowery, herbaceous, slightly fatty odor reminiscent of irone. One of its outstanding characteristics is its diffusing odor. Thus it is readily perceived and recognizable even from a distance. Its lasting power as determined from a smelling blotter is limited to 7 to 8 hours. Accordingly, it can be used advantageously to give top-notes to perfume compositions. It blends well with other perfume components to improve and give character to the total odor-profile.

The hexyloxyacetonitrile may be used in a great variety of perfume compositions such as rose, jasmine, chypre, lavender and phantasy-bouquet. In such compositions it may be used in various concentrations to achieve a desired effect. Consequently, hexyloxyacetonitrile is a valuable addition to the arsenal of aroma chemicals.

The hexyloxyacetonitrile may be produced by the method described in U.S. Patent 3,132,179 which employs the hexyloxyacetonitrile as a pharmaceutical intermediate.

U.S. Patent 3,132,179 describes a procedure for the production of hexyloxyacetonitrile involving a two-step method; chloromethylating n-hexanol to chloromethyl-n-hexyl ether, and reacting the ether with cuprous cyanide in accordance with the following reaction scheme:

(I) $CH_3(CH_2)_5OH + HCHO + HCl \rightarrow CH_3(CH_2)_5OCH_2Cl$

(II) $CH_3(CH_2)_5OCH_2Cl + CuCN \rightarrow CH_3(CH_2)_5OCH_2CN + CuCl$

Alternatively, the hexyloxyacetonitrile may be produced by liberating hexyloxyacetaldehyde from one of its acetals such as its dimethylacetal, reacting the hexyloxyacetaldehyde with hydroxylamine to obtain hexyloxyacetaldehyde oxime, and dehydrating hexyloxyacetaldehyde oxime, for example, with acetic anhydride to obtain the hexyloxyacetonitrile.

Example 1: *Muguet Perfume Composition* – A muguet perfume composition is prepared by mixing together the following:

	Parts by weight
Benzyl acetate	25
Linalool	30
Dimethyl benzyl carbinol	50
Linalyl acetate	20
Citronellyl acetate	20
Phenylethyl alcohol	50
Citronellol	50
Heliotropin	40
Ylang Ylang oil	10
Cinnamyl acetate	100
Hydroxycitronellal	475
Cyclamal	75
Hexyloxyacetaldehyde dimethyl acetal	40
Hexyloxyacetonitrile (50% solution in dipropylene glycol)	10
Tetramethyl ethyl nitrile tetralin [33% solution in dipropylene glycol (Nitrile Musk)]	5
Total	1,000

Example 2: *Chypre Perfume Composition* – A chypre perfume composition is prepared by mixing together the following:

	Parts by weight
Linalyl acetate	180
2,4-Dihydroxy-3-methylbenzaldehyde [20% solution in dipropylene glycol (Oak Moss aldehyde)]	300
Patchonli oil	30

(continued)

	Parts by weight
Phenylethyl alcohol	30
Vetiver oil	40
Clary sage oil	50
Methyl ionone	50
Coumarin	180
Labdanum resinoid	100
Eugenol	20
Hexyloxyacetaldehyde dimethyl acetal	10
Hexyloxyacetonitrile (50% solution in dipropylene glycol)	10
Total	1,000

Cyclopentanone Derivatives

C. Celli; U.S. Patent 4,092,362; May 30, 1978 describes a process for the production of 2-n-pentyl-3-(2-oxopropyl)-1-cyclopentanone and also perfume formulations containing this substance. 2-n-pentyl-3-(2-oxopropyl)-1-cyclopentanone is a valuable odoriferous substance possessing a particularly fragrant magnolialike aroma. The production of 2-n-pentyl-3-(2-oxopropyl)-1-cyclopentanone is accomplished by hydrolyzing and decarboxylating a compound of the formula

(I) [cyclopentanone ring bearing $CH(COOR)-CO-CH_3$ and C_5H_{11} substituents]

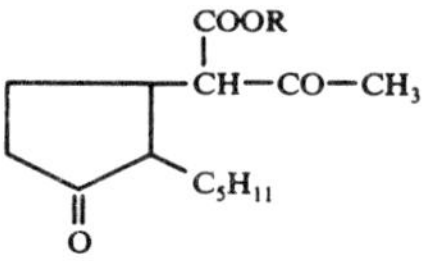

wherein R represents a hydrocarbyl group, with water, under pressure and in an initially substantially neutral medium. Pure stereoisomers may be obtained from mixtures of isomers by conventional separation methods.

The hydrolysis and decarboxylation is conveniently carried out at a temperature of from 120° to 300°C, preferably from 140° to 250°C, under substantially neutral conditions.

The reaction is conveniently effected using the same weight of water as of 2-n-pentyl-3-(2-oxopropyl)-1-cyclopentanone. The reaction is normally effected in an autoclave from which the air is first purged.

The starting substance of formula

(I) [cyclopentanone ring bearing $CH(COOR)-CO-CH_3$ and C_5H_{11} substituents]

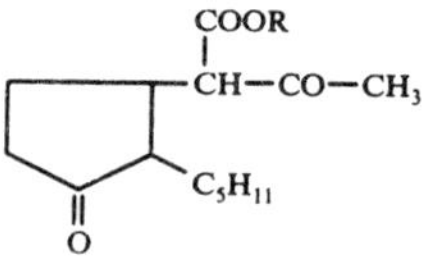

wherein R represents a hydrocarbyl group, may be prepared by condensing 2-n-pentyl-1-cyclopent-2-en-1-one with a keto ester of the formula:

(II) $CO_2R-CH_2-CO-CH_3$

The reaction conditions for this step are those of a conventional Michael condensation. The group R is conveniently an alkyl group and preferably a lower alkyl group having from 1 to 5 carbon atoms, for example, methyl or ethyl. 2-n-pentyl-3-(2-oxopropyl)-1-cyclopentanone and its stereoisomers have a very strong odor of magnolia. Thus, the compound can be used for the preparation of perfumes as well as for the preparation of perfumed products, for example solid and liquid detergents, synthetic washing agents, aerosols or cosmetic products of all kinds. These odorant compositions may conveniently contain from 1 to 20% by weight, preferably 5 to 10% by weight, of 2-n-pentyl-3-(2-oxopropyl)-1-cyclopentanone.

Example 1: (a) One liter of absolute ethanol was introduced into a 4 liter flask. 23 g of sodium were added to it. 650 g of ethyl acetoacetate were added to the resulting sodium ethylate solution, then 456 g of 2-n-pentyl-cyclopent-2-en-1-one were added dropwise over one hour. The solution thus obtained was heated over a period of two hours up to the reflux temperature (87°C), held at this temperature for three hours, then allowed to cool and stand at the ambient temperature for 16 hours.

The resulting product was neutralized with 65 g of acetic acid and subsequently poured onto 3,300 ml of a 10% aqueous sodium chloride solution. The aqueous layer was then decanted off and extracted three times with the aid of 300 ml of toluene. The toluene extracts were washed to neutrality, the toluene was removed by distillation and the remainder fractionally distilled. There was thus obtained 513 g of 2-n-pentyl-3-(1-carbethoxy-2-oxopropyl)-1-cyclopentanone. Yield = 60%.

(b) 513 g of 2-n-pentyl-3-(1-carbethoxy-2-oxopropyl)-1-cyclopentanone were mixed in an autoclave with an equal weight of water. The autoclave was then purged of the air which it contained, closed and heated up to a temperature of 140° to 150°C for 2 hours. After cooling, extraction, washing and distillation, there was obtained 332 g of 2-n-pentyl-3-(2-oxopropyl)-1-cyclopentanone (yield = 90.6%) having n_D^{20} = 1.432 and d_4^{20} = 0.961.

Example 2: A perfumery composition containing 2-n-pentyl-3-(2-oxopropyl)-1-cyclopentanone was prepared as follows:

	Parts by weight
Bergamot peel oil extra	60
Benzyl acetate	140
Phenylethyl alcohol	150
Jasmine absolute	50
Linalol	50
Methylnonylacetaldehyde, 10% in ethylphthalate (E.P.)	60
C_{11} aldehyde 10% in E.P.	20
Hydroxydihydrocitronellal	140
α-Ionone	20
Geranium oil Africa	60
Civet absolute 10% in E.P.	10
Hyacinth absolute	10
Vetiver oil bourbon	35
Cyclopentadecanolide	20
Tonka beans absolute	20

(continued)

	Parts by weight
Rose oil Eastern	15
Musk ketone	30
α-Amylcinnamaldehyde	30
2-n-pentyl-3-(2-oxopropyl)-1-cyclopentanone	80
Total	1,000

2-Ethyl-6,6-Dimethyl-2-Cyclohexene-1-Carboxylic Acid Ethyl Ester

H. Schenk; U.S. Patent 4,113,663; September 12, 1978; assigned to Givaudan Corporation describes processes for producing odorant compositions containing 2-ethyl-6,6-dimethyl-2-cyclohexene-1-carboxylic acid ethyl ester of the formula

(I) $COOC_2H_5$

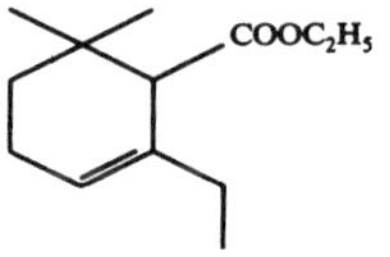

The ester of formula I is manufactured by either (a) cyclizing 3-ethyl-7-methyl-2,6-octadienoic acid ethyl ester, or (b) esterifying 2-ethyl-6,6-dimethyl-2-cyclohexene-1-carboxylic acid.

Processes (a) and (b) can be carried out by known methods for the synthesis of cyclogeranoyl derivatives.

The ester of formula I has particular odorant properties. Thus it provides, for example, an extremely diffuse rose note of very remarkable radiance which is accompanied by honeylike, warm-spicy and fruity berrylike side-notes. The woody, lightly flowery base note with fruity berrylike side-notes reminiscent of dry fruits is also worthy of mention.

The ester of formula I, which can be manufactured easily and cheaply, provides odorant properties which come quite close to the highly sought but expensive damascone of formula II and accordingly can serve as a cheap substitute for damascone.

(II) O

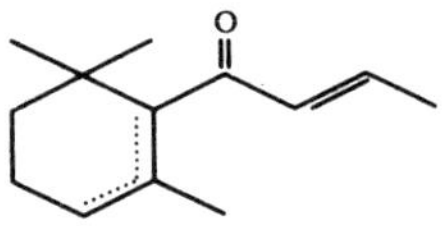

Because of its interesting olfactory properties, the ester of formula I can be used as an odorant; for example, in perfumery for the production of odorant compositions such as perfumes or for perfuming products of all types such as soaps, washing agents, solid and liquid detergents, aerosols or other cosmetic products (e.g., creams, cleansing milk, make-up, lipsticks, bath salts, bath oils, etc.).

Because of its very natural notes, the ester of formula I is also particularly suitable for modifying known compositions (e.g., those of a flowery type and in particular those having rose bases). Its organoleptic character enhances, above all, base notes (e.g., chypre). It is very well suited in combination with wood notes (e.g., those from p-tert-butylcyclohexyl acetate, sandalwood oil, patchouli oil, cedryl acetate and methyl ionone).

Depending on the intended use, the concentration of the ester of formula I can vary within wide limits; for example, between about 1 wt % in the case of detergents and about 15 wt % in the case of alcoholic solutions. In perfume bases or concentrates the concentrations can, of course, also be higher.

Example 1: 30 g (0.143 mol) of cis,trans-3-ethyl-7-methyl-2,6-octadienoic acid ethyl ester are carefully added to a solution, cooled to 5° to 10°C, of 280 ml of concentrated formic acid and 20 ml of concentrated sulfuric acid. After the dropwise addition is complete, the mixture is allowed to attain room temperature. It is then further stirred at this temperature for one hour. The mixture is poured onto ice and extracted three times with hexane.

The combined hexane solutions are washed once with water, twice with sodium bicarbonate solution and once again with water, dried over sodium sulfate and evaporated. The crude product (27.5 g) is fractionally distilled. There are obtained 21 g (70%) of pure 2-ethyl-6,6-dimethyl-2-cyclohexene-1-carboxylic acid ethyl ester of boiling point 102° to 103°C/11 mm Hg; n_D^{20} = 1.4615.

The cis,trans-3-ethyl-7-methyl-2,6-octadienoic acid ethyl ester used as the starting material can be prepared as follows:

A solution of 5.8 g (0.25 g-atom) of sodium in 130 ml of absolute ethanol is placed in a four-necked flask which is fitted with a mechanical stirrer, a dropping funnel, a reflux condenser and a thermometer. A solution of 30 g (0.214 mol) of 7-methyl-6-octen-3-one and 62.5 g (0.279 mol) of carbethoxymethylenediethyl phosphonate in 130 ml of absolute toluene is added dropwise to this at 0° to 10°C over a period of ¾ hour. The mixture is subsequently allowed to attain room temperature and is further stirred at this temperature for two hours. The mixture is poured onto ice water and extracted three times with hexane. The combined hexane solutions are washed three times with water, dried over sodium sulfate and evaporated.

The crude product (41 g) is fractionally distilled. At 55° to 65°C/0.02 mm Hg, there are obtained 29.3 g (65.1%) of pure cis,trans-3-ethyl-7-methyl-2,6-octadienoic acid ethyl ester.

Example 2: *Rose Composition –*

	Parts by weight
3,5-Dimethyl-cyclohex-3-ene-carboxaldehyde, 10% in propylene glycol	2
Decanal, 10% in propylene glycol	3
Geranyl acetate	5
α-Ionone	10
1,1-Dimethyl-4-acetyl-6-tert-butylindane	10
Cinnamic alcohol synth.	50

(continued)

	Parts by weight
Citronellol extra	100
Geraniol extra	200
Phenylethyl alcohol extra	500
Total	880

When 120 parts by weight of ethyl 2-ethyl-6,6-dimethyl-2-cyclohexene-1-carboxylate are added to the foregoing base which has a rose character, then the special note of tea roses becomes more pronounced. The resulting composition is fresher and more natural.

Example 3: *Fruity Character Composition –*

	Parts by weight
Osmanthus absolute	5
Ethylene brassylate	10
γ-undecalactone	15
Palmarosa oil	20
2-Ethoxycarboxymethyl-2-methyl-1,3-dioxolane	30
Allyl ionone	40
Dimethylbenzene carbinyl butyrate	50
α-ionone	80
Propylene glycol	700
Total	950

When 50 parts by weight of ethyl 2-ethyl-6,6-dimethyl-2-cyclohexene-1-carboxylate are added to the foregoing fruity base, the resulting composition is much more diffusive and rounded. It is more natural and the velvet apricot tone is especially good.

α-Oxy(Oxo) Sulfides

W.J. Evers, H.H. Heinsohn, Jr., E.J. Shuster and F.L. Schmitt; U.S. Patents 4,094,824; June 13, 1978; 4,097,396; June 27, 1978; and 4,151,103; April 24, 1979; all assigned to International Flavors & Fragrances, Inc. have found that certain α-oxy(oxo) sulfides impart to perfumed articles such as colognes, soaps, and detergents, sweet, green, floral, herbal, vegetative, basil-like, minty, melony, grapefruit, fruity and alliaceous aromas with yara, neroli and/or verdimalike nuances.

These compounds have the structure:

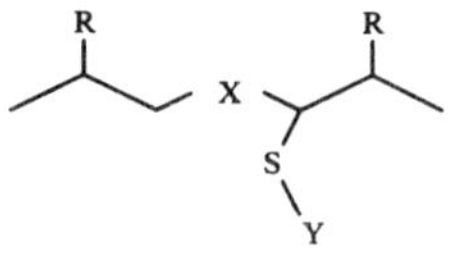

R R X S Y

wherein X is either >C=O or >CHOH; R is methyl or hydrogen and Y is either methyl, acetyl, 1-propyl, 2-methyl-1-propyl or methallyl having the structure:

Such α-oxy(oxo) sulfides are obtained by reacting an alkanone with SO_2Cl_2 to form an α-chloroalkanone; reacting the α-chloroalkanone with an alkali metal mercaptide to form an α-substituted mercaptoalkanone which can be used for its perfumery properties; or, if desired, reacting the α-substituted mercaptoalkanone with a reducing agent such as an alkali metal borohydride in order to obtain an α-hydroxy-substituted mercaptoalkane.

The table below shows specific examples of α-oxy(oxo) sulfides and their perfume characteristics:

COMPOUND	STRUCTURE	AROMA
3-methylthio-4-heptanol		At 1% in food grade ethanol, a sweet, green, floral, herbal, vegetative note.
3-methylthio-4-heptanone		At 1% in food grade ethanol a green, minty, herbaceous note with vegetative basil notes
3-propylthio-4-heptanol		Fatty, cucumber, onion (scallion, shallot) aroma with some green melon and floral nuances.
3-thioacetyl-2,6-dimethyl-4-heptanone		At 1% in food grade ethanol, a sweet, meaty, vegetable aroma with somewhat of a grapefruit topnote.
3-isobutylthio-4-heptanone		Evaluated at 1% in food grade ethanol, a meaty, onion aroma with a green, spicey and peppery nuance and an underlying Bergamot note.
3-isobutylthio-2,6-dimethyl-4-heptanol		Evaluated at 1% in food grade ethanol, a vegetable, green, horseradish, somewhat rubbery, onion-like aroma containing notes of hyacinth and narcisse.
3-methylthio-2,6-dimethyl-4-heptanone		Evaluated at 1% in food grade ethanol, a sweet, sulfurous, slightly floral and woody aroma with a fruity and berry nuance.
3-(methyllylthio)-2,6-dimethyl-4-heptanone		At 1% in ethanol, a fruity, grapefruit, somewhat floral aroma with underlying yara neroli notes and bready, vegetative nuances.

Example 1: *Preparation of 3-Methylthio-4-Heptanone –*

O

Cl

$+ NaSCH_3 \xrightarrow{CH_3OH}$

$HSCH_3 + NaOCH_3$

O

SCH_3

Into a 50 ml, three-necked, round-bottom flask equipped with magnetic stirrer, dry ice condenser, pot thermometer, cold water bath, reflux condenser with nitrogen inlet tube and nitrogen bubbler, is placed a solution of 0.54 g of sodium methoxide in 6 ml anhydrous methanol (0.01 mol sodium methoxide). The sodium methoxide solution is then cooled using the cold water bath to a temperature of 25°C. The nitrogen flow is ceased and methyl mercaptan in methanol (0.48 g methyl mercaptan in 6 ml anhydrous methanol, 0.01 mols methyl mercaptan) is added to the reaction mass while maintaining same at 24°C.

At 24°C, a solution of 1.49 g of 3-chloro-4-heptanone in 2 ml anhydrous methanol (0.01 mol 3-chloro-4-heptanone) is added to the reaction mass. The reaction mass is maintained, with stirring, at 25°C for a period of 1¼ hours. At the end of this period, the reaction mass is flushed with nitrogen. The reaction mass is then concentrated on a rotary evaporator using a water aspirator vacuum to approximately 5 ml.

15 ml of distilled water is then added to the concentrated reaction mixture whereupon the reaction mixture formed into two phases; an oil phase and an aqueous phase. The pH of the aqueous phase is in the range of 5 to 6. The oil phase is then extracted with two 12 ml portions of n-hexane and the phases are separated. The hexane extracts are combined, washed with 5 ml of water, dried over anhydrous sodium sulfate, gravity filtered and concentrated on a rotary evaporator to a weight of 1.29 g. The resulting product contains 90.1% 3-methylthio-4-heptanone by GLC, as confirmed by NMR, IR and mass spectral analyses.

Example 2: *Preparation of 3-Methylthio-4-Heptanol –*

O

SCH_3

$+ NaBH_4 \xrightarrow{\text{anhydrous EtOH}}$

OH

SCH_3

Into a 25 ml, three-necked, round-bottom flask equipped with magnetic stirrer, nitrogen inlet tube, reflux condenser, pot thermometer and cold water bath, is added a solution of 0.10 g sodium borohydride ($NaBH_4$) dissolved in 4 ml anhydrous ethyl alcohol (0.00265 mol sodium borohydride).

While maintaining the pot temperature at 25°C, a solution of 0.8 g of 3-thiomethyl-4-heptanone in 3.5 ml anhydrous ethyl alcohol is added to the sodium borohydride-ethanol solution over a one minute period. The reaction mass then warms up to about 30°C and is maintained at a temperature of between 25° and 30°C for a period of about 1½ hours. At the end of this period another

0.05 g (0.00133 mol) of sodium borohydride and 2 ml ethanol is added.

After 10 minutes of stirring while maintaining the reaction mass at 25°C, the reaction mass is then worked up as follows: The reaction mixture is concentrated to about 4 ml of a thick slurry using water aspirator vacuum. The resulting thick slurry is then combined with 12 ml water thereby causing the solid to dissolve, and the reaction mass to exist in two phases; an aqueous phase and an organic phase. The aqueous phase is acidified to a pH of 2 to 3 using 10% HCl solution. The organic phase is extracted with two 12 ml portions of methylene chloride.

The extracts are then combined, washed with 8 ml water, dried over anhydrous sodium sulfate, gravity filtered, and then concentrated on a rotary evaporator (using water aspirator vacuum) to a weight of 0.58 g. The desired product is trapped out on an 8' x ¼" SE-30 GLC column; and MS, NMR and IR analyses confirm that the resulting compound is 3-thiomethyl-4-heptanol.

Example 3: *Otto of Rose Perfume Formulation* – The following mixture is prepared:

	Parts by weight
Phenyl acetic acid	5
Hydroxycitronellal	10
Geraniol	125
Citronellol	150
Phenylethyl alcohol	50
Phenylethyl acetate	4
Ethyl phenylacetate	5
Citronellyl formate	20
Geranyl acetate	25
Linalool	15
Terpineol	10
Eugenol	3
Phenyl acetaldehyde dimethyl acetal	5
Benzyl acetate	3
Guaiacwood oil	5
3-Methylthio-4-heptanone produced according to the process of Example 1	10

The 3-methylthio-4-heptanone, produced according to the process of Example 1 imparts the green, fruity, spicy top note so characteristic of rose otto to this formulation.

Example 4: *Narcisse Formulation* – The following mixture is prepared:

	Parts by weight
Benzyl alcohol	50
Benzyl benzoate	25
Terpineol	30
Nerol	15
Phenylethyl alcohol	50
Geraniol	40
Linalool	50

(continued)

	Parts by weight
Para-cresyl phenylacetate	10
Benzyl acetate	6
Acetyl isoeugenol	20
Heliotropin	30
Ylang extra	5
Para-cresol	1
3-Methylthio-4-heptanol, produced according to the process of Example 2	20

The 3-methylthio-4-heptanol, produced according to the process of Example 2 imparts the green, floral, heady tobaccolike oriental middle and undertone necessary for narcisse.

Example 5: *Perfumed Liquid Detergents* – Concentrated liquid detergents with rich, pleasant aromas are prepared containing 0.10%, 0.15% and 0.20% of an α-oxy(oxo) sulfide as set forth in the table below. They are prepared by adding and homogeneously admixing the appropriate quantity of α-oxy(oxo) sulfide in the liquid detergent.

The liquid detergents are all produced using anionic detergents containing a 50:50 mixture of sodium lauroyl sarcosinate and potassium N-methyl lauroyl tauride. The detergents all possess a pleasant fragrance as defined in the table below, the intensity increasing with greater concentration of α-oxy(oxo) sulfide.

Compound	Aroma
3-Methylthio-4-heptanol	A sweet, green, floral, herbal, vegetative aroma with an underlying verdima nuance.
3-Methylthio-4-heptanone	A green, minty, herbaceous aroma with basil nuances.
3-Propylthio-4-heptanol	A green aroma with floral nuances.
3-Thioacetyl-2,6-dimethyl-4-heptanone	A grapefruit oil-like aroma.
3-Isobutylthio-4-heptanone	A green, spicy and peppery aroma with an underlying bergamot note.
3-Isobutylthio-2,6-dimethyl-4-heptanol	A green aroma containing notes of hyacinth and narcisse.
3-Methylthio-2,6-dimethyl-4-heptanone	A sweet, slightly floral and woody aroma with fruity and berry nuances.
3-Methallylthio-2,6-dimethyl-4-heptanone	A fruity, grapefruit, somewhat floral aroma with underlying yara and neroli notes and bready, vegetative nuances.

Aliphatic Dibasic Acid Diesters

M. Katada, K. Hayashi, H. Katsura, K. Tomita, H. Morohoshi, K. Uehara and H. Tanaka; U.S. Patent 4,110,626; Aug. 29, 1978; assigned to Shiseido Co., Ltd., Japan have developed a method for improving the quality of the fragrance of a perfume by balancing the fragrance and maintaining the tenacity of the fragrance of the perfume by incorporating at least one perfume controlling agent in the perfume. Aliphatic dibasic acid diesters of the general formula:

$$R_1OCOR_2COOR_3$$

wherein R_1 and R_3, which can be the same or different, each represents a saturated branched chain alkyl group containing 4 or 5 carbon atoms, and R_2 represents a saturated straight chain alkylene group containing 4 carbon atoms, have been found to be colorless, transparent, odorless, tasteless, low viscosity liquids which serve as superior diluents for various odoriferous substances.

When at least one of these compounds is added to a perfume, the fragrance characteristics and fragrance stability of the perfume can be increased greatly.

Suitable examples of saturated branched chain alkyl groups represented by R_1 and R_3 include an isobutyl group and an isoamyl group. A suitable example of the alkylene group represented by R_2 includes a butylene group.

The perfume controlling agents therefore include diisobutyl adipate (DIBA), diisoamyl adipate (DIAA) and isobutyl isoamyl adipate. Of these, DIBA and DIAA are most preferred. The perfume controlling agent concurrently possesses the effects of a stabilizer, a diluent and fixative, and is odorless.

Example 1: *Synthesis of DIBA* – a three-necked flask equipped with a stirring device and a condenser was charged with 3 mols of isobutanol, 1 mol of adipic acid and 0.01 mol of sulfuric acid, and the reaction was performed under reflux for three hours. Then, under reduced pressure, water formed in the reaction and the unreacted isobutanol were evaporated off. 3 mols of isobutanol was then freshly added, and the reaction was carried out in the same way until the reaction was completed. Again, water formed in the reaction and the excess of isobutanol were evaporated off under reduced pressure.

The crude diisobutyl adipate was washed with a 5.0% by weight aqueous solution of sodium carbonate to neutralize the catalyst and remove precipitate. After removing the residue, the product was washed several times with water. The washed product was dried and distilled at reduced pressure to obtain 0.88 mol of a main fraction having a boiling point of 113° to 114°C/2 mm Hg. The main fraction was deodorized using steam to obtain colorless, transparent, tasteless and odorless diisobutyl adipate having a purity, as determined by gas chromatography, of 99.9%.

Example 2: *Floral-Bouquet Type Formulation* – A typical perfume formulation whose fragrance has been improved in quality by the use of such compounds is given below.

	Percent by weight
Linalool	4.0
Bergamot oil	4.0
Linalyl acetate	1.5
Absolute jasmine oil	5.0
Essential rose oil	0.5
Ylang Ylang oil	4.0
Eugenol	1.5
Isoeugenol	2.0
Vetiveryl acetate	2.5

(continued)

	Percent by weight
Hydroxycitronellal	9.0
α-Isomethyl ionone	3.5
Benzyl salicylate	10.0
Phenylethyl alcohol	9.0
Musk ketone	6.5
Lemon oil	1.0
Decanal, 1% ethyl alcohol solution	1.0
Dodecanal, 1% ethyl alcohol solution	1.0
Geraniol	4.0
Vertofix coeur	3.0
Heliotropin	2.0
Santalex	1.0
DIBA	24.0
Total	100.0

3-(10-Undecenyloxy)Propionitrile

M. Plattier and P.J. Teisseire; U.S. Patent 4,115,326; September 19, 1978; assigned to Societe Anonyme Roure Bertrand Dupont, France have found that 3-(10-undecenyloxy)propionitrile, having the formula, $CH_2=CH(CH_2)_9O(CH_2)_2CN$, has a fruity odor reminiscent of pineapple. The odor also has exceptional tenacity, and may be used to improve the fragrance of certain synthetic products. Example 1 describes its synthesis, and Examples 2 and 3 formulations containing it.

Example 1: A 2 liter reactor provided with a stirrer, a heater, a reflux condenser and a 1 liter collecting flask was charged with 850 g of 10-undecen-1-ol and then 2 g of sodium and heated to 90° to 95°C. When the sodium had completely reacted, the reaction mass was cooled to 80°C and 132.5 g of acrylonitrile were added at 80°C in the course of ½ hour. The mass was maintained at this temperature for a further hour and then cooled to 25°C and 20 ml of acetic acid and then 1 liter of benzene were added. The organic phase was decanted off and then washed 3 times with 250 ml of water. The solvent was finally distilled off in a water bath under a pressure of 30 mm Hg.

1,006 g of crude 3-(10-undecenyloxy)propionitrile were obtained, which were fractionated under a pressure of 0.5 mm Hg. 432.7 g of pure nitrile were collected which represents a theoretical yield of 77.6% with respect to acrylonitrile, 38.8% with respect to undecenol employed and 81.1% with respect to undecenol consumed.

Example 2: *Fern Perfume Formulation –*

	Parts by weight, g
40% Lavender	20
Coumarin	20
Patchouli	20
Vetiver bourbon	10
Heliotropin	10
Vanillal containing 10% ethyl phthalate	20
Amyl salicylate	40

(continued)

	Parts by weight, g
Resinoid oak moss A No. 1 (50% solution in ethyl phthalate)	40
12-oxahexadecanolide	10
Geranium bourbon	10
50% Resinoid benzoin No. 1 syrup	40
3-(10-undecenyloxy)propionitrile	8
Total	248

Example 3: *Liquid Detergent Formulation* – A concentrated liquid detergent was prepared composed of the following:

	Parts by weight, g
Sodium alkylbenzenesulfonate	600.0
Sodium lauryl sulfonate	30.0
Water	170.0
Fern composition of Example 2	1.6

This detergent had a particularly pleasant woody, coumarin, fruity odor.

Tricyclodecane-Methylol and Derivatives

W. Skorianetz, K.P. Dastur and H. Strickler; U.S. Patent 4,123,394; October 31, 1978; assigned to Firmenich SA, Switzerland have found that tricyclo[5.2.1.$0^{2,6}$] decane-methylol and its esters or ethers of formula

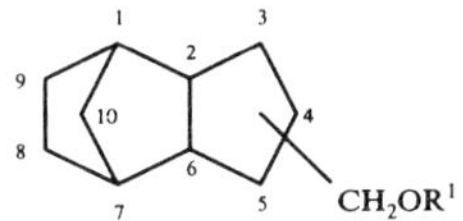

wherein R^1 represents a lower saturated or unsaturated, linear or branched alkyl radical, having from 1 to 6 carbon atoms, an acyl radical containing from 1 to 6 carbon atoms or a hydrogen atom, and wherein the side-chain is bound to the cyclopentane ring at positions 4 or 5, possess very useful perfuming properties, and consequently, can be used advantageously for improving, enhancing, or modifying the fragrance character of perfumes, perfume compositions and perfumed articles.

Compounds of particular utility include tricyclo [5.2.1.$0^{2,6}$] decane-methylol and the formate, acetate, propionate and butyrate esters of this alcohol. The alcohol possesses a powerful scent having a flowery, fresh character. The corresponding acetate develops a pleasant woody, green smell reminiscent of certain nuances of rhubarb. The corresponding formate develops a woody note with an irislike character.

Tricyclo[5.2.1.$0^{2,6}$] decane-methylol is commercially available and can be prepared by the process described in German Patent 934,889. The esters can be prepared using known esterification procedures. It will be understood that in preparing these compounds mixtures of isomers in which the methylol group is attached to the ring at the 4 and 5 positions will be obtained. It is, in general, unnecessary to separate the isomers.

Example 1: A base perfume composition of lavender type fragrance destined to be incorporated into toilet soaps was prepared by mixing together the following ingredients:

	Parts by weight
Lavandin oil	250
Linalyl acetate	100
Linalol	100
10% trans-epoxyocimene	100
Diisobutylcarbinyl acetate	60
Geraniol	50
Coumarin	40
Cedar wood oil	30
Terpineol	30
10% β-Damascone	20
10% Trimethylcyclohexenyl-carbaldehyde*	10
1-(3,3-Dimethyl-cyclohex-6-en-1-yl)-pent-4-en-1-one 1%	10
Total	800

*In diethyl phthalate

By adding to 80 g of the above base composition 20 g of tricyclo[5.2.1.$0^{2,6}$] decylmethyl acetate, there was obtained a composition which possessed an enhanced lavender scent, the character of which was more natural and elegant than that of the base composition.

Example 2: A base perfume composition was prepared by admixing the following ingredients:

	Parts by weight
Phenylethanol	300
Tricyclo[5.2.1.$0^{2,6}$] decane-methylol	200
Heptylacetate	80
p-tert-Butylcyclohexylacetate	80
Dimethyloctanol	60
Benzyl salicylate	50
1% 1-(3,3-Dimethyl-cyclohex-6-en-1-yl)-pent-4-en-1-one	50
Diisobutylcarbinol	40
4-Isopropyl-cyclohexylmethanol	40
Dimethylhydroquinone	20
10% Geranonitrile*	80
Total	1,000

*In diethyl phthalate

Tricyclo[5.2.1.$0^{2,6}$] decane-methylol confers to the flowery composition given above a Ylang-Ylang character. The obtained composition is particularly useful for perfuming soaps and detergents.

2-Methyl-2-Alkyl-Alkanoic Acid Esters

J. Conrad, U.-A. Schaper and K. Bruns; U.S. Patent 4,126,585; November 21, 1978; assigned to Henkel Kommanditgesellschaft auf Aktien (Henkel KGaA),

Germany have found that 2-methyl-2-alkyl-alkanoic acid esters of the general formula

$$R_3-CH_2-CH_2-\underset{\displaystyle R_2}{\overset{\displaystyle CH_3}{\overset{|}{\underset{|}{C}}}}-COOR_1$$

wherein R_1 is a saturated or unsaturated, straight- or branched-chain aliphatic hydrocarbon radical having 1 to 5 carbon atoms, and R_2 and R_3 are an alkyl radical having 1 to 4 carbon atoms, are valuable perfumes having a very natural and complex scent.

The esters can be used in amounts of 0.05% to 2% by weight relative to the total product as perfumes or to perfume cosmetics, toiletries, soaps, etc.

Example 1: *Synthesis of Ethyl 2-Methyl-2-Ethyl-Hexanoate* – (A) Production of 2-methyl-2-ethyl-hexanoyl chloride: 316 g (2 mols) of 2-methyl-2-ethyl-hexanoic acid were heated to boiling with 357 g (3 mols) of thionyl chloride under agitation until the evolution of gas had ended. The surplus thionyl chloride was distilled off and the residue was fractionated in vacuum. 320 g (90% of theory) of 2-methyl-2-ethyl-hexanoyl chloride of boiling point 78°C at 13 mm Hg were obtained.

(B) Production of ethyl 2-methyl-2-ethyl-hexanoate: 11.5 g (0.5 mol) of sodium were dissolved in 150 ml of ethanol. 44 g (0.25 mol) of 2-methyl-2-ethyl-hexanoyl chloride were slowly added under agitation to the solution of sodium ethylate at 0° to 5°C. The mixture was subsequently agitated for 3 hours at room temperature, filtered, absorbed in ether, washed neutral with water, dried, reduced, and fractionated in vacuum. The ethyl 2-methyl-2-ethyl-hexanoate was obtained in the form of a colorless oil having a fruity/fresh odor with a fragrance of apple and camomile, and has the following characteristics: boiling point, 85°C at 14 mm Hg; and refractive index, n_D^{20} = 1.4210.

Example 2: Methyl-2-methyl-2-ethyl-hexanoate was obtained in an analogous manner to that set forth above, using 2-methyl-2-ethyl-hexanoyl chloride and sodium methylate. It had the following characteristics: appearance, colorless oil; odor, fresh, peppermintlike with a menthol fragrance; boiling point, 75°C at 14 mm Hg; and refractive index, n_D^{20} = 1.4228.

Example 3: *Peppermint Base Composition* –

	Parts by weight
Methyl 2-methyl-2-ethyl-hexanoate	300
l-Menthol	300
l-Menthyl acetate	190
Menthofuran	50
Piperitone	25
l-Carvone	15
Pulegone	30
Eucalyptus oil	80
Menthone	10
Total	1,000

Example 4: *Soap Perfume Composition –*

	Parts by weight
Citrenes	450
Ethyl 2-methyl-2-ethyl-hexanoate	325
Methyl anthralinate	100
Indole	5
Bergamot oil	70
Tolu balsam	50
Total	1,000

This soap perfume composition is added to a toilet soap in amounts of from 0.5-1.0% by weight.

Trimethylcyclohexylethyl Ethers

J. Conrad, K. Bruns and P. Meins; U.S. Patent 4,130,509; December 19, 1978; assigned to Henkel Kommanditgesellschaft auf Aktien (Henkel KGaA), Germany have formulated perfumes having 1 to 50% by weight of cis- and trans-trimethylcyclohexylethyl ethers of the formula

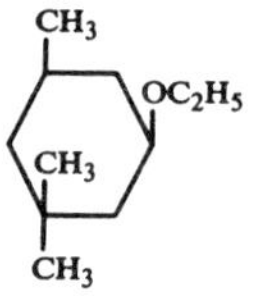

CH_3 OC_2H_5 CH_3 CH_3 and CH_3 CH_3 OC_2H_5 CH_3

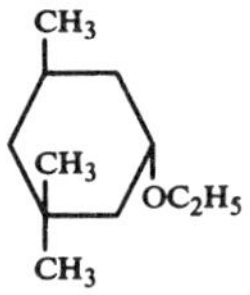

and the remainder customary constituents of perfumery compositions.

They also provide an improved process for making trans-3,3,5-trimethylcyclohexylethyl ether which comprises hydrogenating 3,3,5-trimethylcyclohexenylethyl ether in the presence of a dry nickel catalyst, preferably Girdler nickel 49A, at a temperature of 150° to 200°C, and at 10 to 200 bar hydrogen pressure, and subsequently processing the hydrogenation product in known manner.

These cis- and trans-ethers can also be made by conventional methods, as described in *Helvetica Chimica Acta* 47, 1398 (1964). Both ethers have pleasant fragrances, but the trans-ether is the much stronger and more valuable substance.

Example 1: *Production of trans-3,3,5-Trimethylcyclohexylethyl Ether* – 280 g (2 mols) of 3,3,5-trimethylcyclohexanone and 355 g (2 mols) of triethyl orthoformate were added to 2 g of p-toluenesulfonic acid and stirred for 30 minutes. Subsequently, at normal pressure approximately 200 g of ethyl formate and ethanol were distilled off (bottom 70° to 115°C, top 56° to 77°C), and the residue was fractionated in vacuum. 340 g of enol ether (boiling point$_{17mm}$ 59° to 63°C) were obtained in quantitative yield.

340 g of the enol ether obtained as described above were added to 25 g of Girdler nickel 49A and heated slowly up to 200°C in an autoclave at 50 bar hydrogen pressure. At approximately 170°C hydrogen adsorption commenced. Such adsorption terminated within 15 minutes. The mixture was allowed to react further for one hour at 200°C and at an increased hydrogen pressure of 180 bars. It was then cooled down, and the product was decanted from the catalyst and then fractionated in vacuum.

250 g, i.e., 74% of the theoretical yield, of 3,3,5-trimethylcyclohexylethyl ether were obtained with a boiling point$_{100mm}$ of 112° to 116°C and a refractive index n_D^{20} = 1.4364. The reaction product consisted of 95% trans-3,3,5-trimethylcyclohexylethyl ether and of only 5% cis-3,3,5-trimethylcyclohexylethyl ether.

Example 2: *Artificial Lavender Perfume Composition –*

	Parts by weight
trans-3,3,5-Trimethylcyclohexylethyl ether (95% trans-form; 5% cis-form)	100
Lavandin oil Abrialis	400
Lavender oil acetylated	375
Geranyl acetate	30
Lavender absolute	25
Allyl ionone	20
Citronellol	20
Ethylamyl ketone	15
Sandal (H&R)	10
Coumarin	5
Total	1,000

Cyanoethylidene-Bicyclo-Heptenes

G.W. Shaffer and K.L. Purzycki; U.S. Patent 4,132,677; January 2, 1979; assigned to Givaudan Corporation have developed a class of fragrance compounds having the general formula:

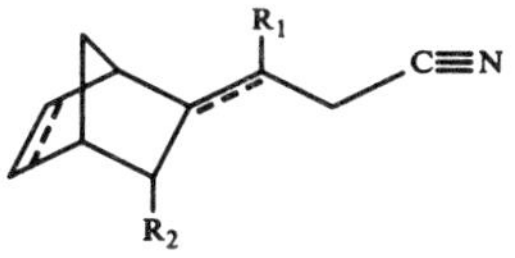

R1 C≡N R2

wherein R_1 and R_2 are alike or different and are chosen from the group consisting of hydrogen or methyl. A dotted line (---) between the two carbons indicates that either a double bond or a single bond may exist between those two carbons.

These compounds have spicy, floral odors and are valuable in fragrance compositions. The compounds can be prepared by Knovenagel reaction and a hydrogenation. These compounds are useful in a variety of odor compositions including rose, jasmine, violet, carnation, galbanum, labdanum, tobacco, leather, cinnamon bark and the like.

Compounds of the general structure shown below are especially preferred; their odor being more intense than the corresponding semihydrogenated and hydrogenated compounds.

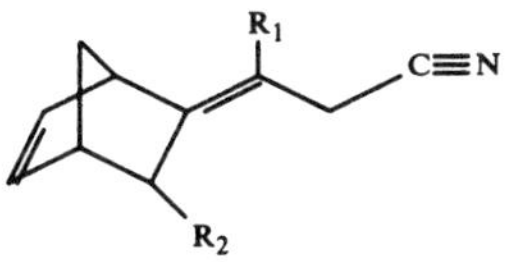

R1 C≡N R2

These aroma chemicals can be used in perfume formulations in a practical range extending from 0.1 to 30%. This will vary, of course, depending upon the type of fragrance formula involved. Higher concentrations above 30% (i.e., 80 to 90%) may be used successfully for special effects.

The compounds can be used to prepare perfumes and toilet waters by adding the usual alcoholic and aqueous diluents thereto; approximately 15 to 20% by weight of base would be used for the former and approximately 3 to 5% by weight would be used for the latter.

Similarly, the base compositions can be used to odorize soaps, detergents, cosmetics, or the like. In these instances, a base concentration of from 0.5 to 2% by weight can be used.

Example 1: *2-(2-Cyanoethylidene)-3-Methylbicyclo[2.2.1] Hept-5-ene* – A stirred solution of 136 g (1.0 mol) of 2-formyl-3-methylbicyclo[2.2.1] hept-5-ene, 91 g (1.05 mols) of cyanoacetic acid, 3 g (0.05 mol) of ammonium hydroxide (58%), 132 ml of dimethylformamide, and 170 ml of benzene were heated to reflux and the water removed with a Dean-Stark trap. The reaction was allowed to continue until the carbon dioxide evolution ceased (approximately 31 hours). Upon completion, the reaction was cooled and the solvent removed under reduced pressure. The residual oil was distilled under vacuum to give 120 g (75.5% yield) of 2-(2-cyanoethylidene)-3-methylbicyclo[2.2.1] hept-5-ene: boiling point 70° to 80°C/1 mm.

The compounds of this process can be used to provide or enhance spicy notes in perfume compositions.

Example 2: *Carnation Type Base Formulation* –

	Parts by weight
Aldehyde C_{11}, 10% in diethyl phthalate	10
Amyl salicylate	100
Baccartol*	50
Benzyl isoeugenol	30
Cinnamic alcohol pure	75
Cinnamon leaf Seychelles	5
Copaiba oil	40
Eugenol USP extra	50
p-Isopropylcyclohexanol	100
Isoeugenol	50
Methyl isoeugenol	20
Methyl undecylenate	10
2,6-Dinitro-3-methoxy-4-t-butyltoluene	15
Nutmeg oil	10
Phenyl ethyl alcohol	100
3,7-Dimethyl-7(6)-octen-1-ol	200
Trichloromethyl phenyl carbinyl acetate	30
7-Acetyl-1,1,4,4-tetramethyl-7-ethyl-1,2,3,4-tetralin	50
Ylang-Ylang #3	50
Compound A	5
Total	1,000

*Givaudan condensation product of citronella oil and acetone.

In the above formulation, Compound A represents either the odorless diethylphthalate or a compound of this process.

The presence of 2-(2-cyanoethylidene)-2-methylbicyclo[2.2.1] hept-5-ene provided the above composition with intensified spiciness, actually changing the top note from fruity to spicy, and made the total odor impression stronger and more rounded at the same time. The overall effect was of a more natural carnation. The other compounds of this process may be used in a similar manner.

Certain Crotonyl-Trimethylcyclohexanes

D.R. De Haan and D.K. Kettenes; U.S. Patent 4,136,066; January 23, 1979; assigned to P.F.W. Beheer B.V., Netherlands have prepared perfume compositions using either the trans- (Compound I) or cis- (Compound II) stereoisomers of E-1-crotonyl-2,2,6-trimethylcyclohexane, which have the formulas:

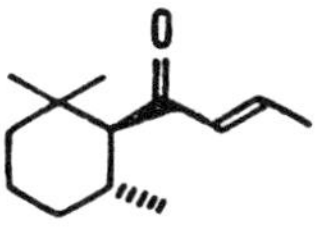

trans E (I) cis E (II)

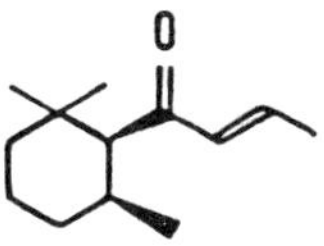

The trans-E-1-crotonoyl-2,2,6-trimethylcyclohexane possesses a powerful odor that can be described as very natural fruity, rich and warm, with an earthy-minty shading that accentuates the natural character. Its tenacity, when smelled on evaporation-blotterstrips, is very good and makes this compound of great interest for use in compositions of floral and/or woody character like iris, mimosa, boronia, tobacco, and fancy bouquets.

The cis-E-isomer has a less fruity character but a stronger earthy-minty aspect that makes it highly useful in perfume compositions. Furthermore, the presence of 10 to 20% of the cis-E-isomer in the trans-E-isomer results in a slightly different shade but not in a negating effect.

The synthesis of the trans-isomer is described in U.S. Patent 4,109,022, in the section of this book on "Flavorings". The preparation of the cis-form, which is not so useful for flavors but is also useful for perfumery, is described in Examples 1 through 3. The reaction scheme for the synthesis is as follows:

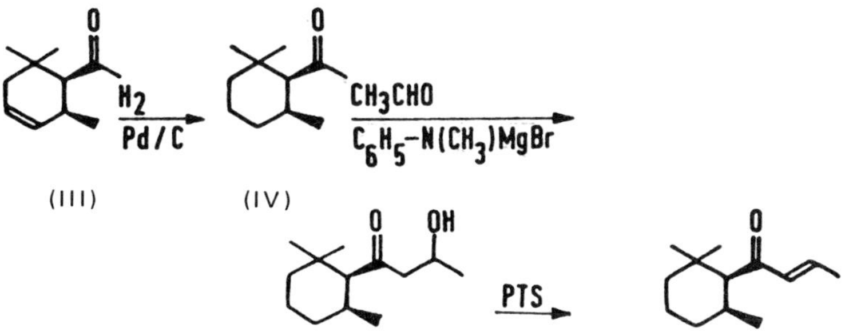

(III) $\xrightarrow[Pd/C]{H_2}$ (IV) $\xrightarrow[C_6H_5-N(CH_3)MgBr]{CH_3CHO}$ (V) $\xrightarrow{PTS}$ (II)

Example 1: *Preparation of cis-1-Acetyl-2,2,6-Trimethylcyclohexane (IV)* – 464 g 1-acetyl-2,2,6-trimethyl-4-cyclohexene (2.8 mols) in 800 ml of ethanol and 3 g 10% palladium on charcoal are hydrogenated in an autoclave under a pressure of 50 atm and at 50°C. After the uptake ceases, the reaction mixture is filtered and concentrated by distillation under reduced pressure. The product cis-1-acetyl-2,2,6-trimethylcyclohexane (IV, containing about 10% of the trans-isomer) is collected at 83° to 88°C/12 mm; yield, 420 g (90%).

Example 2: *Preparation of cis-1-(2,2,6-Trimethylcyclohexyl)-3-Butanolone-1 (V)* – To 750 ml of an ethereal solution of ethylmagnesium bromide [prepared from 302 g (2.7 mols) of ethyl bromide and 58.7 g (2.4 mols) magnesium turnings in 550 ml of ether] is added with cooling and stirring a solution of 238 g (2.2 mols) of freshly distilled dry N-methylaniline in 700 ml of dry benzene (N_2 atmosphere).

To the above freshly prepared solution of N-methylanilinomagnesium bromide is added during 25 to 30 minutes a solution of 370 g (2.2 mols) of 1-acetyl-2,2,6-trimethylcyclohexane (IV) in 370 ml of dry benzene while keeping the temperature at 15°C. After the addition of the ketone, the solution is allowed to stand for 30 minutes. A solution of 145.5 g (3.3 mols) of freshly distilled acetaldehyde in 150 ml of dry benzene is then added during 45 minutes, keeping the temperature at -13° to -10°C.

After the addition of the aldehyde, the solution is allowed to stand for 90 minutes at the same temperature. 2,000 ml of 3 N hydrochloric acid is then added with stirring and cooling. The organic layer is separated and washed 6 times with 1,000 ml 3 N hydrochloric acid (to remove the N-methylaniline) and finally with 500 ml of water, 500 ml of sodium bicarbonate solution, and then water. The combined organic layers are dried over anhydrous sodium sulfate, and the solvents are removed by distillation at reduced pressure. The residue is distilled through a short Vigreux column. The product, cis-1-(2,2,6-trimethylcyclohexyl)-3-butanolone-1 (containing about 10% of the trans-isomer) is collected at 98° to 104°C/1 mm; yield, 320 g (70%).

Example 3: *Preparation of cis-E-1-Crotonoyl-2,2,6-Trimethylcyclohexane (II)* – A solution of 320 g (1.5 mols) of the ketol prepared in Example 2 in 1,000 ml of benzene containing 2 g p-toluene-sulfonic acid is heated in an apparatus for azeotropical water separation (Dean-Stark) until no more water separates. The reaction mixture is washed with a solution of sodium bicarbonate until neutral and dried over anhydrous sodium sulfate.

The solvent is removed by distillation at reduced pressure and the residue is distilled through a short Vigreux column. The product, cis-E-1-crotonoyl-2,2,6-trimethylcyclohexane (II, contains about 10% of the trans-isomer) is collected at 76°C/1 mm; yield, 273 g (94%). Recrystallization from pentane gives cis-E-1-crotonoyl-2,2,6-trimethylcyclohexane in high purity (melting point 27° to 28°C).

Examples 4 and 5 show perfumery compositions, one of which uses the cis- and the other the trans-stereoisomer.

Example 4: A perfume composition concentrate or base, suitable for admixture with other perfume ingredients, was prepared by mixing together the following ingredients:

	Parts by weight
δ-Undecalactone	250
Dihydrojasmone	100
Methyl-2-n-hexyl-3-oxo-cyclopentane-carboxylate (PFW)	50
Allyl heptanoate	100
Eugenol	50
Geranyl acetate	50
Dimethylbenzylcarbinyl butyrate	50
Isocyclocitral	5
cis-3-Hexenyl acetate (10% solution in diethylphthalate)	15
cis-3-Hexenol (10% solution in diethylphthalate)	15
Methyl ethyl phenyl glycidate	5
Linalyl isobutyrate	60
cis-E-1-crotonoyl-2,2,6-trimethylcyclohexane	250
Total	1,000

This composition has a very strong mixed fruity character: apple, prune, apricot-gooseberry, in which the cis-E-isomer of 1-crotonoyl-2,2,6-trimethylcyclohexane plays a dominant role.

Example 5: A perfume composition or base, suitable for admixture with other perfume ingredients, was prepared by mixing together the following ingredients:

	Parts by weight
Methyl octinecarboxylate	10
Heliotropine	50
Hydroxycitronellal	200
Linalool	200
Ylang-Ylang premier	100
δ-Dodecalactone	10
Schiff's base of 4-(4-methyl-4-hydroxypentyl)-3-cyclohexenecarboxaldehyde (IFF) with methyl anthranilate	100
Methyl anthranilate	10
4-(p-hydroxyphenyl)-butanone-2 (IFF)	10
8,8-di(3H-indolyl-3)-2,6-dimethyl-2-octanol (Roure Bertrand)	10
cis-3-Hexenyl benzoate	25
cis-3-Hexenyl salicylate	25
trans-E-1-crotonyl-2,2,6-trimethylcyclohexane	250
Total	1,000

This composition has a strong tuberose character with a very long lasting dry-out. The trans-E-isomer of 1-crotonoyl-2,2,6-trimethylcyclohexane is one of the pushing factors.

Ketone Acetals

S.A.R. Dewaele; U.S. Patents 4,092,331; May 30, 1978; and 4,144,249; March 13, 1979; both assigned to SA Texaco Belgium NV, Belgium provides a process for the preparation of ketone acetals from polyhydric phenols containing at least two pairs of hydroxy groups in ortho positions which comprises reacting a polyhydric phenol having at least four OH groups, including at least two pairs of OH groups in ortho positions, with a ketone, or a dialkyl acetal of such a ketone, in the presence of a nonaqueous solvent and a strong Lewis acid catalyst.

Preferably, the polyhydric phenol is a compound of the formula

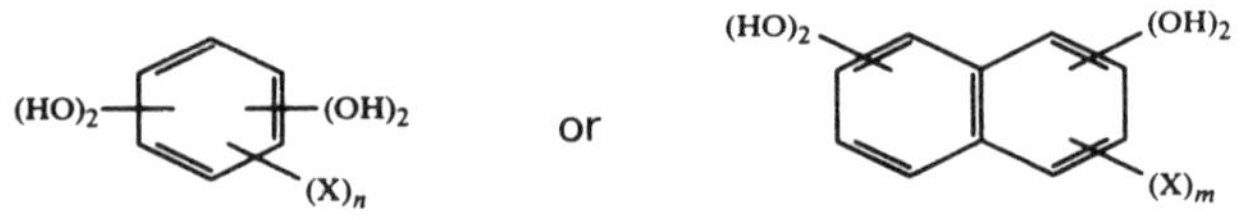

$(HO)_2$–[benzene ring]–$(OH)_2$, $(X)_n$ or $(HO)_2$–[naphthalene ring]–$(OH)_2$, $(X)_m$

in which the OH groups in each pair are in positions ortho- to one another, X represents a halogen atom, an OH group or a group of the formula R, OR or SR in which R represents a hydrocarbon or substituted hydrocarbon group, m represents 0, or an integer from 1 to 4, and n represents 0, 1 or 2.

Preferably, m and n are 0, and the most preferred polyhydric phenol is 1,2,4,5-tetrahydroxy benzene.

The preferred ketones are acetone, hexan-2,5-dione, cyclohexan-1,4-dione or 1,4-diacetylbenzene. Acetone is the most preferred ketone. The preferred ketone acetal is the dimethyl acetal of acetone, i.e., 2,2-dimethoxypropane.

The process is carried out in a nonaqueous solvent and in the presence of a strong Lewis acid catalyst. Examples of suitable catalysts are phosphorus pentoxide, toluene sulfonic acid, hydrochloric acid, sulfuric acid, boron trifluoride, zinc chloride and aluminum chloride.

Examples of suitable solvents include tetrahydrofuran, dioxan dimethoxyethane, and diethylene glycol dimethyl ether. Where the ketone is acetone, it can also constitute the solvent, although only dilute solutions of the polyhydric phenols can be obtained in acetone. Reaction can be carried out at room temperature or at elevated temperature up to the boiling point of the solvent; for example, in diethylene glycol dimethyl ether, reaction can conveniently be carried out at a temperature of about 110°C. When a dialkyl acetal of the ketone is employed, the alcohol formed during the course of the reaction (methanol in the case of 2,2-dimethoxypropane) should be distilled off during the reaction to prevent it from interfering with the desired reaction.

Polymeric acetals formed from diketones are of high heat stability.

The acetals formed by the process are of interest in perfumery. For example, 2,2,6,6-tetramethylbenzobisdioxole, the compound of Example 1, has an odor providing a green top note and may be employed in perfumes and fragrances to replace or supplement other materials possessing green top notes, such as phenyl acetaldehyde, hydrotropic aldehyde, acetals of these aldehydes, extract of violet leaves, galbanum oil, methyl heptine carbonate or methyl octine carbonate.

The process is further illustrated by the following examples, in which the following abbreviations are employed: THB: 1,2,4,5-tetrahydroxybenzene; THF: tetrahydrofuran; and DMP: 2,2-dimethoxypropane.

Example 1: THB (926 g) and acetone (28.3 g) are dissolved in tetrahydrofuran (500 ml) in a flask protected from moisture by $CaCl_2$ tubes. Phosphorus pentoxide (84 g) is added to this mixture whereafter it is heated at 65°C for 4 hours. The reaction mixture is allowed to cool, the liquid is decanted from the solid mass, and solid, dry potassium carbonate (5 g) is added. After stirring for 1 hour the liquid is decanted again and the solvent (THF) is distilled off leaving a brown solid residue. The latter is extracted with petroleum ether (100 to 200 ml) and filtered and the petroleum ether is distilled off.

The residue is 2,2,6,6-tetramethylbenzobisdioxole (11.5 g). Yield, 28%. Melting point after crystallization from petroleum ether, 120° to 121°C. Infrared spectrum shows peaks at 6.7, 7.2, 8.0, 10.2 and 11.9 μ. The 60 Mc proton NMR spectrum shows peaks at 1.6 ppm (H atoms of the methyl groups) and 6.2 ppm (aromatic H atoms). Elemental analysis–Calculated for $C_{12}H_{14}O_4$: C, 64.8%; H, 6.4%. Found: C, 65.0%; H, 6.3%.

Example 2: THB (10 g), dimethoxypropane (14.8 g) and p-toluene sulfonic acid (0.025 g) are heated in THF (100 ml) for 120 hours. During the heating period a liquid of boiling point 60° to 65°C is distilled off through a distillation column and the volume in the flask is maintained at its original level by the addition of THF. In this way methanol is distilled off as it is formed.

A mixture of THF and methanol was distilled off during reaction (using a 60 cm Vigreux column) and the level in the flask was maintained by addition of THF. After 120 hours all DMP had disappeared. The solvent is then removed by distillation, the residue is dissolved in diethyl ether (100 ml) and washed with sodium hydroxide solution (2 N, 100 ml). The two layers are separated, and the diethyl ether is distilled from the organic layer leaving 2,2,6,6-tetramethyl benzobisdioxole (5.3 g). Yield, 34%. Melting point after recrystallization from petroleum ether: 120° to 121°C.

Example 3: A parma violet fragrance suitable for incorporation in violet-scented perfumes and cosmetics was produced having the following composition:

	Parts by weight
Benzyl acetate	100
Bergamol	100
Methylheptine carbonate	5
2,2,6,6-tetramethylbenzobisdioxole	5
Orris conc.	20
Methyl ionone	500
Violet leaf absolute	20
Ionone-α	150
Benzyl isoeugenol	40
Ylang Ylang	20
Jasmine absolute	20
Cassia absolute	20
Total	1,000

Certain Allyl Ethers

K.P. Dastur and J.J. Becker; U.S. Patent 4,150,000; April 17, 1979; assigned to Firmenich SA, Switzerland have synthesized allyl ethers corresponding to formula I in which the dotted lines signify that the methyl groups may be attached at either position 1 or position 2.

(I)

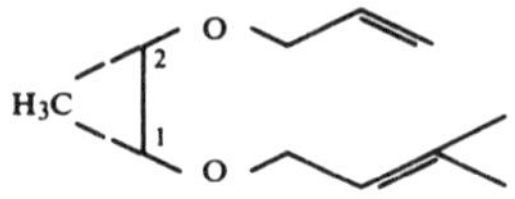

In the described synthesis, the reaction product, which will be referred to as Reaction Product I, is a mixture of the two isomers 1-[3'-methyl-but-2'-en-1'-oxy]-2-[prop-2'-en-1'-oxy]-propane and 1-[prop-2'-en-1'-oxy]-2-[3'-methyl-but-2'-en-1'-oxy]-propane.

Reaction Product I is fully useful in perfume applications without separation of the isomers, although this can be accomplished, if desired, by conventional techniques such as vapor phase chromatography.

Reaction Product I has a natural fugacious scent of green, slightly flowery-aromatic character and can be used in a large variety of applications, namely for the manufacture of flowery compositions to which they confer a much appreciated naturalness. The product is particularly adapted to the perfuming of household products, such as soaps, detergents, air fresheners as well as cosmetic materials.

Example 1: (A) A mixture of 200 g of propylene glycol, 214 g of 2-methyl-but-3-en-2-ol and 20 g of acidic diatomaceous earth was kept under stirring during 20 hours at about 90°C. After cooling and filtration, an extraction was effected with 1 liter of diethyl ether, whereupon the combined organic extracts were washed, dried over anhydrous sodium sulfate and finally concentrated. The obtained residue was distilled to yield 114 g (yield 31.8%) of 3-[3'-methyl-but-2'-en-1'-oxy]-propan-2-ol and 2-[3'-methyl-but-2'-en-1'-oxy]-propan-1-ol; boiling point 77° to 83°C/12 torrs.

(B) A solution of the alcohol obtained as indicated above (70 g) in 500 ml of anhydrous tetrahydrofuran was added dropwise while stirring for 20 minutes to a suspension of 12 g of sodium hydride in 500 ml of anhydrous tetrahydrofuran. The reaction mixture was kept under stirring at room temperature for 1½ hours, then 60 g of allyl bromide in 120 ml of tetrahydrofuran were added thereto and stirring was carried on for 16 supplementary hours while the mixture was heated to 70°C. After cooling, the excess of sodium hydride was decomposed with 5 ml of methanol and the mixture was evaporated to dryness. The obtained residue was extracted with 250 ml of diethyl ether and the organic extracts were washed with 200 ml of water. After drying over Na_2SO_4, evaporation and distillation, 54 g of the desired products (yield 60.3%) were collected at boiling point 40° to 48°C/0.2 torr.

It is clear that since in step (A) of the process the alcohol produced is a mixture of positional isomers the ether obtained from this alcohol in step (B) will be a corresponding mixture of isomers. If isomer separation is desired it is better accomplished with the alcohol than with the ether.

Example 2: A base perfume composition of flowery type was prepared by admixing the following ingredients:

	Parts by weight
Phenethylol	150
Terpineol	100
tert-Butylcyclohexyl acetate	100
Cinnamic alcohol	80
Hydratropic alcohol	60
Rosewood oil of Brazil	60
Cyclohexylethyl acetate	60
2,4-Dimethylcyclohex-3-en-1-yl carbaldehyde 10%*	60
Benzyl salicylate	60
Synthetic lily-of-the-valley oil	50
Cyclamen aldehyde	20
Hexyl salicylate	20
Isobutyl benzoate	20
Galbanum oil 10%*	20
Isononyl aldehyde 10%*	20
Isopropylcyclohexylmethanol**	10
Methylnonylaldehyde 10%*	10
Total	900

*In diethyl phthalate
**Mayol, see Swiss patent 578,312

By adding to 90 g of the above base 10 g of Reaction Product I, there was obtained a composition whose odorous characters were more natural and aromatic than those of the base composition. The composition possessed, moreover, an enhanced top note.

Nonanols, Nonenols and Their Monocarboxylic Acid Esters

The perfume compositions described by *K. Kulka; U.S. Patent 4,168,248; September 18, 1979; assigned to Fritzsche Dodge & Olcott Inc.* include at least 1% by weight in the formulation of one or more of 3-methylnonan-3-ol, 3-methyl-1-nonen-3-ol, 3-methylnonan-1-ol, 3-methyl-2-nonen-1-ol or monocarboxylic acid esters of these alcohols, plus at least 9% by weight of one or more perfume components, the odoriferous properties of which are enhanced or modified, without chemical reaction, by the specified alcohols or certain of their monocarboxylic acid esters.

The monocarboxylic acid esters of the specified alcohols are produced by reaction of the specified alcohol with one of the following monocarboxylic acids: saturated 1 to 7 carbon atoms aliphatic monocarboxylic acids, unsaturated 3 to 7 carbon atoms aliphatic monocarboxylic acids having one double bond, cyclopentane carboxylic acid, cyclohexane carboxylic acid, salicylic acid, cinnamic acid, hexahydrobenzoic acid or p-toluic acid.

Examples of saturated aliphatic monocarboxylic acids are acetic, propionic, butyric, pentanoic, hexanoic and heptanoic acids.

Examples of unsaturated aliphatic monocarboxylic acids having one double bond are acrylic, crotonic, tiglic acids; a pentenoic, hexenoic or heptenoic acid.

The 3-methylnonan-3-ol, 3-methyl-1-nonen-3-ol, 3-methylnonan-1-ol, 3-methyl-

2-nonen-1-ol, or the monocarboxylic acid ester of any of the foregoing alcohols is thoroughly mixed with at least 9% by weight of the perfume component or components. Desirably, the specified alcohol or specified ester comprises 1 to 91% by weight, more advantageously 5 to 85% by weight of the perfume composition.

The monocarboxylic acid esters of 3-methylnonan-3-ol or 3-methyl-1-nonen-3-ol may be produced by mixing the required alcohol and the anhydride of the required monocarboxylic acid with a suitable solvent, such as toluene. The formed acid is removed as an azeotrope with the solvent by distillation through a column such as a 14" Vigreux column. Any excess anhydride and the solvent are removed by distillation under vacuum at steam bath temperature. A solvent, such as benzene, is added to the crude reaction product, and the mixture is neutralized to remove any remaining acid or acid anhydride. The mixture is then subjected to fractional distillation through a column such as a 14" Vigreux column to obtain the desired ester.

Alternatively, an acid halide of the required carboxylic acid, such as the acid chloride, may be employed. In such case, the required alcohol is mixed with a tertiary base, such as pyridine, and preferably a solvent, such as toluene or benzene.

To this solution is gradually added, with agitation, the required acid chloride. The reaction mixture is then heated to 50° to 60°C to complete the reaction. The formed tertiary base hydrochloride is removed by washing with water and the crude desired ester is rectified by fractionation in vacuum.

The 3-methylnonan-3-ol and the 3-methyl-1-nonen-3-ol utilized as components in these perfume compositions and for the production of the monocarboxylic acid esters also employed as components of the compositions, are known compounds. A convenient starting material for both of these alcohols is the acetylenic alcohol 3-methyl-1-nonyne-3-ol. This acetylenic alcohol may be prepared according to the procedures described in U.S. Patent 2,385,547. The 3-methylnonan-3-ol is produced by the hydrogenation of 3-methyl-1-nonyne-3-ol and the 3-methyl-1-nonen-3-ol may be prepared by partial hydrogenation of 3-methyl-1-nonyn-3-ol.

The 3-methyl-2-nonen-1-ol employed in these perfume compositions and as a reactant in the production of monocarboxylic acid ester of that alcohol is produced by employing 3-methyl-1-nonen-3-ol as the starting material. The starting alcohol is reacted in the absence of solvents with a monocarboxylic acid such as acetic acid together with sulfuric acid and maintained at a temperature of about 50° to 80°C.

The progress of the reaction is followed by IR spectroscopy. Under these conditions there is dehydration of the tertiary alcohol, 3-methyl-1-nonen-3-ol to a hydrocarbon, allylic rearrangement of the tertiary alcohol to the primary alcohol, 3-methyl-2-nonen-1-ol and partial esterification with formation of a monocarboxylic acid ester of 3-methyl-2-nonen-1-ol. A solution of an alkali metal hydroxide, such as sodium hydroxide, and a suitable solvent such as methanol is added to the reaction product and the solution refluxed for three to six hours to saponify the esters. The solvent is removed by distillation under a slight vacuum.

The remaining reaction product is then washed successively with water, an alkali metal bicarbonate solution and water. Fractional distillation under vacuum is conducted to obtain a fraction containing the 3-methyl-2-nonen-1-ol. The fraction distilling at 10 mm and a temperature of 110° to 111°C is recovered and contains primarily the 3-methyl-2-nonen-1-ol. The corresponding saturated alcohol, 3-methylnonan-1-ol is produced by hydrogenation of the unsaturated alcohol 3-methyl-2-nonen-1-ol in the presence of Raney nickel or other suitable catalyst.

The specified alcohols or their esters enhance the olfactory properties of the perfume component or components of the perfume compositions, giving a more aesthetic physical impact. This result does not stem from a chemical reaction which would change the structure of the perfume components.

A whole range of perfume compositions—rose, lavender, gardenia, jasmine, lilac, for example—may result from varying the types and amounts of the specified alcohols or esters employed. Perfume compositions are prepared in the following examples by mixing together the components as given.

Example 1: *Gardenia –*

	Parts by weight
3-Methylnonan-3-yl-acetate	2.0
Aldehyde C_{10}	0.2
Aldehyde C_{11}	0.3
Oil sandalwood E.I.	1.5
Styrolyl acetate	1.5
Coumarin	1.5
Resinoid labdanum absolute	1.5
Musk ketone	2.5
Isoeugenol	4.0
F.B. synthetic violet parma type	8.0
Oil bergamot	8.0
F.B. synthetic otto rose	12.5
F.B. synthetic lilac	16.4
3-Methyl-1-nonen-3-yl-isobutyrate	40.1
Total	100.0

Example 2: *Jasmine –*

	Parts by weight
Fritzbro jasmine provence	10.0
3-Methyl-1-nonen-3-yl-isovalerate	90.0
Total	100.0

Example 3: *Lilac –*

	Parts by weight
Jasolea base	10.0
3-Methylnonan-3-yl-cyclohexanecarboxylate	90.0
Total	100.0

Example 4: *Woody Bouquet –*

	Parts by weight
Oil geranium bourbon	4.0
10% solution methyl heptine carbonate in D.E.P.	8.0
α-Ionone	8.0
Isoeugenol	12.0
Oil bois de rose	12.0
Phenylethyl alcohol	12.0
Cinnamic alcohol	12.0
Methyl ionone	32.0
3-Methyl-1-nonen-3-yl-cinnamate	900.0
Total	1,000.0

FRUITY FRAGRANCES

α-Oxy(Oxo)Mercaptans

W.J. Evers, H.H. Heinsohn, Jr. and F.L. Schmitt; U.S. Patent 4,070,308; Jan. 24, 1978; assigned to International Flavors & Fragrances Inc. have developed perfumes and perfumed articles containing α-oxy(oxo)mercaptans having the structure:

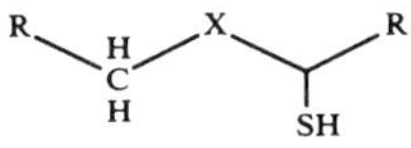

R H C H X R SH

where X is >CHOH or >C=O, and R is ethyl, 1-propyl, 2-propyl or 1-butyl.

Such α-oxy(oxo)mercaptans are obtained by reacting an alkanone with SO_2Cl_2 to form an α-chloroalkanone; reacting the α-chloroalkanone with an alkali metal hydrosulfide to form an α-mercaptoalkanone which can be used for its perfumery properties; and, if desired, reacting the α-mercaptoalkanone with a reducing agent such as an alkali metal borohydride in order to obtain an α-hydroxymercaptoalkane. Specific examples of such mercaptans and their perfumery properties are shown in the following table.

COMPOUND	STRUCTURE	AROMA
3-mercapto-4-heptanone	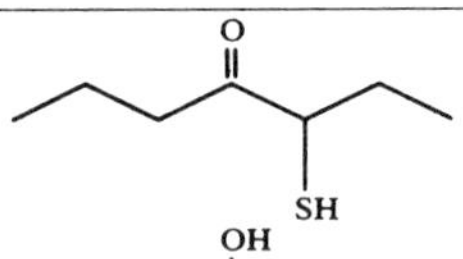 O, SH	Strong, green, buchu, grapefruit character with a cassis note.
3-mercapto-4-heptanol	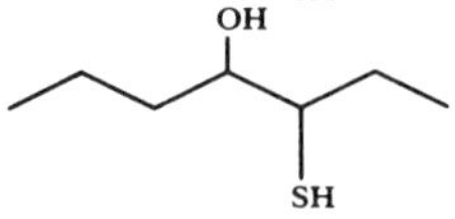 OH, SH	Strong, buchu, grapefruit character.
4-mercapto-5-nonanone	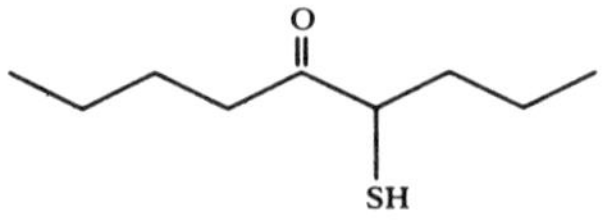 O, SH	Green, fruity, grapefruit aroma with minty and leafy nuances.

(continued)

COMPOUND	STRUCTURE	AROMA
4-mercapto-5-nonanol	OH / SH	Grapefruit aroma with green pepper nuance.
5-mercapto-6-undecanone	O / SH	Grapefruit aroma with vetiver nuances.
5-mercapto-6-undecanol	OH / SH	Grapefruit, buchu oil-like aroma with minty nuances.
3-mercapto-2,6-dimethyl-4-heptanone	O / SH	Powerful green, tart grapefruit aroma with concord grape nuance.
3-mercapto-2,6-dimethyl-4-heptanol.	OH / SH	Green, fruity aroma having concord grape and grapefruit oil-like nuances.

Example 1: *Preparation of 3-Chloro-4-Heptanone* – Into a 3,000 ml, three-necked, round-bottom flask, equipped with mechanical stirrer, 500 ml addition funnel, Y-tube, pot thermometer and gas outlet tube with rubber tubing leading over a stirring solution of 10% sodium hydroxide is added 1,000 g 4-heptanone. Addition of 434 g of SO_2Cl_2 dropwise into the 4-heptanone is commenced while maintaining the pot temperature in the range of 22° to 34°C and is continued over a period of two hours. A water aspirator vacuum is applied to the reaction mass in order to pull the acidic gases, sulfur dioxide and hydrogen chloride, over the NaOH solution. The reaction mass is periodically sampled using GLC analysis until such time as about 25% monochlorinated ketone product is found to be present.

While maintaining the reaction mass at 15°C, 1,000 ml saturated sodium chloride is added to the mixture, and the mixture is then stirred for a period of 10 minutes. The reaction mass is then transferred to a 5 liter separatory funnel and shaken well, whereupon the organic and aqueous phases separate. The lower aqueous phase (approximately 1,000 ml) has a pH of about 1. The upper organic phase is washed with 700 ml saturated sodium bicarbonate solution to a pH of 6 to 7. The organic phase is then dried over 50 g anhydrous sodium sulfate and filtered yielding a yellow oil weighing 1,063 g.

The organic layer is determined to contain 24.94% chlorinated ketone and 68.12% original ketone starting material. This material is then vacuum distilled by first adding it to a 2,000 ml, three-necked, round-bottom flask equipped with a 2.5 x 60 cm vacuum jacketed column packed with 6 mm Raschig rings, and equipped with an automatic reflux head, a pot thermometer, a heating mantle, a vacuum pump and a dry ice trap. Twelve cuts are made by the fractional distillation. Cuts 8, 9 and 10 are blended and were found by GLC to contain 93.89% 3-chloro-4-heptanone.

Example 2: *Preparation of 3-Mercapto-4-Heptanone* – Into a 50 ml, three-necked, round-bottom flask, equipped with magnetic stirrer, pot thermometer, 6" distillation column with gas outlet at top attached to rubber tubing leading above stirring solution of 10% sodium hydroxide solution, gas inlet tube (for hydrogen sulfide bubbling), gas bubbler, empty trap between hydrogen sulfide cylinder and bubbler, hydrogen sulfide cylinder and isopropanol/dry ice bath is added a solution of 1.62 g sodium methoxide dissolved in 13.5 ml anhydrous methanol.

The sodium methoxide solution is cooled to -10°C and the hydrogen sulfide bubbling is commenced below the surface of the sodium methoxide solution. The reaction is maintained at a temperature of -5° to -10°C while continuing the hydrogen sulfide bubbling and stirring the reaction mass for a period of 1½ hours. At this point, 5 ml of the cold sodium hydrosulfide solution is transferred to a 25 ml Erlenmeyer flask equipped with magnetic stirrer, dry nitrogen flow, pot thermometer, and isopropanol/dry ice bath. At -4° to 0°C, 0.75 g (0.005 mol) of 3-chloro-4-heptanone is added dropwise over one minute using a pipette. After all of the chlorinated ketone is added, a heavy solid precipitate forms which is stirred at 0°C for 15 minutes, then allowed to warm to 23°C over an additional 50 minutes.

About 4 ml of 10% sodium hydroxide solution is then added to the reaction mass while stirring under a nitrogen blanket. Unreacted chloroketone is extracted with 7 ml of methylene chloride and separated. The basic aqueous phase is acidified to a pH of 2 with 10% aqueous hydrochloric acid. The oil out is extracted twice with 10 ml methylene chloride. The methylene chloride extracts are combined, washed with saturated sodium chloride solution, dried and concentrated to yield 0.55 g of product. GLC, IR and NMR analyses of trapped product yield the information that the product is 3-mercapto-4-heptanone.

Example 3: *Preparation of 3-Mercapto-4-Heptanol* – Into a 25 ml, round-bottom flask, equipped with magnetic stirrer, nitrogen inlet tube, gas outlet tube, dry ice/acetone bath and reflux condenser is added 2.5 ml of a 95% ethanolic solution containing 0.06 g of sodium borohydride (0.0015 mol). While maintaining the reaction mass at a temperature of between 25° and 35°C over a period of about 5 minutes, 0.44 g (0.003 mol) of 3-mercapto-4-heptanone in 95% ethanol (2.5 ml) is added to the sodium borohydride solution. During this time, the reaction mass is stirred under a blanket of dry nitrogen.

The reaction mass is then continued to be stirred for a period of 3 hours at which time the reaction mixture is concentrated on a rotary evaporator using water aspirator vacuum to 3 ml of a thick slurry. To the slurry is added 10 ml water with stirring, and the solid then dissolves. The aqueous solution is then acidified to a pH of 6 with 4% aqueous hydrochloric acid, at which time the reaction mass exists in two phases; an aqueous phase and an organic phase. The organic phase is extracted with two 10 ml portions of methylene chloride. The extracts are combined, dried over anhydrous sodium sulfate, gravity filtered and concentrated on a rotary evaporator to yield a yellow oil weighing 0.3 g. GLC analysis indicates 96.3% 3-mercapto-4-heptanol.

Example 4: Both 3-mercapto-4-heptanone and 3-mercapto-4-heptanol are useful in creating a synthetic grapefruit oil as follows (where 3-mercapto-4-heptanone is added at a concentration of 10 ppm and 3-mercapto-4-heptanol is added at a concentration of 20 ppm):

Ingredients	. . .Parts by weight. . .	
Orange oil Florida	98.0	97.0
Nootkatone (1% in Limonene)	1.0	1.0
3-Mercapto-4-heptanone (0.1% in Limonene)	1.0	–
3-Mercapto-4-heptanol (0.1% in Limonene)	–	2.0

At the levels demonstrated, these powerful aroma chemicals twist the odor of orange oil to the fresh bitter grapefruit character. Even with the Nootkatone as the only additive to the orange oil, the grapefruit character does not come alive until the addition of either of the above compounds, the 3-mercapto-4-heptanone and the 3-mercapto-4-heptanol. The recommended preferred use level of 3-mercapto-4-heptanone and 3-mercapto-4-heptanol is in the range of from 0.1 ppm up to 50 ppm.

Example 5: The following intense, long lasting buchu-type essence having grapefruit nuances is prepared:

Ingredients	Parts by weight
3-Mercapto-4-heptanone	0.08
α-pinene	0.10
Myrcene	0.15
Limonene	1.00
Menthone	1.40
Isomenthone	2.60
Pulegone	0.80
Pulegyl acetate	0.15
α-Terpineol	0.10
Geraniol	0.04
Methyleugenol	0.10
Cedryl acetate	0.05
Eucalyptol	0.30
Terpinen-4-ol	0.15

The 3-mercapto-4-heptanone is responsible for adding the pleasant highly valuable grapefruit/buchu nuance to this otherwise bland essence.

cis-Oct-6-en-1-al

F. Näf; U.S. Patent 4,132,675; January 2, 1979; assigned to Firmenich SA, Switzerland have found that cis-oct-6-en-1-al, an unsaturated aliphatic aldehyde, develops perfuming properties which considerably differ from those developed by the prior known aldehydic derivatives. In its pure state cis-oct-6-en-1-al possesses a fresh and elegant fruity character reminiscent of the fragrance of melon associated with the freshness developed by cucumber. Consequently, the aldehyde finds a particular utility in perfume compositions destined to be incorporated in cosmetic articles such as shampoos, deodorizers, toilet soaps and bath preparations, as well as in detergents. In these applications cis-oct-6-en-1-al improves, enhances or modifies the natural, fresh, green, aldehydic and fruity fragrance.

The compound may also be used for enhancing, improving or modifying the flavor properties of foodstuffs, beverages, pharmaceutical preparations and tobacco, by adding to these materials a small but flavor modifying amount of cis-oct-6-en-1-al.

The aldehyde can be used in its isolated form or, more frequently, in association with other perfuming coingredients, perfume bases, diluents, excipients and carriers. Its power is such that minute amounts of it can already achieve satisfactory results. Concentrations as small as 0.01 to 0.1% by weight of the total weight of the composition into which the aldehyde is incorporated, may be used in most applications. These values, however, can be increased up to 1 or 2% whenever special effects are desired.

When cis-oct-6-en-1-al is used as a flavor modifier, it develops fatty, soapy, slightly green and fruity gustative characters. Consequently, this compound finds a particularly suitable application for the aromatization of beverages such as fruit juices, puddings, pastries and desserts in general.

Its proportions can vary within a wide range; preferably, however, concentrations of between 0.0005 and 0.05 ppm by weight of the aldehyde, based on the total weight of the article into which it is incorporated, achieve the best results.

cis-Oct-6-en-1-al can be synthesized starting from deca-cis,cis-2,8-diene via a process which consists in ozonizing the olefin and reducing then the obtained ozonide according to known methods.

Example: A base perfume composition was prepared by mixing together the following ingredients:

	Parts by weight
Artificial bergamot oil	4.0
Hydroxycitronellal	1.0
Methyl dihydrojasmonate	1.0
Phenylethanol	1.0
Synthetic jasmine oil	1.0
Vetiveryl acetate	1.0

By adding to 4.5 g of the base composition 0.05 g of cis-oct-6-en-1-al, there was obtained a perfume composition possessing, by comparison with the base composition, an improved fruity character which was particularly pleasant and reminiscent of the odor of melon.

Safranic Acid Esters

Increasing attention is being devoted to the preparation and utilization of artificial perfuming and odor modifying agents in perfumes and perfumed products, and of artificial flavoring and taste modifying agents in foodstuffs, beverages, pharmaceuticals and tobacco. This attention has been stimulated not only because of the inadequate quantity of natural perfume and flavoring materials available, but, perhaps even more importantly, because of the need for materials which can combine several natural nuances, will blend better with other perfuming or flavoring compositions, and will give perfumed or flavored products which can be specifically tailored to a given use and can be duplicated at will.

This latter factor confers a major advantage to artificial perfuming and food flavoring agents, since natural products, such as essential oils, extracts, concentrates, and the like are subject to wide variation because of changes in the quality, type or treatment of the raw material.

H.J. Wille, W.M.B. Konst and J. Kos; U.S. Patent 4,144,199; March 13, 1979; assigned to Naarden International NV, Netherlands have found that the safranic acid esters and some homologues thereof possess useful organoleptic properties and can be used for flavoring and perfuming a wide range of products. The compounds may be used, combined with other flavoring agents, diluents or carriers, for flavoring foodstuffs, beverages or tobacco products; they may be compounded with other odoriferous compounds to make perfumery compositions, in the manner conventional in the perfumery art.

The odoriferous esters of this process have the general formula:

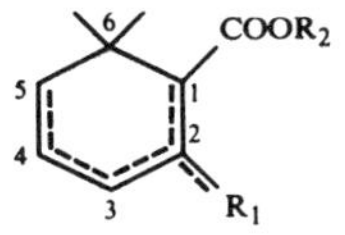

wherein the dotted lines represent two conjugated double bonds in the positions 2 (endocyclic) and 4 (α-isomer), 1 and 3 (β-isomer) or 2 (exocyclic) and 3 (γ-isomer); R_1 represents an alkyl group, or, in the γ-isomers an alkylidene group, with 1 to 3 carbon atoms; and R_2 represents an alkyl or alkenyl group with 1 to 4 carbon atoms.

These compounds may be prepared by the method given in Dutch Patent Application 73.01451, pp 11-13, where the α-, β-, and γ-safranic acid ethyl esters are prepared from mesityl oxide and ethyl acetoacetate, as intermediates in a synthesis of damascenone analogues, or modifications of this method.

These compounds were found to blend particularly well with damascones and damascenones. When the compounds were dissolved in odorless diethyl phthalate they gave odors of the following types:

Code	Compound	Odor Description
A	ethyl safranate (mixture, α:β:γ = 20:60:20)	general fruity, but not apple-like
B	ethyl 6,6-dimethyl-2-ethylidene-3-cyclohexenecarboxylate	apple-like
C	α-damascone	apple-like
D	β-damascone	apple-like
E	damascenone homologue	Naar threshold: fruity, in higher concentration coffee-like
F	damascone homologue	apple-like
G	A + E (5:2)	distinctly apple-like
H	B + E (5:2)	apple-like
I	A + D (5:2)	apple-like

A combination of the compounds with codes A and E has a distince applelike Idor, whereas the individual components have a general fruity odor, with no reminiscence of the odor of apples.

Example 1: *Perfume Compositions of the Pine-Fantasy Type* – Three pine-fantasy type compositions were prepared by mixing the following ingredients.

Compound	.... Composition		
	A	B	C
Olibanum resinoid	20	10	10
Elemi oleoresin	20	25	10
Bornyl acetate liquid	550	580	600
Turpentine oil rectified	130	100	110
Orange oil terpenes	150	120	135
Rosemary oil French	5	20	20
Guaiyl acetate	15	10	10
Cistus absolute*	10	–	10
Methylnonylacetaldehyde*	30	40	25
Ligustral (N) (Naarden)*	20	15	5
Juniper berry oil	20	30	20
Oak moss absolute*	20	25	10
Thyme oil*	5	–	–
Serpolet oil	–	20	–
Origanum oil	–	–	10
Synthetic geranium oil	–	–	5
Lavender oil French	–	–	5
Laurel leaf oil	–	–	5
Total	995	995	990

*10% solution in diethyl phthalate.

The effect of adding 5 g of a 1% solution in diethyl phthalate of ethyl safranate to composition A (995 g) is a considerable increase in brilliancy and natural richness of this basic composition.

Adding 5 g of a 1% solution in diethyl phthalate of ethyl 6,6-dimethyl-2-ethylidene-3-cyclohexenecarboxylate to 995 g of composition B results in a fuller, warmer character of this composition, a greater brilliancy and accentuation of the warm herbal notes.

By adding to 990 g of mixture C, 10 g of a 1% solution in diethyl phthalate of ethyl 6,6-dimethyl-2-propylidene-3-cyclohexenecarboxylate a composition with an enhanced, full bodied natural character is obtained.

Example 2: *Imitation Apple Flavor* – An imitation apple flavor composition was prepared by mixing the following ingredients.

	Parts by weight
Amyl acetate	35
Ethyl butyrate	15
Hexyl acetate	3
Amyl propionate	1
Orange oil Florida	0.4
Ethyl alcohol 96%	945.6
Total	1,000.0

An apple beverage was prepared using 0.3 g of this flavor composition for each liter of finished beverage. Two modifications of the apple flavor composition were made by adding to the composition 25 or 50 ppm of ethyl safranate.

The apple beverages made from these two modifications were compared with the unmodified beverage by a panel of experienced tasters. All members of this panel had a strong preference for the beverages with the addition of ethyl safranate, these beverages being more naturally applelike in odor and taste.

Maltyl-2-Methyl Alkenoates

C.J. Mussinan, B.D. Mookherjee, F.L. Schmitt and E.J. Shuster; U.S. Patent 4,151,181; April 24, 1979; assigned to International Flavors & Fragrances Inc. describe maltyl-2-methyl alkenoates having the structure:

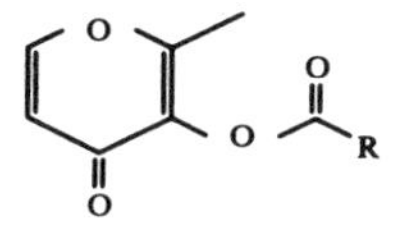

wherein R is an alkenyl moiety having one of the structures:

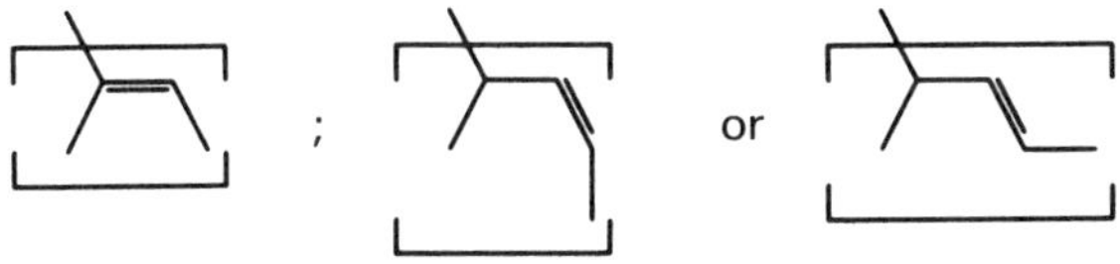

; or

as well as the uses of these compounds in perfumery.

Perfumes and perfumed articles may be made by utilizing these maltyl-2-methyl alkenoates, to give strawberry, pineapple, sweet, fruity and jammy aromas with green floral nuances.

Example 1: *Preparation of Maltyl-2-Methyl-cis-3-Pentenoate Mixture –*

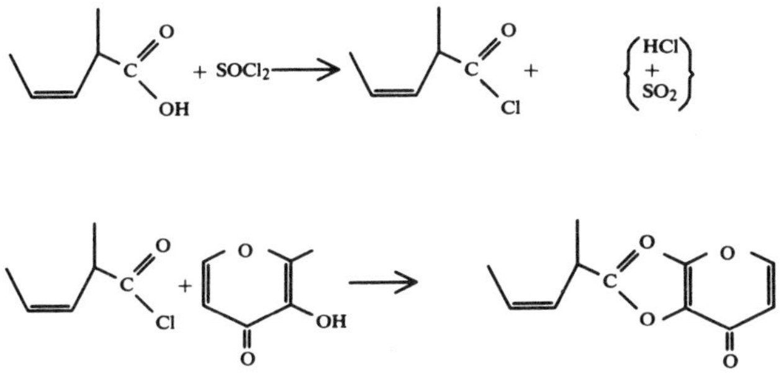

+ $SOCl_2$ → + $\{HCl + SO_2\}$

+ →

Into a 250 ml two-necked reaction flask equipped with mechanical stirrer, Friedrich's condenser and gas trap, 12 g of a mixture containing a high proportion of 2-methyl-cis-3-pentenoic acid prepared according to U.S. Patent 3,984,579 in 50 ml benzene is added. Fredrich's condenser side arm is equipped in such a way as to vent the gases released through a trap into a beaker containing commercial preparation of sodium hypochlorite (Clorox). With vigorous stirring, 11.9 g of freshly distilled thionyl chloride are added to the reaction mixture. The reaction mixture is then stirred and heated until no more hydrogen chloride or sulfur dioxide gas is released (over a period of 45 minutes).

The reaction mass is then allowed to cool and 12.6 g of maltol and 50 ml of benzene are added thereto. The mixture is again heated until no more gas evolves (about 45 minutes). The benzene solvent is then removed on a rotary evaporator. The resulting reaction product, maltyl-2-methyl-3-pentenoate containing a high proportion (80%) of maltyl-2-methyl-cis-3-pentenoate and 20% maltyl-2-methyl-trans-3-pentenoate is trapped on a preparative GLC column. Conditions: 12' x ⅜" 20% SE-52 column programmed at 100° to 190°C at 8°C per minute. A total of 0.55 g of maltyl-2-methyl-3-pentenoate is collected having a purity of greater than 99%.

Example 2: *Strawberry Fragrance* – The following mixture is prepared:

	Parts by weight
Cuminic acetate	15
Ethyl acetoacetate	3
Ethyl laurate	30
Cinnamyl isobutyrate	15
Cinnamyl decylate	20
Diacetyl (10% in 95% aqueous ethanol)	2
Ethyl pelargonate	5
γ-Undecalactone	20
Ethyl isobutyrate	110
Ethyl isovalerate	60
Ethyl heptanoate	12
Dulcinyl	5
2 (para-hydroxyphenyl)-3-butanone	2
Ethyl acetate	5
β-Ionone	5
Palatone	3
Vanillin	10
Ethyl vanillin	10
Ethyl-3-methyl-3-phenyl glycidate	70
Maltyl-2-methyl-3-pentenoate prepared according to process of Example 1	10

The mixture containing the maltyl-2-methyl-3-pentenoate imparts the berry undertone necessary to complete and round out the strawberry aroma.

Example 3: *Preparation of a Cologne and Handkerchief Perfume* – The composition of Example 2 is incorporated in a cologne at a concentration of 2.5% in 85% aqueous ethanol; and into a handkerchief perfume at a concentration of 20% (in 95% aqueous ethanol). The use of the mixture containing maltyl-2-methyl-3-pentenoate in the composition of Example 2 affords a distinct and definite strong strawberry aroma with a fruity note to the handkerchief perfume and cologne.

OTHER FRAGRANCES

MUSK AROMAS

Isomeric Hydroxymethyl Tricyclodecanes

The odor of musk has been highly valued for a long time. The musk pods (dried glands of the musk deer) are the most important animal source. Because this natural product is extremely expensive, repeated attempts were made in the laboratory to duplicate this odor. The fragrance of the synthetic musks varies considerably from that of the natural musk. Even very expensive synthetic products do not adequately possess the required fragrance.

J. Weber and H. Grau; U.S. Patent 4,146,505; March 27, 1979; assigned to Ruhrchemie AG and Gebrüder Grau & Co. KG, both of Germany have provided a product hydroxymethylformyltricyclo[5.2.1.$0^{2,6}$]-decane of the formula:

wherein R_1 and R_2 are the hydroxymethyl group CH_2OH or the formyl group CHO and R_1 represents CH_2OH when R_2 is CHO and vice versa.

An isomeric mixture of such hydroxymethyl formulas exhibits a strongly adherent animal odor, similar to Tonkin musk.

The odor of this product can be mistaken for natural animal musk odor. It has an odor much closer to natural musk than the odor of even very expensive synthetic musks.

An unexpected increase in the odor qualities and odor strength is attained when the hydroxymethylformyltricyclo[5.2.1.$0^{2,6}$]-decanes are dissolved in isomeric tricyclo[5.2.1.$0^{2,6}$]-decane-3(4,5),8(9)-dimethylol. Beside its fixative properties, the mentioned decane and dimethylol interact in a synergistic manner. Especially favorable musk bases are obtained by mixing 0.5 to 20 wt % of the mentioned decanes with the corresponding 80 to 99.5 wt % dimethylol.

Example 1: 1,000 g dicyclopentadiene, mixed with toluene in the weight ratio 1:1, are hydroformylated in a 5 liter high pressure steel vessel in the presence of rhodium-2-ethylhexanoate with a rhodium content of 5 mg at a pressure of 270 bars and a temperature of 90°C. The synthesis gas used had a volume ratio of CO/H_2 of 1:1. The conversion rate is monitored by continuous determination of the iodine number and the carbonyl number. When the iodine number falls below 190 the reaction is stopped by releasing and cooling the pressure vessel.

The diacetal is synthesized by adding 500 ml methanol and acid treatment with paratoluene sulfonic acid to the reaction product of the first stage. After neutralization (pH 7) the second double bond is hydroformylated in a second hydroformylation step at a temperature of 130°C and a pressure of 270 bars with synthesis gas of a volume ratio of 1:1 in the presence of 100 mg rhodium as a 2-ethylhexanoate. When the iodine number falls below 5, the reaction is stopped. Thereafter the whole reaction product is hydrogenated for three hours in the presence of 100 g of a nickel-hydrogenation catalyst consisting of 55 wt % nickel, 4.4 wt % MgO, 33 wt % kieselguhr and oxygen (part of the nickel is oxidized) in a steel pressure vessel at a temperature of 100°C and at a pressure of 100 bars.

After separating the rhodium and nickel catalyst by filtration, the deacetalization is performed in acidic medium (para-toluenesulfonic acid) by distilling the methanol. After the subsequent distillative separation of the toluene, the isomeric mixture of the hydroxymethyltricyclo[5.2.1.$0^{2,6}$]-decane is obtained as product.

The reaction product is characterized by determination of the hydroxyl number and the carbonyl number. The product is highly viscous and has a strong musk odor.

Example 2: A perfume composition is made with the following components:

	Grams
Laudanum absolue	100
Vetiver oil	50
Patchouli oil	10
Bergamot oil	50
Rose absolue	30
Ambrette musk	50
Ketone musk	50
Vanillin	40
Gamma-methylionone	100
Scarlet sage oil	40
East Indian sandalwood oil	80
Cypress oil	20
Tube rose absolue	20
Decolored oak moss	30
Ambergris tincture (3% in ethanol)	150
Iris concret	30
Sweet orange oil	50

(continued)

	Grams
Tricyclo[5.2.1.$0^{2,6}$]-decane-3(4,5),8(9)-dimethylol (isomeric mixture)	23
Hydroxymethylformyltricyclo[5.2.1.$0^{2,6}$]-decane (isomeric mixture)	2
Ethanol	75

The composition has a distinctive amber odor and has a more lasting effect on the skin than the same composition without the addition of the isomeric hydroxymethylformyltricyclo[5.2.1.$0^{2,6}$]-decane mixture.

Nitrated Synthetic Musk plus a Modifying Agent

Nitrated synthetic musks, such as musk xylene (1-tert-butyl-3,5-dimethyl-2,4,6-trinitrobenzene), musk ketone (4-tert-butyl-2-methyl-3,6-dinitroacetophenone), musk ambrette (6-tert-butyl-3-methyl-2,4-dinitroanisole), musk tibetene (2,6-dinitro-3,4,5-trimethyl-tert-butylbenzene), and musk moskene (1,1,3,3,5-pentamethyl-4,6-dinitroindane) have been known and used for a long time. However, there is some difficulty in their stability when used in certain perfumery compositions and they are occasionally mentioned in the literature as being potentially dangerous, i.e., explosive.

L. Dürr and F. Legrand; U.S. Patent 4,126,221; July 24, 1979; assigned to Societe des Produits Chimiques et Matieres Colorantes de Mulhouse, France have found that satisfactorily stable nitrated synthetic musk compositions can be prepared by using from 5 to 10% by weight of the total composition of a modifying agent, which must be (1) soluble in the lower alcohols; (2) practically free of odor; and (3) non-film-forming. It has been found that the most satisfactory compounds for use as modifying agents are sugars, particularly saccharose, maltose, glucose and lactose.

Example 1: In a double cone 500 liter mixer, 150 kg of pure musk xylene in powder or in crystal form and 15 kg of pure glucose are introduced. It is allowed to rotate for one hour until complete homogenization.

In this way, 165 kg are obtained of a composition whose fragrance properties, wettability with water and solution capacities in solvents are identical with those of pure musk xylene.

Example 2: The same operational conditions as those described in Example 1 were used, but replacing the glucose by lactose. Identical results were obtained.

Example 3: In a heatable double cone 500 liter mixer capable of being placed under vacuum and containing 92.5% of alcohol and 20 kg of pure powder glucose, 217 kg of musk xylene (derived from recrystallization in isopropyl alcohol) are introduced. The apparatus is placed under vacuum of 500 mm of mercury and heated to 60°C. It is allowed to rotate for 2 hours. In this way, 220 kg were obtained of a mixture possessing the properties described in Example 1.

Example 4: The same operational conditions as described in Example 3 were used, but replacing the glucose by lactose. Identical results were obtained.

Example 5: Examples 1 to 4 above were reproduced replacing the musk xylene by musk ketone and musk ambrette, respectively. An equivalent result was obtained.

Example 6: 194 kg of crude musk xylene were dissolved in 600 kg of isopropyl alcohol under reflux. The solution was filtered. Into this solution, 23 kg of pure glucose was added with stirring. The mixture was cooled, still with slow stirring, until complete crystallization of the musk. The crystals were filtered off. After drying, 200 kg were obtained of a mixture having the same properties as that indicated in Example 1.

Isochromans

M.A. Sprecker and E.T. Theimer; U.S. Patents 4,162,256; July 24, 1979; and 4,178,311; December 11, 1979; both assigned to International Flavors & Fragrances Inc. has provided a simplified, economical process for producing synthetic musks such as isochromans and acylated indanes and acylated tetrahydronaphthalenes using an economical process involving the use of one reactor and one physical step which comprises intimately admixing an indane derivative such as a pentamethylindane (Compound A) or a hexahydrotetramethylnaphthalene (Compound B), both of which compounds conform to the following structures,

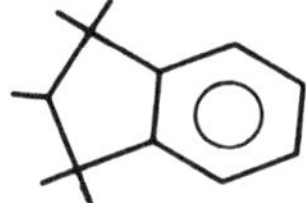

Compound A

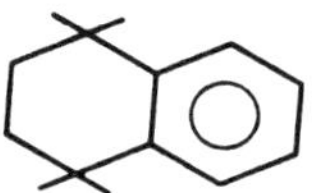

Compound B

with aluminum chloride and an alkylene oxide such as propylene oxide in the presence of an alkane or alkane mixture comprising C_{5-10} hydrocarbons such as isooctane (2,2,4-trimethylpentane), n-hexane or n-octane. This part of the reaction occurs at temperatures from -20° up to -5°C with the alkane:pentamethylindane weight ratio being from 1:10 up to 10:1 with a weight ratio of 0.2:1 up to 0.3:1 being preferred. A mol ratio of aluminum chloride:alkylene oxide (e.g., propylene oxide) may vary from 1:2 up to about 2:1 with a mol ratio of 1:1 up to 1.02:1 being preferred. The mol ratio of pentamethylindane:alkylene oxide may vary from 1:1 up to 10:1 with a mol ratio of 2:1 being preferred.

In the second part of the reaction, in the same reactor, a lower alkanol (e.g., isobutanol, ethanol or isopropanol) is added to the reaction mass in order to deactivate the aluminum chloride, preferably at temperatures below 0°C, followed by the addition of a formaldehyde precursor such as paraformaldehyde, diisopropylformal or dimethoxymethane. The temperature at this point of the process may vary from 20° to 80°C with a temperature of 40°C being preferred. Such a temperature gives rise to a convenient rate of reaction yet still permits a high yield of isochroman to be formed. The mol ratio of lower alkanol such as isopropanol to aluminum chloride is preferably 1:1. The mol ratio of the formaldehyde precursor:propylene oxide is preferably 1:1.

Products of the reaction are such compounds as Galaxolide (Compound C) or "Musk 89" (Compound D) which have the structures as shown on the following page.

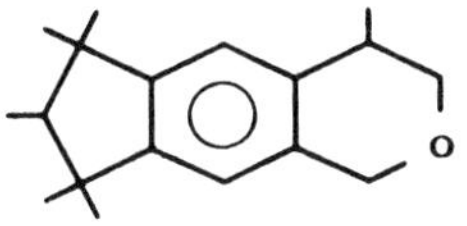

Compound C

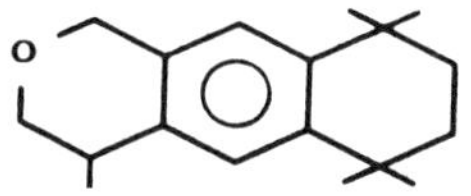

Compound D

The Galaxolide produced according to this process is from an organoleptic standpoint, a product superior to that produced according to the methods of the prior art in that it contains a major odoriferous isomer which causes the resulting product to be at least 20% stronger than any prior art Galaxolide products. This can be determined from a comparison of the major peak of the GLC profiles. The major peak comprises the major odoriferous component in each case.

The following examples illustrate the preparation of isochromans.

Example 1: *4,5,5,6,7,7-Hexamethyl-1H-Indano[2,3-c] Pyran (Galaxolide) –*

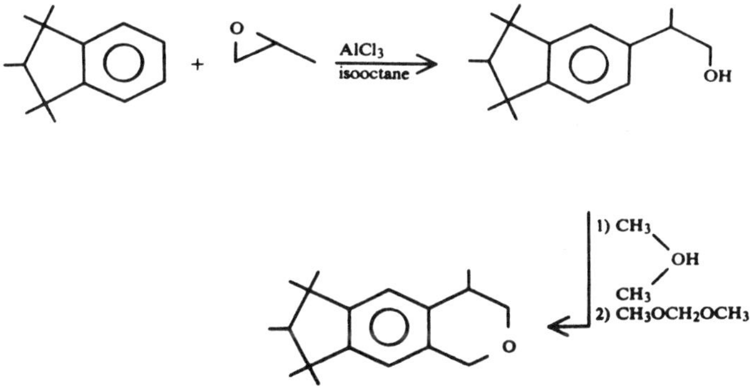

$AlCl_3$, isooctane

OH

1) CH_3, OH, CH_3
2) $CH_3OCH_2OCH_3$

A solution of 388 g of 1,1,2,3,3-pentamethylindane and 152 g of propylene oxide is added over a 2½ hour period to a stirred slurry of 720 g of 1,1,2,3,3-pentamethylindane, 360 g of aluminum chloride, and 242 g of isooctane at -10° to -5°C. At the end of this period, 194 g of isopropyl alcohol is added to the reaction mixture over a ten minute period while retaining a temperature between -5° and 0°C. External cooling is removed, and 251 g of dimethoxymethane is added to the reaction mass.

The reaction mass is then heated up to 50°C and maintained at that temperature for a period of 2 hours. The reaction mass is then poured onto 4 liters of crushed ice with 2 liters of water and stirred for a period of 15 minutes. The reaction mass is then washed with a 2 liter volume of water followed by a 1 liter portion of dilute sodium hydroxide (5%). The reaction mass is then distilled after adding thereto 30 g Primol, yielding 239 g of recovered isooctane (vapor temperature 32° to 45°C at 35 mm vacuum); 875 g of recovered 1,1,2,3,3-pentamethylindane (vapor temperature 70° to 110°C at vacuum of 3 mm); and 304 g of the product Galaxolide (vapor temperature 127° to 136°C at vacuum of 2 mm).

The yield of product based on propylene oxide is 44%. The yield based on consumed 1,1,2,3,3-pentamethylindane is 95%.

Example 2: *3,4,6,7,8,9-Hexahydro-4,6,6,9,9,-Pentamethyl-1H-Naphthalene[2,3-c]-Pyran (Musk 89)* –

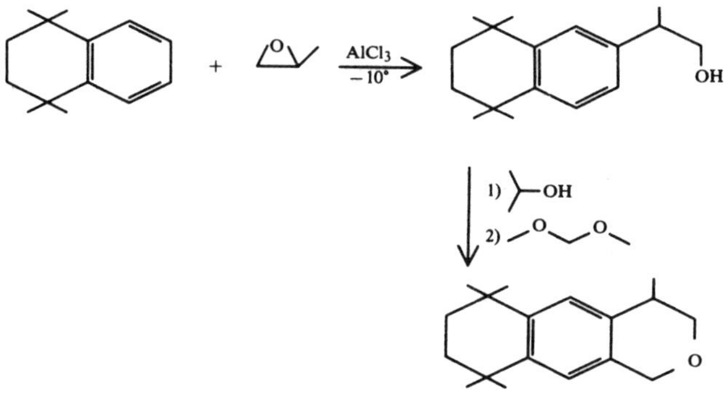

A solution of 388 g of 1,1,4,4-tetrahydro-1,1,4,4-tetramethylnaphthalene and 152 g of propylene oxide is added over a 2½ hour period to a stirred slurry of 360 g of aluminum chloride, 720 g of 1,1,4,4-tetrahydro-1,1,4,4-tetramethylnaphthalene, and 399 g of isooctane at 70°C. At the end of this period, 162 g of isopropanol are added at -10°C. External cooling is removed and 206 g of dimethoxymethane is added. The reaction mass is heated to 40°C for 2 hours. At the end of this time, the reaction mass is poured into ice water with stirring. The organic layer is washed with water and then washed with dilute base.

Distillation of the organic solution affords recovered solvent, 1,029 g of recovered 1,1,4,4-tetrahydro-1,1,4,4-tetramethylnaphthalene (BP 122° to 134°C, vacuum 3 mm) and 226 g of product (BP 140° to 175°C, 1.5 mm).

The yield of product, Musk 89, is 33½% based on propylene oxide and 91% based on consumed 1,1,4,4-tetrahydro-1,1,4,4-tetramethylnaphthalene.

AMBERLIKE

Trimethylacetyl Octalins

D. Helmlinger and P. Naegeli; U.S. Patent 4,162,266; July 24, 1979; assigned to Givaudan Corporation describe fragrant mixtures of trimethylacetyl octalins and a process for preparing such mixtures.

The compounds in these mixtures correspond to the general Formula 1.

(1)

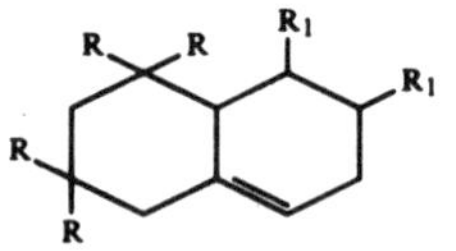

wherein three of the R symbols represent methyl groups and the fourth R symbol represents a hydrogen atom and one of the R_1 symbols represents a hydrogen atom and the other R_1 symbol represents an acetyl group.

Formula (1) includes compounds which are designated as (1a), (1b), (1c) and (1d) and the mixtures described here contain compounds corresponding to all four of these formulas.

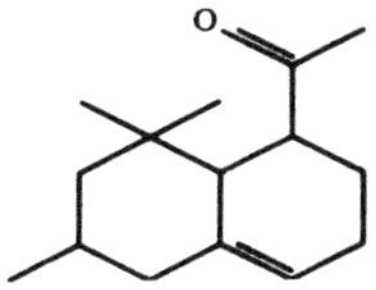

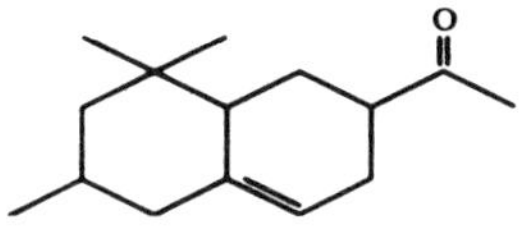

(1a) (1b)

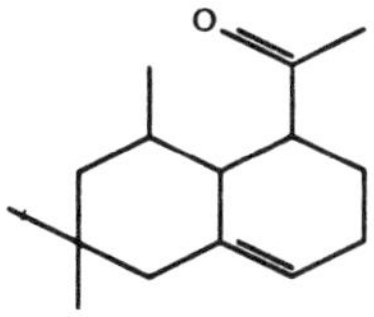

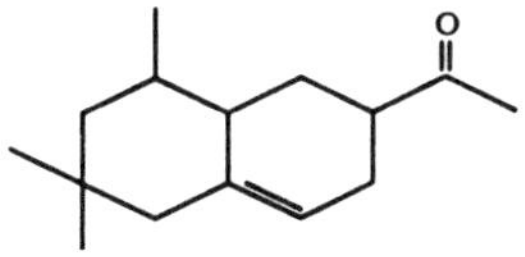

(1c) (1d)

Furthermore, because three asymmetric centers are present, the compounds of the four formulas (1a) to (1d) can each occur in 8 enantiomeric forms or can each occur in the form of 4 enantiomer pairs.

These odorant mixtures are manufactured by reacting a mixture of compounds of Formula 2.

(2)

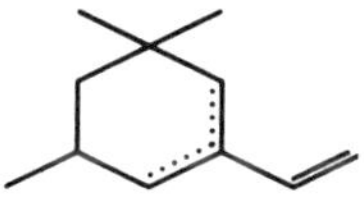

wherein one of the two lines denoted by dots represents an additional bond, with methyl vinyl ketone.

The reaction of the mixture of compounds of Formula 2 with methyl vinyl ketone can be carried out under the conditions of a Diels-Alder reaction according to known methods.

These odorant mixtures provide a pronounced, warm, tobacco-like and amberlike fragrance of very good tenacity. They can be incorporated, for example, into odorant compositions of the tobacco, chypre, lavender, amber, woody (especially spicy, rootlike) types. Such compositions are thereby modified in an advantageous manner in that they act fuller, softer, rounder and with more harmony.

The high stability of odorant mixtures and their tenacity enables them to be used as components of the most diverse compositions, especially since they are inexpensive to produce.

Example 1: A Grignard compound is prepared from 80 g of vinyl bromide and 20 g of magnesium in 1 liter of tetrahydrofuran. To this Grignard compound there are added dropwise at room temperature 100 g of 3,3,5-trimethylcyclohexanone (dihydroisophorone) in 100 ml of tetrahydrofuran. The mixture is held at reflux temperature for 12 hours. The mixture is then cooled, treated with saturated ammonium chloride solution and extracted with ether. The organic phase is washed neutral with saturated sodium chloride solution, dried and evaporated. The crude product is distilled in a high vacuum (55°C/0.1 mm Hg). There are obtained 88.9 g of 1-vinyl-1-hydroxy-3,3,5-trimethylcyclohexane.

A mixture of 75 g of 1-vinyl-1-hydroxy-3,3,5-trimethylcyclohexane and 2.3 g of p-toluenesulfonic acid in 350 ml of benzene is held at reflux temperature for 2 hours under a water separator. The mixture is cooled, washed neutral with water, dried and evaporated. The crude product is distilled in a high vacuum (73°C/10 mm Hg). There are thus obtained 28 g of 1-vinyl-3,3,5-(and 3,5,5-)trimethyl-cyclohex-1-ene. 34 g of 1-vinyl-3,3,5-(and 3,5,5-)trimethyl-cyclohex-1-ene and 17 g of methyl vinyl ketone are held at 180°C for 12 hours under an argon atmosphere in an autoclave while adding a few crystals of hydroquinone monomethyl ether and pyrogallol. After cooling, the product is distilled and there are thus obtained 19.3 g of 1,1,3-(and 1,3,3-)trimethyl-7-(and 8)acetyl-$\Delta^{5,10}$-octalin of boiling point 75° to 100°C/0.2 mm Hg. The fragrance of the product is warm, amber-like, tobacco-like, cedar-dry, earthy and spicy.

Example 2: An illustrative composition of the chypre type is as follows:

	Parts by Weight
1,1,3-(and 1,3,3-)trimethyl-7-(and 8-)acetyl-$\Delta^{5,10}$-octalin (compounds of formula 1)	70
γ-Methylionone	200
Ambrette musk	100
Hydroxycitronellol	80
Bergamotte oil	80
Phenethyl alcohol	70
Lilial (p-tert-butyl-α-methylhydrocinnamaldehyde)	70
α-Hexylcinnamaldehyde	60
Citronellol extra	50
Wood moss absolute	50
Vetiveryl acetate Bourbon	40
Patchouli oil	40
Eugenol extra	40
Gardenol	20
Citral-base	10
Artemisia oil	10
	1,000

Preparation of Dihydroionones

The process developed by *W.L. Schreiber, J.N. Siano and J.B. Hall; U.S. Patent 4,129,569; December 12, 1978; assigned to International Flavors & Fragrances Inc.* is one for the preparation of dihydro-gamma-ionone and mixtures containing dihydro-gamma-ionone, dihydro-alpha-ionone and dihydro-beta-ionone and homologues thereof which are useful ingredients for foodstuff flavorants, foodstuff flavor enhancers, tobacco flavorant, tobacco flavor modifiers, perfumes, perfumed articles and cosmetics. This three-step process renders the preparation of such compounds commercially feasible and inexpensive.

This process involves, firstly, the reaction of geranyl acetone or a homologue of geranyl acetone with an amino alcohol thereby forming a cyclic oxazo derivative of the geranyl acetone whereby the ketone moiety of the geranyl acetone is protected.

The second step of the process concerns the reaction of the cyclic oxazo derivative of geranyl acetone or one of its homologues whereby a cyclization takes place thereby causing the formulation of the mixture of dihydro-gamma-ionone cyclic oxazo derivative, dihydro-alpha-ionone cyclic oxazo derivative and dihydro-beta-ionone cyclic oxazo derivative or homologues thereof. This cyclization is accomplished using a Lewis acid catalyst, preferably a boron trifluoride complex, for example boron trifluoride diethyl etherate of BF_3 itself.

The third reaction of the process involves the hydrolysis of the cyclic oxazo derivatives of the dihydro-gamma, alpha and beta-ionone mixture whereby the dihydro-gamma, alpha and beta-ionone mixture is produced; or it involves the hydrolysis of the mixture of cyclic oxazo derivatives of the dihydro-gamma-ionone homologue, the dihydro-alpha-ionone homologue and the dihydro-beta-ionone homologue.

The reaction, using geranyl acetone and ethanol amine as starting materials can be illustrated by the following sequence:

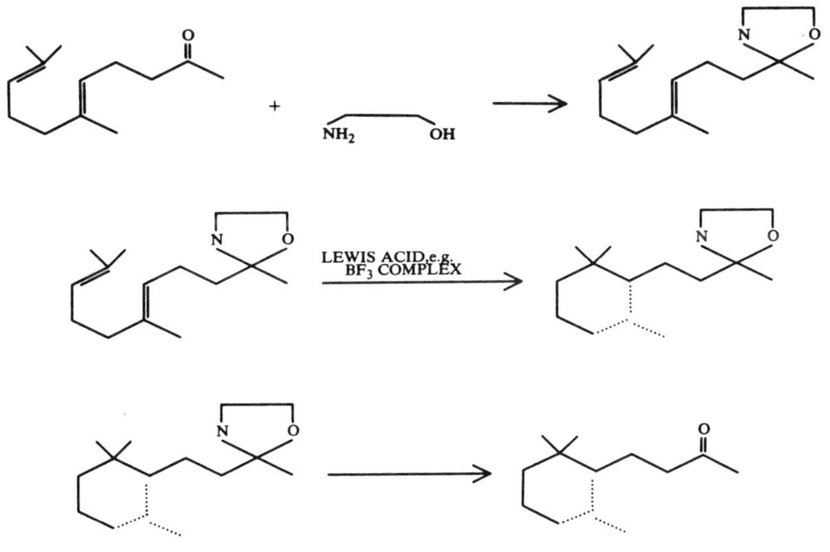

where one of the dashed lines is a C–C double bond, and the other two are single bonds. The BF_3 complex is preferably BF_3 diethyl etherate.

The reaction product mixtures containing the dihydro-gamma-ionone, dihydro-beta-ionone and dihydro-alpha-ionone as well as the specific ionone isomers produced according to the process are clear liquids with intense and persistent warm, woody and earthy notes important in creating amber aromas. These materials are particularly suited to use as perfume materials in the preparation of perfume compositions. They are very suited to perfumery wherein an amber or woody-amber aroma is required. To make such an amber or woody-amber type of perfume, the materials can be combined with auxiliary perfume adjuvants including one or more of many types of odor materials such as bergamot oil, vertivert oil, patchouli oil, sandalwood oil, oakmoss and floral musk.

Example 1: To a 22 liter, 3 neck flask fitted with thermometer, stirrer, Dean-Stark trap and condenser is added 4,268 g (22.0 mols) of geranyl acetone, 1,610 g (26.4 mols) ethanolamine (Aldrich Chemical Co.) and 4,200 ml toluene. The mixture is heated to rapid reflux with water separation for 8 hours. The mixture is then cooled to 20° to 25°C as 5,467 g (40.0 mols) boron trifluoride etherate is added. The mass is aged for 20 hours at room temperature. A slurry of 3,280 g sodium acetate and 3,280 ml water is added to this reaction mixture while it is stirred at 15° to 25°C. More water is added to dissolve salts and the aqueous layer is separated.

The organic layer is washed with water and then with aqueous sodium bicarbonate. The organic layer is rapidly distilled to remove solvent and free the product from residue. The material is then fractionally distilled through a 2' x 1½" Goodloe packed column to give 1,678 g of product, BP 89° to 96°C (2.3 to 2.9 mm), which contains (by GC area normalization) 66% dihydro-gamma-ionone, 26% dihydro-alpha-ionone and 5.5% dihydro-beta-ionone.

Example 2: The reaction product of Example 1 is incorporated into a cologne having a concentration of 2.5% in 80% aqueous ethanol; and into a handkerchief perfume in a concentration of 20% (in 95% ethanol). The use of the ionone affords a distinct and definite warm, woody, earthy aroma to the handkerchief perfume and to the cologne.

In other perfume formulations, the addition of a mixture of the dihydro-gamma, alpha, and beta-ionones gives a high degree of richness and persistence in a natural amber quality.

CASTOREUM REPLACEMENT

In perfumery many natural materials are used. However, these materials suffer from the disadvantage of not always being available in sufficient quantities. Moreover they are often expensive and not of consistent quality. The use of certain animal materials also may be objected to because of the danger of extinction of the animal species under consideration, or because collection of the product is annoying to the animal. So it is advantageous to prepare or compound synthetic perfume materials whose odor properties approach those of the natural materials as closely as possible. Castoreum is an animal material derived from beavers. It is very popular for use in perfumery but is expensive and hard to obtain for the reasons set forth above.

H. Boelens and H.J. Wobben; U.S. Patents 4,147,671; April 3, 1979; and 4,124,771; November 7, 1978; both assigned to Naarden International, NV, Netherlands found that a very natural and satisfactory castoreum odor can be imparted to perfume compositions, without using any natural castoreum, by adding o-hydroxybenzyl ethyl ether of the formula (1) and/or 8-allyl-8-hydroxytricyclo[5.2.1.$0^{2,6}$]decane of the formula (2) to other components commonly used in preparing perfume compositions. The formulas are shown on the following page.

(1) 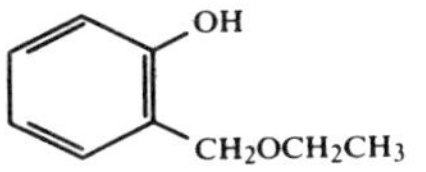 [structure: OH, $CH_2OCH_2CH_3$] (2) 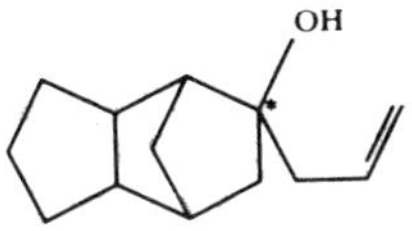 [structure: OH]

The compound 8-allyl-8-hydroxytricyclo[5.2.1.0^{2,6}]decane is prepared from tricyclo[5.2.1.0^{2,6}]decane-8-one and an allylmagnesium halogenide. Both stereoisomeric ketones (endo and exo) will undergo this Grignard reaction and both will yield two new stereoisomers because of the introduction of a new asymmetric center (indicated by *). It is possible to separate the four stereoisomers by methods known per se, but this is not necessary. For economical reasons the use of the mixture of isomers is preferred.

Although the odors of o-hydroxybenzyl ethyl ether and 8-allyl-8-hydroxytricyclo[5.2.1.0^{2,6}]decane are very different, they are both clearly reminiscent of castoreum. The compounds may be used either separately or together in preparing a synthetic castoreum, if desired, in conjunction with compounds known to be components of natural castoreum (e.g., benzoic acid, salicylic acid, their methyl and ethyl esters, cresol and other phenolic compounds, acetophenone, etc.). The addition of o-hydroxybenzyl ethyl ether accentuates the phenolic character and improves the odor of a synthetic castoreum when used in an amount of 1,000 ppm by weight or more.

On the other hand, 8-allyl-8-hydroxytricyclo[5.2.1.0^{2,6}]decane has a sweeter animal character and improves the odor of a synthetic castoreum when used in amounts of 100 ppm by weight or more. The two make it possible to obtain a very near approach to the odor of natural castoreum. With this combination a synthetic castoreum may be prepared which is so close to nature that it is able to substitute natural castoreum in any desired application.

The weight ratio of o-hydroxybenzyl ethyl ether to 8-allyl-8-hydroxytricyclo-[5.2.1.0^{2,6}]decane to be used depends on the desired effect. In general it will be between 300:1 and 1:5. Preferably a ratio between 30:1 and 1:1 is used.

Example 1: *Preparation of 8-Allyl-8-Hydroxytricyclo[5.2.1.0^{2,6}] Decane* – In a 1 liter reaction vessel, equipped with a stirrer, a cooler and a dropping funnel, 26.4 g magnesium, 100 g tetrahydrofuran and 1 g allyl chloride were heated to about 40°C to start the reaction. A mixture of 76.5 g allyl chloride, 250 g tetrahydrofuran and 150 g tricyclo[5.2.1.0^{2,6}]decane-8-one was added in four hours, keeping the temperature at 35° to 40°C. The reaction mixture was stirred for an additional 30 minutes and subsequently poured into a mixture of 600 g of ice and 80 g of acetic acid.

The layers were separated and the water layer extracted twice with toluene. The combined organic layers were washed with 5% soda solution until alkaline and subsequently with water until neutral. The organic solvent was removed by distillation under reduced pressure. The residue was distilled on a Vigreux head at 2 mm Hg. The fraction boiling between 93° and 95°C was collected. Yield: 80%, n_D^{20}: 1.5154.

Example 2: Synthetic castoreums were prepared by mixing the following ingredients.

Mixture A

	Parts by Weight
Benzoic acid	738
Farnesol	50
Farnesyl acetate	20
Farnesyl isobutyrate	10
o-Cresol	15
p-Cresol	5
m-Cresol	4
Salicylic acid	6
Borneol	2
Eugenylphenyl acetate	1
Ethyl benzoate	25
Methyl benzoate	10
Methylphenylcarbinol	6
Acetophenone	4
Pentanoic acid	2
Butanoic acid	2
	900

Mixture B

	Parts by Weight
Mixture A	900
o-Hydroxybenzyl ethyl ether	85
	985

Mixture C

	Parts by Weight
Mixture A	900
8-allyl-8-hydroxytricyclo-[5.2.1.$0^{2,6}$] decane	15
	915

Mixture D

	Parts by Weight
Mixture A	900
o-Hydroxybenzyl ethyl ether	85
8-allyl-8-hydroxytricyclo-[5.2.1.$0^{2,6}$] decane	15
	1,000

Mixtures A, B, C and D were compared by seven trained perfumers. Mixtures B and C were unanimously preferred over Mixture A. Two experts preferred Mixture C over Mixture B, the others preferred Mixture B over Mixture C. However, Mixture D was judged unanimously as the mixture with the best and most natural castoreum odor.

Example 3: By mixing the following ingredients a men's cologne concentrate was prepared using the synthetic castoreum Mixture D of Example 2.

	Parts by Weight
Musk ketone	15
Musk R1* (11-oxa-hexadecanolide)	15
Tonka absolute	10
Heliotropine	10
Benzylisoeugenol	20
Mousse absolute	15
Galbanum resin	10
Benzoin resin Siam	15
Lemon oil Italian	75
Bergamot oil	200
Vervain oil	10
Sandalwood oil EI	100
α-Isomethylionone	100
Cedarwood oil Virginia	50
Clove bud oil	25
Rosana NB 131*	50
Jasmin NB 133*	50
Vetiveryl acetate	50
Ylang ylang I	25
Geranium oil Bourbon	10
Basil oil	10
Angelica root oil	10
Clary sage oil	10
Lavender oil, 45 to 47%	50
Lauric aldehyde, 10% in diethyl phthalate	30
Civet absolute, 10% in ethanol	10
Mixture D of Example 2, 20% in benzyl alcohol	25
	1,000

*Naarden International

GREEN AND HERBLIKE ODORS

1,5-Alkadienols and Their Esters

Fragrance compounds with a naturally green odor are highly esteemed in the perfume industry. Such fragrance compounds are mostly natural in origin and thus a sufficient supply is not always available. Moreover, the quality of these naturally occurring fragrance compounds tend to vary. Accordingly a growing demand exists for synthetic fragrance compounds which can replace these natural compounds.

H.J. Wille; U.S. Patent 4,104,202; August 1, 1978; assigned to Naarden International NV, Netherlands has found that 1,5-alkadienols and the esters derived therefrom, having the general structural formula (1), are good fragrance compounds

$$R_1{-}CH_2CH_2CH{=}CH{-}\underset{\displaystyle OR_2}{\underset{|}{C}H}{-}CH_2CH{=}CH_2 \qquad (1)$$

In the formula on the preceding page R_1 is an alkyl group having from 1 to 4 carbon atoms and R_2 is selected from the group consisting of hydrogen (dienols), formyl radicals and alkyl carbonyl radicals, the alkyl portion of which has from 1 or 2 carbon atoms (dienol esters).

Of these compounds, those where R_1 is a straight chain (unbranched) alkyl group having from 1 to 3 carbon atoms, are preferred. Especially important among the compounds of Formula (1) are the alcohols (R_2 = H) and acetates (R_2 = $COCH_3$). One of these compounds is 1,5-nonadien-4-ol. This compound has a green, slightly earthy and mushroomlike odor. Another specific compound of the process is 1,5-undecadien-4-ol and the corresponding acetate; both have a green and slightly fatty odor. The odor of this acetate is particularly suggestive of galbanum.

The 1,5-alkadien-4-ols may be prepared by reacting an α,β-unsaturated aldehyde having the structural formula:

$$R_1CH_2CH_2CH{=}CHCHO \qquad (2)$$

where R_1 has the same meaning as R_1 in Formula (1) above, with allyl magnesium chloride

$$ClMgCH_2CH{=}CH_2 \qquad (3)$$

The reaction scheme is as follows:

$$\underset{(2)}{R_1CH_2CH_2CH{=}CHCHO} + \underset{(3)}{ClMgCH_2CH{=}CH_2} \longrightarrow \underset{(4)}{R_1CH_2CH_2CH{=}CH\overset{OH}{\overset{|}{C}}HCH_2CH{=}CH_2}$$

The alcohols thus obtained may be converted into esters in accordance with esterification procedures well known in the art.

These compounds are strong and stable fragrance compounds with a long lasting odor. As such, they may be incorporated successfully in perfume compositions.

Example 1: *Synthesis of 1,5-Undecadien-4-ol* – A mixture of 300 ml of tetrahydrofuran, 21 g of magnesium, 0.07 g of iodine and 2 g of allyl chloride in a nitrogen atmosphere is carefully heated to 35°C under constant stirring. When the reaction has started, a mixture of 700 ml of tetrahydrofuran, 58.2 g of allyl chloride and 100.8 g of 2-octenal is carefully added in 3 hours. During this addition the temperature is kept at 30° to 35°C, if necessary by cooling with a water bath. The stirring of the reaction mixture at this temperature is continued for another 1.5 hours. After cooling the reaction mixture to room temperature, it is poured in a mixture of 80 g of acetic acid, 800 g of ice and 200 ml of toluene.

The total mixture is stirred vigorously; the water layer is separated and extracted with 100 ml of toluene. The combined organic layers are washed with water, 5% soda solution and again with water. The solvent is evaporated and the residue is distilled under reduced pressure. 80.6 g of 1,5-undecadien-4-ol is obtained (64%); BP 75° to 78°C/2 torrs; n_D^{20} = 1.4568.

Example 2: A perfume composition of the men's cologne type was prepared according to the following recipe:

	Parts by Weight
Bergamot oil, bergaptene-free	285
Lemon oil	180
Vetiveryl acetate	50
α-Amylcinnamaldehyde	50
Jasmin NB 133*	30
Methyl dihydrojasmonate	25
Lavender concrete	50
Musk R_1**	10
Oakmoss absolute	10
Acetylcedrene	30
Sandalwood oil	30
Hydroxycitronellal	40
α-Isomethylionone	30
Mysorol***	30
Tincture of civet	50
Orange oil Florida	70
1,5-undecadien-4-yl acetate	30
	1,000

*Jasmin perfume base of Naarden International
**Perfume compound of Naarden International
***Sandalwood perfume compound of Naarden International

Ethyl Dichloroisothiazolecarboxylate

J.A. Virgilio; U.S. Patent 4,132,676; January 2, 1979; assigned to Givaudan Corporation has found that ethyl 3,4-dichloro-5-isothiazolecarboxylate is valuable in perfumery because of its mild minty anise odor which is particularly useful as a light miscible note in a variety of perfumes.

The compound has the following structure:

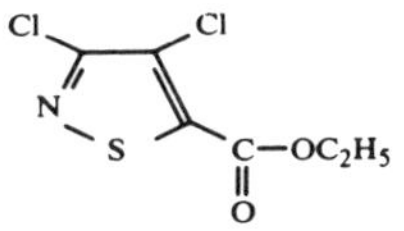

Its preparation is accomplished by reacting ethanol and the corresponding acid under esterifying conditions.

Example 1: *Preparation of Ethyl 3,4-Dichloro-5-Isothiazolecarboxylate* – 3,4-dichloro-5-isothiazolecarboxylic acid (55 g, 0.278 mol) was dissolved in 500 ml of ethanol. The solution was saturated with hydrochloric acid and then refluxed for 3 hours. The ethanol was distilled on a rotary evaporator leaving a residue which was then charged into 200 ml of water and neutralized with $NaHCO_3$. The water layer was extracted with 2 x 200 ml of CH_2Cl_2. The extracts were combined, dried over $MgSO_4$, filtered and the solvent distilled. The remaining liquid was distilled to yield 52.3 g (83%) of ethyl 3,4-dichloroisothiazole-5-carboxylate by 92°C (3.0 mm).

Example 2: *Use of Ethyl 3,4-Dichloro-5-Isothiazolecarboxylate as an Odorant* – The following perfume formulation was made up:

Woody Base	Parts per Thousand
Methylionone	50
p-tert-Butylcyclohexyl acetate	50
Methyl dihydrojasmonate	70
Cedryl acetate	100
Sandalwood oil EI	200
Patchouli oil	200
Bergamot oil	200
2-ethyl-6,6-dimethyl-2-cyclohexen-1-carboxylic acid ethyl ester	100
Estragole	5
Ethyl 3,4-dichloro-5-isothiazolecarboxylate	25

The ethyl 3,4-dichloro-5-isothiazolecarboxylate enhances the woodiness, giving a more blended woody base.

Replacement for Coumarin

Coumarin is a natural constituent of certain plants such as e.g., tonka beans, lavender, woodruff and sweet clover, all vegetable species containing coumarin at a relatively high concentration. This latter possesses a bitter gustative note which modifies itself when the compound is tasted in a diluted state, then becoming sweetish and herbal. Its use in perfumery is rather extensive wherein it is used to support herbaceous odors, such as lavender, lavandin, rosemary and citrus oils and as a fixative in various perfume compositions.

E. Sundt and R. Aschiero; U.S. Patents 4,144,200; March 13, 1979; and 4,173,550; November 6, 1979; both assigned to Firmenich SA, Switzerland report that 3-phenyl-cyclopent-2-en-1-one, having the formula

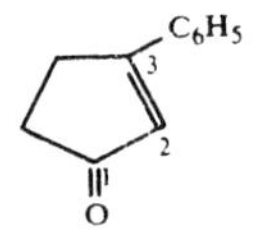

has a spicy, lactonic odor reminiscent of coumarin and can be used as a partial replacement for coumarin in various perfume compositions. Due to its stability towards current perfume ingredients, diluents and supports, this compound can be used in a large variety of applications both in the field of fine perfumery and in the manufacture of technical compositions and cosmetics.

Depending on the nature of the perfumed materials or on the effect desired, the proportions used may vary within a wide range and may be, for example, of the order of about 1 to 10% by weight, based on the total weight of the perfumed materials. Preferred proportions are of about 5%. Lower concentrations may be utilized in the manufacture of perfumed products such as soaps, cosmetics, detergents and household materials in general.

The compound can be prepared in various ways, one of which, for instance, consists in reacting bromo-benzene with 3-ethoxy-cyclopent-2-en-1-one as indicated by the reaction scheme on the following page.

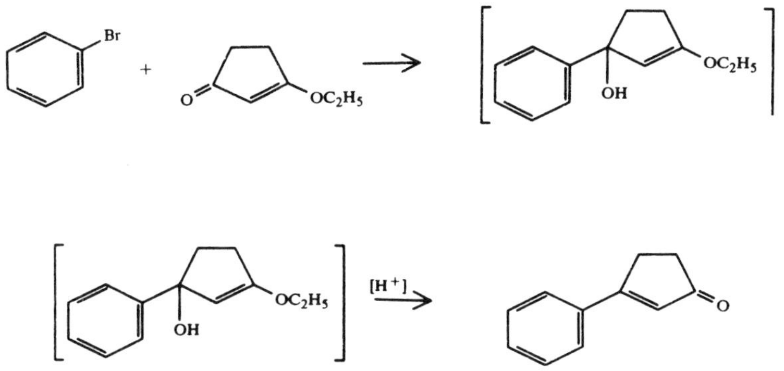

The compound of the formula is a crystalline solid having a melting point of 81.5° to 83°C; its purification was effected by several successive crystallizations with petroleum ether.

Example: A base perfume composition was prepared by admixing the following ingredients.

	Parts by Weight
Lavandin oil	200
African geranium oil	100
Cedrene	80
Oakmoss concrete, 50%*	60
Benzyl salicylate	60
Terpenyl acetate	60
Linalol	60
Amyl salicylate	40
Myrcenyl acetate	40
Synthetic oriental sandalwood	40
Isobutyl benzoate	30
Terpineol	20
1,1-Dimethyl-4-acetyl-6-tert-butyl indane	20
Cyclopentadecanolide	20
Geraniol	20
Galbanum resinoid	10
Patchouli oil	10
α-Isomethylionone	10
	880

*In diethyl phthalate

By adding to 90 g of this perfume base, 6 g of 3-phenyl-cyclopent-2-en-1-one, there is obtained a perfume composition which possessed an improved "fougere' character and a deeper, more tenacious and spicy smell than the base composition.

GENERAL PROCESSES FOR PREPARING PERFUMERY INGREDIENTS

ALDEHYDES

Phenylpropanals

W. Aquila, W. Hoffmann, W. Himmele and H. Siegel; U.S. Patent 4,113,781; September 12, 1978; assigned to BASF AG, Germany have found that phenylpropanals of the formula (1)

(1) 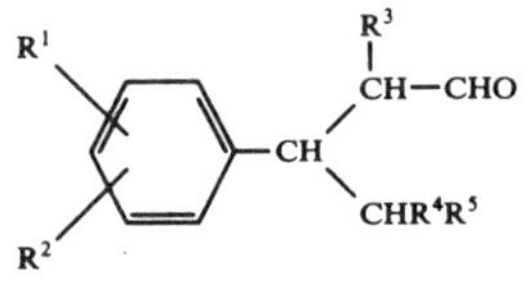 R^1, R^2, R^3, CH—CHO, CH, CHR^4R^5

in which R^1 and R^2 are identical or different and each is hydrogen, halogen, alkyl of 1 to 7 carbon atoms, cycloalkyl of 3 to 8 carbon atoms or alkoxy of 1 to 4 carbon atoms, and R^3, R^4 and R^5 are identical or different and each is hydrogen or alkyl of 1 to 7 carbon atoms, with the proviso that at least one of the substituents R^1 to R^5 is not hydrogen,

are obtained in an advantageous manner when compounds of the formula (2)

(2) 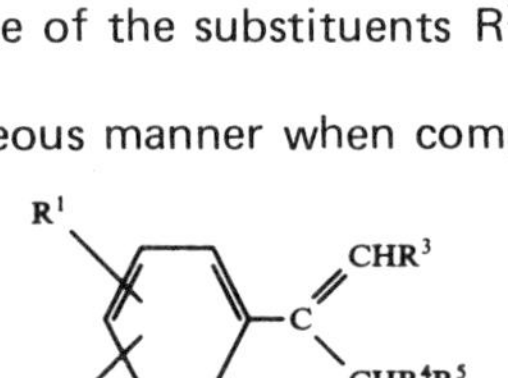 R^1, R^2, CHR^3, C, CHR^4R^5

in which R^1 to R^5 have the above meanings,

are reacted with carbon monoxide and hydrogen at 50° to 180°C and under pressures of 20 to 1,500 atm gauge in the presence of rhodium carbonyl complexes.

The process is of particular importance for the manufacture of phenyl propanals of formula (1) in which R^1 is alkyl of 1 to 4 carbons in the para position and R^2 to R^5 are hydrogen, since these phenylpropanals, especially 3-(4-isopropylphenyl)-butanal-1 and 3-(4-tert-butyl-phenyl)-butanal-1, have outstanding scents.

Carbon monoxide and hydrogen are preferably employed in a volume ratio of 2-1:1-2.

The reaction is preferably carried out at temperatures from 60° to 130°C with particularly good results obtained under pressures from 100 to 700 atm gauge.

The reaction is carried out in the presence of rhodium carbonyl complexes. In general, from 0.001 ppm to 0.05% by weight of rhodium, calculated as metal and based on the amount of the compounds of formula (2) employed, is used. From 0.1 to 100 ppm of rhodium has proved to be a particularly advantageous amount. The carbonyl complexes of rhodium can be prepared separately before the reaction, or the corresponding starting materials, namely the halides, oxides, chelates, or fatty acid salts of rhodium, can be fed separately to the reaction. The catalyst then forms in situ under the reaction conditions. Particularly advantageous starting materials are square-planar rhodium(1) complexes, which give a homogeneous solution in the reaction mixture, such as dimeric rhodium carbonyl chloride, dimeric cyclooctadien-1,5-ylrhodium chloride or rhodium carbonyl acetylacetonate.

Particularly advantageous results are obtained when tertiary organic phosphines are additionally used as catalyst modifiers. Triaryl phophines derived from benzene, e.g., triphenylphosphine, tritolylphosphine and mixed substituted phosphines, such as methyldiphenylphosphine, are especially suitable. The phosphines used as modifiers are employed in amounts of from 0.25 to 30 mols per gram atom of rhodium.

The compounds of the formula (1) possess pleasant scents, and 3-(4-isopropylphenyl)-butanal-1 and 3-(4-tert-butyl-phenyl)-butanal-1 in particular have a suitable type of odor for the preparation of flower oil such as lavender, lily of the valley, linden, iris, lily, lilac or cyclamen.

The examples which follow illustrate the manufacture of the compounds of the formula (1). The abbreviation COD represents cyclooctadienyl-1,5.

Example 1: 3,000 g of 2-(4-isopropyl-phenyl)-propene-1 and 75 mg of dimeric cyclooctadien-1,5-ylrhodium chloride $[Rh(Cl)COD]_2$ in 3,000 ml of benzene as the solvent are heated to 110°C in a high-pressure autoclave of 10-liter capacity and reacted with a mixture of carbon monoxide and hydrogen (volume ratio 55:45) under a pressure of 600 atm gauge. The pressure is maintained for 10 hours by injection of further gas mixture. After completion of the reaction, the autoclave is cooled under pressure, and the pressure is then released. The reaction mixture is fractionally distilled in a column, the main fraction passing over at 103°C per 1.0 mm Hg. In total, 2,918 g of 3-(4-isopropyl-phenyl)-butanal-1 are obtained. The 2,4-dinitrophenylhydrazone of the aldehyde melts at from 108° to 109°C.

Example 2: 150 g of 3-phenylpentene-2 and 50 ppm of rhodium as $[Rh(Cl)COD]_2$, in 200 ml of benzene as the solvent, are heated to 110°C in a high-pressure vessel of 0.8 liter capacity and reacted with carbon monoxide and hydrogen (volume ratio $CO:H_2$ = 1:1) under a pressure of 600 atm gauge. The pressure is maintained constant by injection of further gas mixture. Gas corresponding to a pressure drop of 220 atm gauge is consumed in the course of 4 hours. The autoclave is then cooled under pressure, the pressure is released and the mixture is worked up by distillation. 143 g of 2-methyl-3-phenyl-pentanal-1 of boiling point 121° to 123°C/14 mm Hg are obtained.

Cyclohexane Derivatives

R.T. Gray and A.J. DeJong; U.S. Patent 4,122,121; October 24, 1978; assigned to Shell Oil Company have found a class of cyclohexane derivatives ring substituted with aliphatic carboxyaldehyde groups and hydroxy or epoxy moieties which possess distinctive floral odors. This class of cyclohexane derivatives is represented by the general formula (1)

(1)

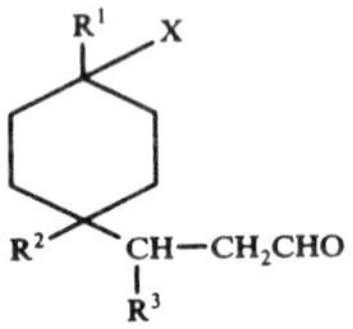

R^1 X

R^2 $CH-CH_2CHO$

R^3

wherein R^1 represents an alkyl group; R^2 and R^3 each represents a hydrogen atom or an alkyl group; and X represents an epoxy or hydroxy group.

Examples of preferred compounds according to formula (1) above are: 3-(3,4-epoxy-4-methylcyclohexyl)butanal, 3-(4-hydroxy-4-methylcyclohexyl)butanal, 3-(1,4-dimethyl-3,4-epoxycyclohexyl)propanal, and 3-(1,4-dimethyl-4-hydroxy-cyclohexyl)propanal. These compounds may be prepared by a process which comprises hydroformylating an olefin of formula (2)

(2)

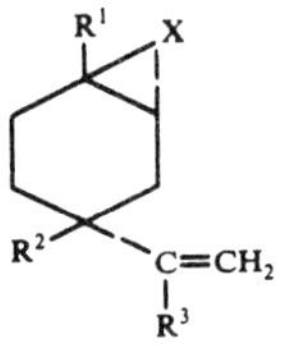

R^1 X

R^2 $C=CH_2$

R^3

in the presence of a metal carbonyl catalyst. The catalyst may be a cobalt or rhodium carbonyl catalyst but is preferably a rhodium carbonyl complex, in particular a complex which also contains hydride and/or phosphine groups. The catalyst may be a homogeneous catalyst such as the compound $HRh(CO)[P(C_6H_5)_3]_3$, or a heterogeneous catalyst obtained, e.g., by incorporating such a compound onto a solid carrier such as silica. The temperature of the hydroformylation is preferably from 50° to 200°C and the total pressure of the carbon monoxide and hydrogen used is preferably up to 200 atm. The process may be carried out in an organic solvent, e.g., an aliphatic, cycloaliphatic or aromatic hydrocarbon.

The olefinic starting materials of formula (2) which are useful in preparing these compounds may be obtained from a variety of natural or synthetic sources. For example, crude terpineols obtained from commercial pine oils typically contain significant quantities of suitable alkenyl-hydroxy cyclohexane starting materials. Further, substituted vinylcyclohexenes obtained from synthetic sources such as isoprene dimerization can be epoxidized by known techniques, e.g., reaction with an organic hydroperoxide, such as tertiary-butyl hydroperoxide over a titanium dioxide/silica catalyst or hydroxylated, for example, by reaction with formic acid and subsequent saponification, to provide the vinyl-epoxy or hydroxy cyclohexane starting materials.

The term "hydroformylation" is used to mean a reaction in which a compound containing a $>C=CH_2$ group is reacted with a mixture of hydrogen and carbon monoxide in the presence of a catalyst to form a compound containing a $>CH-CH_2-CHO$ grouping.

Example: The hydroformylation catalyst used was a heterogeneous catalyst which was prepared by incorporating the homogeneous catalyst $HRh(CO)[P(C_6H_5)_3]_3$ onto a solid silica carrier.

Crude terpineol [20 g containing 20 to 25% by weight 2-(4-hydroxy-4-methylcyclohexyl)propene], the hydroformylation catalyst (2.0 g), benzene (5 ml) and cyclohexane (20 ml) were mixed in a 100-ml stainless steel autoclave. The autoclave was then pressurized to 50 to 55 atm with a mixture of equal volumes of carbon monoxide and hydrogen. The autoclave contents were then stirred at 100°C for 4.5 hours. After cooling to room temperature the reaction mixture was filtered and distilled. The desired product 3-(4-hydroxy-4-methylcyclohexyl)-butanal was obtained in a yield of 3.8 g, boiling point 106° to 109°C at 0.6 mm Hg. The compound possesses a mild floral odor reminiscent of lilac and muguet.

ALCOHOLS

Tertiary Alcohols

In another patent by *R.T. Gray and A.J. DeJong; U.S. Patent 4,125,484; Nov. 14, 1978; also assigned to Shell Oil Company*, a class of tertiary alcohols which have distinctive aroma properties, primarily woody, has been synthesized. These compounds have the general formula (1):

$$(1) \qquad \begin{matrix} R_1 \\ \end{matrix} \!\!\!\! \searrow CH-Y-\overset{R_3}{\overset{|}{C}}-R_4 \qquad \text{with } R_2 \text{ on CH, OH on C}$$

$$\begin{array}{c} R_1 \\ \; \end{array}\!\!>\!CH-Y-\underset{OH}{\underset{|}{\overset{R_3}{\overset{|}{C}}}}-R_4 \quad (R_2 \text{ on lower branch})$$

in which R_1 represents an alkenyl or a cycloalkyl or cycloalkenyl group of up to 8 carbon atoms optionally substituted by one or more alkyl groups; R_2 represents a hydrogen atom or an alkyl group, or R_1 and R_2, together with the carbon atom to which they are linked, from a bicyclo group of up to 10 carbon atoms optionally substituted by up to two alkyl groups; R_3 and R_4 each represents an alkyl group; and Y represents a divalent aliphatic group of up to 4 carbon atoms.

Specific examples of preferred tertiary alcohols of formula (1) are: 5-(2,2-dimethylbicyclo[2.2.1]hept-3-yl)-2-methylpent-3-en-2-ol, 5-(6,6-dimethylbicyclo-[3.1.1]hept-2-yl)-2-methylpent-3-en-2-ol, 5-(1,4-dimethylcyclohex-1-en-4-yl)-2-methylhex-3-en-2-ol, 5-(1-methylcyclohex-1-en-4-yl)-2-methylhept-3-en-2-ol, 2,7,11-trimethyldodeca-3,10-dien-2-ol, 3-(2,5-dimethylbicyclo[3.3.1]non-2-en-8-yl)-2-methylpropan-2-ol and 5-(6,6-dimethylbicyclo[3.1.1]hept-2-en-2-yl)-2-methylpentan-2-ol.

These tertiary alcohols may be prepared by a process which comprises reacting a ketone of formula:

$$\begin{matrix} R_1 \\ & \rangle CH-Y-\underset{\underset{O}{\|}}{C}-R_3 \\ R_2 \end{matrix}$$

wherein R_1, R_2, R_3 and Y are as defined for formula (1) above, with a metal alkyl derivative, preferably an alkyllithium of formula R_3Li or an alkyl Grignard reagent, and hydrolyzing the adduct thus formed. The process is preferably carried out in an organic solvent, for example, an ether such as diethyl ether.

The compound 5-(6,6-dimethylbicyclo[3.1.1]hept-2-yl)-2-methylpent-3-en-2-ol is particularly preferred because of its odor of sandalwood.

Example: (a) 6,6-Dimethylbicyclo[3.1.1]hepten-2-yl acetaldehyde was prepared by hydroformylating 6,6-dimethyl-2-methylenebicyclo[3.1.1]heptane (β-pinene) in the presence of a rhodium carbonyl catalyst. Then 10 g substituted acetaldehyde, 100 g acetone, and 4.0 g barium hydroxide were heated together under reflux for 15 hours. The mixture was then filtered and the excess acetone was removed under reduced pressure. The residue was then fractionally distilled to give 5-(6,6-dimethylbicyclo[2.1.1]hept-2-yl)pent-3-en-2-one, boiling point 110°C at 0.4 mm Hg.

(b) 1.0 g substituted pentenone prepared as in (a) was dissolved in 1.0 g dry diethyl ether, and the solution was cooled to 0°C. A 2 M solution of methyllithium in 5 ml diethyl ether was then added and the mixture was stirred at 20°C for 1 hour. The mixture was cooled again to 0°C and 20 ml water, followed by 10 ml diethyl ether were than added. The ether phase was separated, washed and dried and the solvent was then removed under reduced pressure. Fractional distillation of the residue gave the desired unsaturated alcohol 5-(6,6-dimethylbicyclo[3.1.1]hept-2-yl)-2-methylpent-3-en-2-ol, boiling point 140°C at 0.4 mm Hg. The aroma was woody, like that of sandalwood.

Cis Unsaturated

A number of unsaturated alcohols possess desirable organoleptic properties which make them useful in the preparation of flavor and fragrance intermediates. One such compound is the cis isomer of 3-hexen-1-ol or "leaf alcohol" which possesses a pleasant, green aroma characteristic of new-mown grass. Originally, the compound was extracted from the leaves of green plants, but soon after the structure was determined, a synthetic procedure for the preparation of cis-3-hexen-1-ol was discovered and commercialized. The starting material for the manufacture of cis-3-hexen-1-ol has been butyne-1, which is a very costly starting material.

S.C. Watson, D.B. Malpass and G.S. Yeargin; U.S. Patent 4,069,260; January 17, 1978; assigned to Texas Alkyls, Inc. have described a more economical process for producing these alcohols by reaction of an appropriate predominately cis-1-alkenylaluminum dialkyl intermediate compound with an appropriate epoxide compound. The reaction of the cis-1-alkenylaluminum dialkyl with the epoxide can be conducted at any temperature between -80° and 200°C. The preferred temperature depends on the reactivity of the epoxide with the unsaturated aluminum dialkyl. Very reactive epoxides such as 1,2-epoxyethane and 1,2-epoxy-propane necessitate relatively low temperatures of reaction in order to avoid conditions which would favor the polymerization of these epoxides to polyethers. Less reactive epoxides require prolonged reflux temperatures between 60° and 130°C to complete the reaction. The ideal range is between -80° and 130°C. Following this reaction step, the resulting cis-dialkylaluminum-3-alken-1-oxide is converted to the corresponding cis unsaturated alcohol with a dilute solution of aqueous acid. The resulting unsaturated alcohol will have the following generic formula:

$$HO-\overset{R_2}{\underset{R_3}{C}}-\overset{R_4}{\underset{R_5}{C}}-\overset{H}{C}=\overset{H}{C}-R_1$$

wherein R_1 is a straight-chained or branch-chained alkyl group having from 1 to 8 carbon atoms. R_2, R_3, R_4 and R_5 can be the same or different and can each be selected from a group consisting of hydrogen, straight- or branch-chained alkyl groups having from 1 to 12 carbon atoms, and an aromatic radical having from 6 to 10 carbon atoms, which may be optionally substituted with an alkyl group having from 1 to 6 carbon atoms. It is preferred, however, that R_2 be hydrogen, R_3 be selected from hydrogen and lower alkyl having 1 to 6 carbon atoms, and R_4 and R_5 be selected from hydrogen, alkyl having from 1 to 6 carbon atoms, and phenyl.

The predominately cis-1-alkenylaluminum dialkyl compound used as an intermediate to manufacture the above-defined unsaturated alcohols will have the following formula:

$$\begin{matrix} R \\ R \end{matrix} > Al-\overset{H}{C}=\overset{H}{C}-R_1$$

wherein R is a straight or branch-chained alkyl group having from 1 to 6 carbon atoms, and R_1 is as has been previously defined.

This compound is prepared by the reaction of an appropriate trialkylaluminum compound with acetylene at a pressure of from atmospheric to about 15 psig. The temperature of this reaction should be maintained preferably between about -40° and 40°C.

Preparation of cis unsaturated alcohols in up to 97 to 99% isomeric purity is possible by precoordination of the cis-1-alkenylaluminum dialkyl compounds with at least one molar equivalent of a suitable Lewis base. Especially preferred are the cyclic ethers and diethers, and aliphatic diethers and polyethers selected from tetrahydrofuran, tetrahydropyran, dioxane, 1,2-dimethoxyethane, 1,2-diethoxyethane, and the dimethyl ether of diethylene glycol. While these coordinating agents can also function as solvents, it is preferred to use cosolvents that include pentane, isopentane, petroleum ether, hexane, cyclohexane, heptane, benzene, toluene and xylene.

In addition to ensuring a high cis content in the resultant unsaturated alcohols, the coordinating agents also stabilize the unsaturated cis-1-alkenylaluminum dialkyl compounds to decomposition reactions provided at least a molar equivalent of coordinating agent is used.

The epoxide compounds can be selected from any number of epoxide compounds commercially available. In general, the epoxide compound will have the following generic formula:

$$R_2-\overset{\overset{R_3}{|}}{C}\underset{O}{\overline{\quad\quad}}\overset{\overset{R_4}{|}}{C}-R_5$$

wherein R_2, R_3, R_4 and R_5 are as have been previously defined.

As noted above, the reaction between the cis-1-alkenylaluminum dialkyl compound and the epoxide provides an intermediate oxide which must be hydrolyzed. Thus, the solutions are hydrolyzed with dilute aqueous mineral acids to convert the resultant dialkylaluminum-3-alkenyl-1-oxide to the corresponding unsaturated alcohol. The aqueous mineral acids can be selected from sulfuric acid, nitric acid and phosphoric acid. There is a concomitant production of two mols of an alkane per mol of alcohol, and aluminum salts. The organic layer containing product solvent is separated, the aqueous layer is extracted several times more with solvent and the organic fractions combined. The solvent is removed by flash distillation usually under vacuum.

Example: A one-half gallon stirred autoclave equipped with cooling and heating coils was dried, purged with nitrogen, and charged with 3.0 mols of triethylaluminum. The triethylaluminum was heated to 35°C and rapidly agitated while the reactor was pressured to 13 psig and continually fed with acetylene gas. After a period of 56 hours, a yellow slightly viscous liquid was obtained, which analysis showed to be 88 wt % diethyl-1-butenylaluminum by gas chromatography analysis of the resultant hydrolysis gases. A sample of this material was hydrolyzed with deuterium oxide 24 hours after preparation, and the corresponding 1-deutero-butene-1 was analyzed by proton magnetic resonance spectroscopy. Analysis showed the material to contain greater than 97% cis-1-deuterobutene-1.

To the stirred autoclave containing the 1-butenylaluminum diethyl was immediately added 342 g of benzene solvent and the temperature lowered to 0°C by constant circulation of −10°C silicone fluid through the cooling coils. Gaseous 1,2-epoxyethane was slowly introduced above the surface of the liquid at a rate such that the reaction exotherm never rose above 5°C. After 3 hours of feeding, the reaction exotherm ceased after the addition of approximately 2.9 mols of 1,2-epoxyethane and the reaction was assumed to be complete. The product solution was then slowly hydrolyzed by addition to 2 liters of 10% sulfuric acid. The organic layer separated cleanly and was withdrawn.

The aqueous layer was reextracted three more times with 200-ml portions of ether which were later combined with the original organic layer. The ether and benzene were removed in a flash evaporator and the crude product transferred to a still containing a 3-ft packed distillation column equipped with a partial take-off head. Under vacuum distillation at 27 mm pressure approximately 116 g of 3-hexen-1-ol were obtained boiling between 70° to 73°C. The material was analyzed by gas chromatography and showed a content of 67.8% cis-3-hexen-1-ol. The overall yield of 3-hexen-1-ol was approximately 27% based on cis-1-butenylaluminum diethyl.

Alkoxy-Alkenyl Phenols

K. Bauer, R. Mölleken and F. Exner; U.S. Patent 4,071,564; January 31, 1978; assigned to Haarmann & Reimer GmbH, Germany describe a process for the production of 2-alkoxy-4-alkenyl phenols and discuss their use as clovelike odorants. The 2-alkoxy-4-alkenyl phenols correspond to the general formula:

(1) [structure: benzene ring bearing OH, OR_1 ortho to OH, and para to OH the chain $CH{=}C(R_2){-}CH_2{-}R_3$]

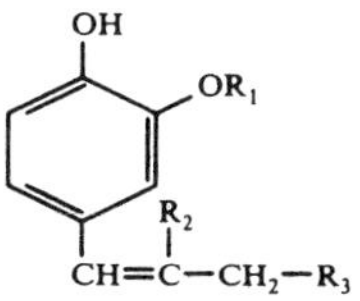

in which R_1 represents a methyl or ethyl radical while R_2 and R_3, which are different from one another, each represent hydrogen or a methyl radical. Compounds falling within the scope of formula (1) include 1-(4-hydroxy-3-ethoxyphenyl)-1-butene and 1-(4-hydroxy-3-methoxyphenyl)-2-methyl-1-propene.

They are prepared by reacting in a first stage phenol derivatives of the general formula (2) with aldehydes of formula (3) in the presence of acid catalysts.

(2) [structure: benzene ring bearing HO and R_1O ortho to each other] (3) $H(O{=})C{-}CH(R_2){-}CH_2{-}R_3$

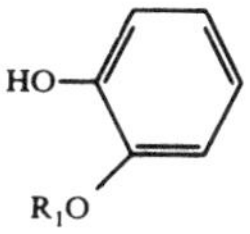

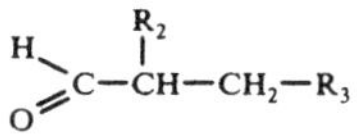

The product of condensation formed in the first stage is split in a second stage by heating in the presence of a preferably basic catalyst to form the required compound of general formula (1).

Suitable starting materials for the first stage of the process according to the process are the phenol derivatives 1-hydroxy-2-methoxy and 1-hydroxy-2-ethoxybenzene, while butyraldehyde and isobutyraldehyde may be used as the aldehydes.

Acid catalysts suitable for use in the condensation reaction include medium-strength to strong proton acids, acidic ion exchangers, acid anhydrides of inorganic acids or Lewis acids, for example, strong to medium-strength inorganic acids, strong organic acids, Lewis acids or inorganic and organic ion exchangers or solid acid anhydrides of strong inorganic, nonoxidizing acids.

The formula (1) compounds are valuable odorants which add a new range of nuances to known clove odorants, for example of the isoeugenol and ethyl isoeugenol type. By comparison with conventional clove odorants, they have smoky, sweet-vanillin-like odor notes reminiscent of oakmoss leather, providing perfume manufacturers with new, interesting possibilities for perfume compositions. For example, the odor of 1-(4-hydroxy-3-ethoxyphenyl)-1-butene and 1-(4-hydroxy-3-ethoxyphenyl)-2-methyl-1-propene, is softer, more flowery and sweeter than that of isoeugenol, so that these compounds may be used with advantage, for example, in perfume compositions where the powdery-spicy clove note of isoeugenol is undesirable. 1-Hydroxy-3-methoxyphenyl)-1-butene is particularly characterized by soft, flowery-sweet and hard, mossy leather-like components.

In 1-(4-hydroxy-3-methoxyphenyl)-2-methyl-1-propene, powdery, spicy, clove-like and smoky, guaiacum-wood-like components are combined in the odor complex of a single compound, which provides for a completely uniform odor combination throughout the entire vaporization process and precludes any differences in the way the complex disintegrates with time. Combinations of this kind may be used with advantage, for example, in men's perfumes.

Example: 72 g (1 mol) of butyraldehyde were added dropwise over a period of 1 hour at 70°C to 2,670 g (19.35 mols) of guaethol (1-hydroxy-2-ethoxybenzene) and 214 g of anhydrous cation exchanger in the H-form (Lewatit SC 102/H^+). The mixture was stirred for a further 3 hours at 70°C, after which the water formed and the unreacted butyraldehyde were distilled off from the reaction mixture at a sump temperature of at most 70°C/10 mm Hg. Another 72 g (1 mol) of butyraldehyde were added to the remaining mixture over a period of 1 hour. After stirring for 2 hours at 70°C, the liquid phase of the reaction mixture was filtered off from the ion exchanger and the unreacted butyraldehyde and excess guaethol were separated off by distillation (110°C/10 mm Hg), leaving 383 g of a condensation product of guaethol and butyraldehyde [consisting essentially of 1,1-bis(4-hydroxy-3-ethoxyphenyl)-butane].

Following the addition of 400 mg of NaOH, 377 g of this condensation product were heated in a distillation apparatus with a 10-cm Vigreux column attachment. A mixture of guaethol and 1-(3-ethoxy-4-hydroxyphenyl)-1-butene distilled off at a sump temperature of 200° to 260°C/10 mm Hg. A total of 270 g of the required product of splitting were obtained, 1-(3-ethoxy-4-hydroxyphenyl)-1-butene being isolated therefrom in pure form by fractional distillation at 142° to 144°C and 10 mm Hg. The yield amounted to 134.8 g (61.5% of the theoretical yield based on the condensation product used). Assessment of odor: clove odorant with a sweet, soft flowery note, less powdery-spicy by comparison with isoeugenol.

Certain Norbornyl Derivatives

The principal object of the process developed by *A.A. Schleppnik; U.S. Patent 4,128,509; December 5, 1978; assigned to Monsanto Company* was to provide a class of aroma chemicals consisting of 1-(norborn-5'-en-2'-yl) or (1-(norborn-2'-yl) substituted 1-alken-3-ols and alkan-3-ols, having a characteristic aroma which can be utilized in the preparation of fragrances and fragrance compositions.

These compounds are characterized by the structural formulas:

(1)

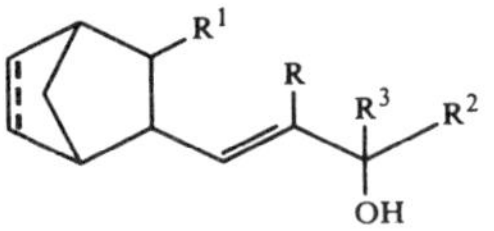

and

(2)

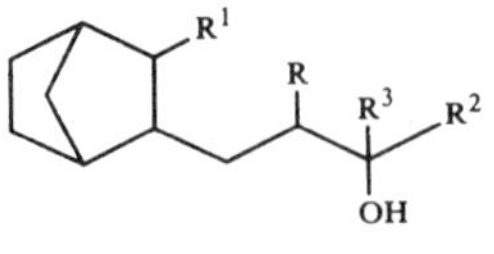

wherein R^1 represents hydrogen or methyl, R represents an alkyl with from 1 to 8 carbon atoms, and R^2 and R^3 represent hydrogen or an alkyl with from 1 to

8 carbon atoms, provided that R and R^2 can be $-(CH_2)_n-$ wherein n represents the integer 2, 3 or 4.

The class of compounds as a whole exhibits a characteristic pleasant, lasting floral-woody aroma. The preparation of representative compounds can be illustrated by the following examples.

Example 1: *1-(Norborn-2'-yl)-2-Methyl-Butan-3-ol* – To a solution of 8.9 g (0.05 mol) of 1-(norborn-2'-yl)-2-methyl-butan-3-one in 50 ml methanol was added, with stirring, a solution of 1 g sodium borohydride in 10 ml 5% aqueous sodium hydroxide solution. An exothermic reaction took place, the temperature rising to 50°C. (Cooling with cold water bath was necessary.) The reaction mixture was left at room temperature for 15 hours and yielded 7.1 g (80% product), boiling point 115° to 118°C/10 mm Hg having a n_D^{25} = 1.4820. The reaction product was found to be 1-(norborn-2'-yl)-2-methyl-butan-3-ol. It was a viscous colorless liquid with a distinct woody aroma.

Example 2: *1-(Norborn-5'-en-2'-yl)-2,3-Dimethyl-But-1-en-3-ol* – The title compound can be prepared by adding a solution of 1-(norborn-5'en-2'-yl-2-methyl-but-1-en-3-one in ether, with stirring and cooling, to one equivalent of methyllithium in ether maintaining the solution formed at room temperature for 15 hours.

Example 3: *1-(Norborn-2'-yl)-2-Methylpentan-3-ol* – The title compound can be obtained by the hydrogenation of 1-(norborn-5'-en-2'-yl)-2-methylpent-1-en-3-one over Raney nickel and ethanol solution at 100 psi H_2 pressure and 80°C.

Example 4: *1-Methyl-2-(Norborn-2'-yl-Methyl)Cyclopentan-1-ol* – The title compound is obtained by reacting 2-(norborn-2'-yl-methyl)cyclopentanone and methyl magnesium bromide in ether.

Allylic Alcohols from Allylic Halides

Allylic halides including terpene allylic halides are available from various sources, and are valuable intermediates in the preparation of allylic esters and alcohols which are useful as perfumery ingredients and flavors. Terpene allylic halides such as myrcene hydrohalides are a source of geraniol and linalol, while allylic alcohols of the spearmint series may be obtained from carvyl chloride, and allylic alcohols of the peppermint series from 5-chloro-3-menthene.

All of the processes suggested for the preparation of the allylic alcohols from allylic halides have a common problem, and that is disposal of the alkali metal carboxylic acid salt after the hydrolysis reaction. Discarding the alkali carboxylate salt solution and starting a new batch with a fresh supply of salt is feasible, but economically unattractive. Besides the direct cost of the salt, which is relatively high, there are problems associated with discarding the solution. How to recover the salt by evaporation of its aqueous solution is well known. To build the necessary installation for this recovery is expensive, and the operation requires labor in handling the solid salt that is recovered.

P.S. Gradeff; U.S. Patent 4,152,530; May 1, 1979; assigned to Rhone-Poulenc Inc. proposes a process which accomplishes the recycling of the salt economically as an integral part of a cyclic process, each cycle of which can be carried out in the same reactor.

In accordance with this process, the allylic alcohols are prepared from allylic halides via an intermediate ester which makes it unnecessary to isolate the ester and realizes the regeneration of the carboxylic acid salt in the same equipment, in a manner obviating costly alternate drying procedures that require special installations and handling operations. The result is that the allylic halide can be converted to allylic alcohol in a minimum of steps, and the only reagent is an aqueous solution of alkali metal hydroxide.

The process comprises:

(1) esterifying an allylic halide having the formula

$$\begin{matrix} R_1 & & R_3 & R_4 & X \\ & \searrow & | & | & \nearrow \\ & & C{=}C & -C & \\ & \nearrow & & & \searrow \\ R_2 & & & & H \end{matrix}$$

(a) R_1, R_2, R_3 and R_4 are selected from the group consisting of hydrogen and alkyl;

(b) one of R_1 and R_2 has from one to about twenty carbon atoms, the remaining one of R_1 and R_2, R_3 and R_4 having from one to about five carbon atoms, and

(c) X is halogen selected from the group consisting of chlorine, bromine and iodine, at a temperature within the range of from 50° to 150°C, under anhydrous conditions and in the presence of a basic nitrogen-containing catalyst with an alkali metal carboxylic acid salt, the allylic halide being in solution in a water-insoluble ketone forming an azeotrope with water, thereby forming an ester of the allylic halide and the carboxylic acid, and an alkali metal halide;

(2) adding water to dissolve the alkali metal halide and then discarding the alkali metal halide aqueous solution;

(3) adding aqueous alkali metal hydroxide to the reaction mixture in an amount to hydrolyze the allylic ester, forming aqueous and organic phases and then heating the reaction mixture at a temperature within the range from 50° to 150°C until allylic alcohol is formed;

(4) separating the aqueous and the organic phases;

(5) distilling off the allylic alcohol from the organic phase, and optionally recycling the ketone that remains;

(6) adding ketone to the aqueous layer from (4) and azeotropically distilling off the water from the aqueous phase, and forming anhydrous alkali metal carboxylic acid salt as a slurry in the ketone; and

(7) recycling the recovered alkali metal carboxylic acid salt and ketone to step (1).

The class of allylic halides to which the process is applicable includes open-chain allylic-type halides including aliphatic allylic halides. Exemplary allylic halides include 1-chloro-3-methyl-2-butene; 1-chloro-3,7-dimethyl-octa-2,6-diene; and 1-chloro-3,7,11-trimethyl-dodecyl-2,6,10-triene.

The alkali metal carboxylic acid salt is preferably a salt of an aliphatic or aromatic carboxylic acid having from 1 to 5 carbon atoms, such as acetic acid, formic acid, propionic acid, butyric acid, and valeric acid; maleic acid; oxalic acid; tartaric acid; gluconic acid; citric acid; succinic acid; malic acid; benzoic acid; terephthalic acid. The alkali metals include sodium and potassium. Preferred salts are sodium acetate and potassium acetate.

The preferred nitrogen catalysts are tertiary amines, such as trimethylamine, triethylamine, triethanolamine, tripropylamine, and tributylamine.

Example: Into a one-liter three-neck flask equipped with a condenser, mechanical stirrer, thermometer, and dry ice trap was charged 180.5 g (2.2 mols) sodium acetate, 161.6 g (100 ml/mol) diisobutyl ketone, and 218.8 g (2.0 mols) 1-chloro-3-methyl-2-butene, with stirring. The mixture was heated to reflux and 1.0 g triethylamine added. The reaction was allowed to proceed while monitoring the amount of 1-chloro-3-methyl-2-butene by gas-liquid chromatographic analysis.

When the reaction was 95% complete, another 1.0 g of triethylamine was added. When the reaction was about 99.5% complete, 340 ml of water was added. The aqueous layer was then separated at above 80°C, and discarded. The ketone layer was then heated to about 120°C, whereupon 208.7 g (2.05 mols) 33.93% aqueous NaOH was charged in about twenty minutes, keeping the pot at reflux throughout the addition. Refluxing was continued for an additional hour, at which time sampling showed virtually quantitative hydrolysis to 3-methyl-1-butanol. The organic layer was separated at 80° to 90°C; it weighed 327.8 g (333.8 grams theory).

The ketone layer was transferred into a 500 ml distilling flask, and distilled under reduced pressure, initially at 130 mm, and 25°C, gradually increasing both temperature and vacuum to 2.5 mm at 77°C vapor temperature, 108°C pot temperature. Gas-liquid chromatographic analysis of the distillate showed:

Product	Percent	Weight, g
3-Methyl-1-butanol	49.7	156.2
3-Methyl-3-butanol	1.1	3.5
2,4-Dichloro-2-methyl-butane	0.9	2.8
Diisobutyl ketone	39.0	126.6
Yield based on 1-chloro-3-methyl-2-butene = 90.5%		

The aqueous layer, 290 g, was then heated to 90°C, neutralized to a pH of 7.5 with glacial acetic acid, diisobutyl ketone added, and the H_2O/diisobutyl ketone azeotrope distilled off. The water was collected and the ketone returned to the pot. When water ceased to distill, the amount of ketone was adjusted to 100 ml/mol and a new cycle begun by adding 218.8 g (2.0 mols) 1-chloro-3-methylbutene-2. The mixture was stirred and heated to 115°C, and then 1.0 g triethylamine was added. When the reaction was 95% complete another 1.0 g triethylamine was added. When the reaction was about 99.5% complete, it was quenched wtih 340 ml H_2O and worked up as before.

The ketone layer was hydrolyzed as before, and 236 g 3-methyl-1-butanol recovered.

KETONES

Certain Tricyclododecene Derivatives

Kitchens, et al, in U.S. Patent 3,678,119 and others, showed that the cis-thujopsene produced a product having a strong woody odor which upon analysis by gas chromatography revealed seven major components which were designated in order of elution as isomers A through G.

The major component constituted approximately half the mixture and possessed a powerful, woody, musk, ambergris odor far greater than that of any of the six other isomers. The structure of this was shown to be 4-acetyl-1,7,7-trimethyltricyclo[6.2.2.$0^{3,8}$]-3-dodecene, (1), (also known as 4-aceto-6,8a-ethano-1,1,6-trimethyl-1,2,3,5,6,7,8,8a-octahydronaphthalene). (This compound will be designated Isomer G from now on, for convenience.)

R.F. Tavares and E. Katten; U.S. Patent 4,067,906; January 10, 1978; assigned to Givaudan Corporation have developed a general process for the preparation of tricyclo[6.2.2.$0^{3,8}$]dodecene derivatives such as Isomer G from 4-methyl-4-p-tolylpentanoic acid. The process comprises:

(a) 1,4-hydrogenating the aromatic ring of the 4-methyl-4-p-tolylpentanoic acid via a Birch reduction to 4-methyl-4-(4-methyl-1,4-cyclohexadien-1-yl)pentanoic acid;

(b) isomerizing the 4-methyl-4-(4-methyl-1,4-cyclohexadien-1-yl)pentanoic acid to 4-methyl-4-(4-methyl-1,3-cyclohexadien-1-yl)pentanoic acid in the presence of a base;

(c) reacting the 4-methyl-4-(4-methyl-1,3-cyclohexadien-1-yl)pentanoic acid with vinyl lithium to form the intermediate 6-methyl-6-(4-methyl-1,3-cyclohexadien-1-yl)hepten-3-one which is converted via an internal Diels-Alder reaction to form the 1,7,7-trimethyltricyclo-[6.2.2.$0^{3,8}$]dodec-9-en-4-one.

This sequence may be illustrated by the steps used in the synthesis of Isomer G as shown below.

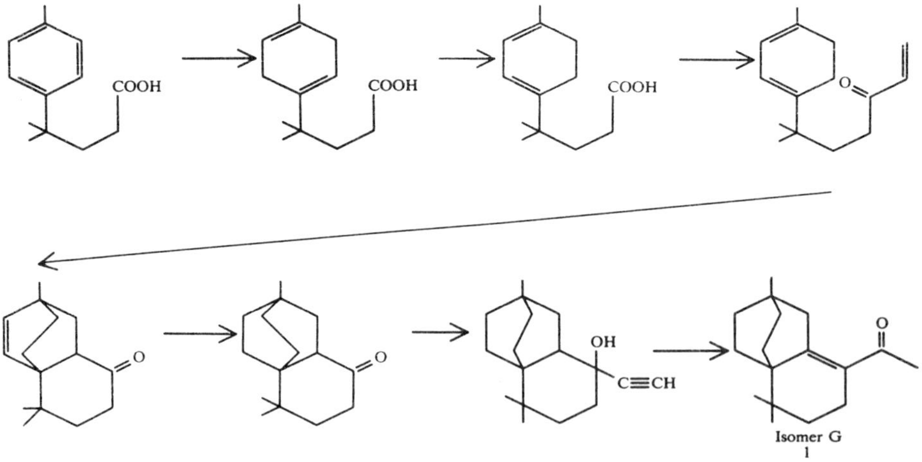

Isomer G
1

Menthone

Menthone is a useful compound for formulating perfumes, and can be converted to menthol by reduction.

Heretofore, synthetic levo-menthol has been produced mainly from the dextro-citronellal present in natural citronella oil in an amount of 40 to 50% by weight. The residual oil remaining after the separation of dextro-citronellal is substantially a mixture of geraniol and citronellol in a ratio of 6:4 by weight; and studies on utilization of the residual oil led *T. Yamanaka and M. Yagi; U.S. Patent 4,134,919; January 16, 1979; assigned to Takasago Perfumery Co., Ltd, Japan* to several processes for the preparation of menthone and menthol.

One process for producing menthone in a high yield comprises dehydrogenating citronellol in a reaction system containing a catalyst at a temperature of about 150° to 260°C in an atmosphere of hydrogen under a gauge pressure of 0 to 5 kg/cm^2. (Pressures stated hereafter should also be understood to be gauge.)

Another process for producing menthol in good yields comprises dehydrogenating citronellol in a reaction system containing a catalyst at a temperature of about 150° to 260°C in an atmosphere of hydrogen under a pressure of 0 to 5 kg/cm^2 to produce menthone, and then, in the same reaction system without separating the catalyst or adding additional catalyst, hydrogenating the menthone produced by decreasing the temperature to 110° to 130°C and increasing the hydrogen pressure to about 10 to 50 kg/cm^2 to produce menthol.

An even further embodiment provides a process for producing menthone which comprises hydrogenating geraniol or a mixture of geraniol and citronellol in a reaction system containing a catalyst at a temperature of about 110° to 180°C in an atmosphere of hydrogen under a pressure of 2 to 10 kg/cm^2 to produce citronellol, and then, in the same reaction system without separating the catalyst or adding additional catalyst, dehydrogenating the citronellol produced by increasing the temperature to about 150° to 260°C and decreasing the hydrogen pressure to 0 to 5 kg/cm^2 to produce menthone.

In a still further embodiment there is provided a process for producing menthol which comprises hydrogenating geraniol or a mixture of geraniol and citronellol in a reaction system containing a catalyst at a temperature of about 110° to 180°C in an atmosphere of hydrogen under a pressure of 2 to 10 kg/cm^2 to produce citronellol, then, in the same reaction system without separating the catalyst or adding additional catalyst, dehydrogenating the citronellol produced by increasing the temperature to 150° to 260°C and decreasing the hydrogen pressure to 0 to 5 kg/cm^2 to produce menthone, and, further in the same reaction system without separating the catalyst or adding additional catalyst, hydrogenating the menthone produced by decreasing the temperature to 110° to 130°C and increasing the hydrogen pressure to 10 to 50 kg/cm^2 to obtain menthol.

In each of the embodiments which have been described above a common step is involved. This common step is demonstrated by the reaction scheme which is shown on the following page.

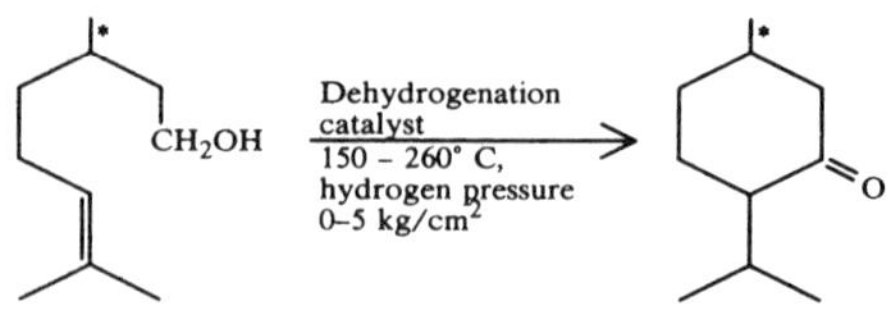

CH₂OH → Dehydrogenation catalyst, 150 - 260° C, hydrogen pressure 0-5 kg/cm²

Since the optical activity on the asymmetric carbon atom shown by the symbol* is maintained in this reaction, dextro-citronellol in citronella oil is advantageously used to produce levo-menthol by the method for producing menthol through menthone.

Example 1: Citronellol (156 g, 1 mol) and 7.8 g of a copper-chromium oxide catalyst ($CuO:Cr_2O_3$ = 50:50 by weight, mainly as $CuCr_2O_4$, less than 200 mesh) were charged into a pressure reactor, and reacted at 230°C under a hydrogen pressure of 2 kg/cm^2. In about 4 hours, the evolution of hydrogen ended. The amount of hydrogen evolved was 24 liters. The catalyst was separated from the reaction product by filtration, and the residue was distilled under reduced pressure to afford 140 g of a fraction having a boiling point of 82° to 90°C/12 mm Hg. This fraction was a mixture of menthone and iso-menthone in a ratio of 6:4 by weight. The theoretical yield was 90%. Also, 7.7 g of thymol boiling at 108° to 110°C/12 mm Hg was obtained.

Example 2: At the end of the reaction in Example 1, the reaction product was further reacted at 120°C under a hydrogen pressure of 25 kg/cm^2 without separating the catalyst from the reaction product. In 5 hours, the absorption of hydrogen ended. Distillation of the reaction mixture under reduced pressure afforded 158 g of an isomeric menthol mixture. By gas chromatographic analysis, the product was found to comprise 35 wt % of neo-menthol, 25 wt % of neo-iso-menthol, 35 wt % of menthol, and 5 wt % of iso-menthol.

Substituted Cyclopropane

J.O. Bledsoe, Jr. and W.E. Johnson, Jr.; U.S. Patent 4,097,531; June 27, 1978; assigned to SCM Corporation have developed a process for the preparation of vinyl cyclopropyl ketone.

These compounds, being ionone analogs, can be useful in perfumery. They have sweet, floral odors.

The process provides a method for making substituted cyclopropanes represented by the following general formulas:

(1) $R{=}C(CH_3){-}CH{<}(CH_2){>}CH{-}C({=}O){-}CH_3$ and (2) $R^1{-}C({=}CH_2){-}CH{<}(CH_2){>}CH{-}C({=}O){-}CH_3$,

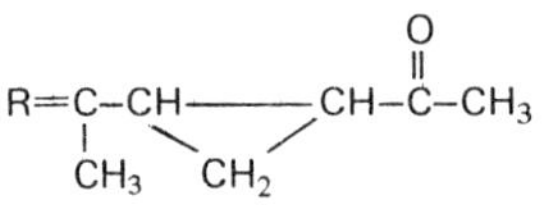

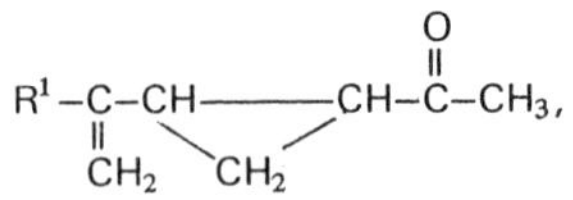

where R is a monovalent alkyl or alkylene radical, preferably a C_{1-6} alkyl or alkylene radical, and R^1 is hydrogen or R. These substituted cyclopropanes (or substituted cyclopropyl ketones) are made by heating at about 100° to 250°C in the presence of an alkaline earth metal carbonate and liquid glycol vehicle, a compound represented by

$$(3) \qquad R^1\text{-}\underset{\displaystyle CH_3}{\underset{|}{C}}\text{=CH-CH}_2\text{-}\underset{\displaystyle X}{\underset{|}{CH}}\text{-}\overset{\displaystyle O}{\overset{||}{C}}\text{-CH}_3$$

where X is a halogen, preferably chlorine, until the product cyclopropane is formed. Reaction times vary from 1 to 20 minutes up to several hours, depending upon reaction temperature and starting material (3). Suitable reaction times of 1 to 20 minutes will suffice at elevated temperatures of 200°C and above, and 4 to 5 hours at about 150° to 160°C, for higher molecular weight starting materials. For lower molecular weight reactants, lower temperatures and longer reaction times often are useful.

Suitable alkaline earth metal carbonates (and bicarbonates) include calcium carbonate, barium carbonate, strontium carbonate, and magnesium carbonate, often provided in pulverulent form. About 1 to 8 equivalents of the carbonate per mol of reactant (3) is useful. Preferably, the carbonate is calcium carbonate, for economy, in a proportion of about 2.5 equivalents.

The liquid glycol (or polyol) can be an alkylene glycol or polyalkylene glycol, such as diethylene glycol, propylene glycol, ethylene glycol, triethylene glycol and the like. Preferably, diethylene glycol is used for atmospheric operation. The glycol vehicle advantageously is 2 to 20 parts per part of reactant, and preferably about 10 parts for best yields and ease of filtration of solids.

The alpha-haloketone reactant most suitable is 3-chlorogeranyl acetone, although other feedstocks such as 3-chloro-6-methyl-5-hepten-2-one and 3-chloro-6-methyl-5-nonen-2-one are suitable. The halogen usually is chlorine, but can also be bromine or iodine.

In the example below all parts are by weight, all percentages are by weight, and all temperatures are in degrees centigrade, unless otherwise expressly noted.

Example: In a 250-ml three-necked glass flask fitted with a stirrer, thermometer, condenser, addition funnel and a nitrogen gas inlet, a slurry of 40 g of calcium carbonate (precipitated chalk) and 100 g of diethylene glycol were stirred and heated to 210° under nitrogen. 30 g of α-chlorogeranyl acetone (Compound A) was added over a five-minute period, and, after an additional 45 minutes, the reaction mixture was cooled to 25° and 300 ml of isopropanol was added. Then solids were filtered off. After removal of the isopropanol on a rotary evaporator at reduced pressure, the diethylene glycol was extracted into an aqueous layer by the addition of water, and the whole resulting mass was extracted with ether. The crude product isolated from the ether weighed 27 g and was distilled to give 12.6 g (49.9% yield) of a mixture of the three cyclopropyl pseudo-ionones, namely Compounds B, C, and D in a ratio of about 50:22:28, respectively.

Fractional distillation on a 2-foot Nester-Faust spinning band column effectively gave pure Compound B, boiling point 62° at 0.2 torr, but Compounds C and D were obtained in only about 80% purity.

Analysis of Compund B gave: C, 81.02%; H, 10.50%. Calculated for $C_{13}H_{20}O$: C, 81.20%; H, 10.48%.

The reaction is shown by the following equation:

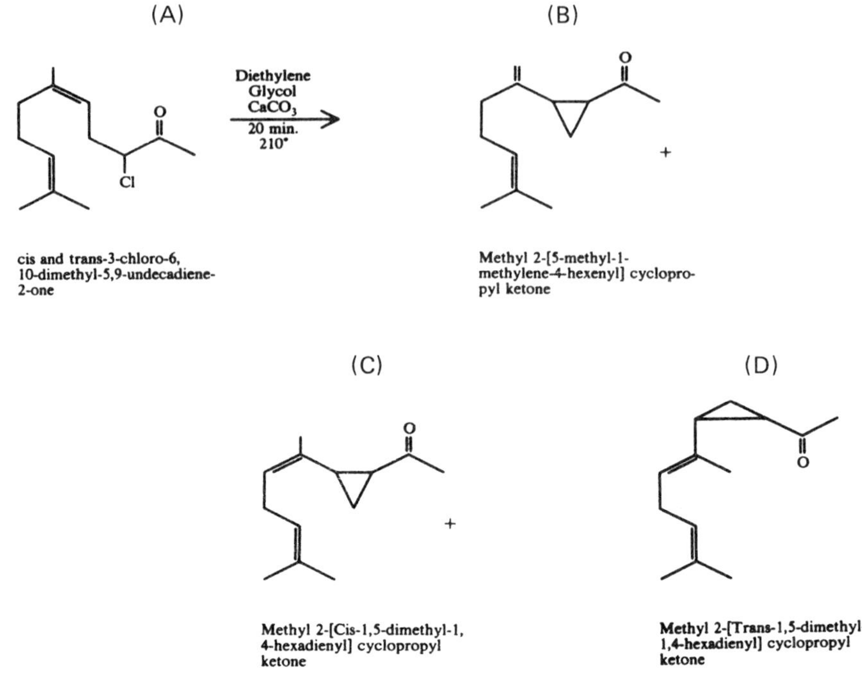

HETEROCYCLIC COMPOUNDS

2,4-Dioxa-Spiro[5.5]Undec-8-ens

J. Conrad and K. Bruns; U.S. Patent 4,113,664; September 12, 1978; assigned to Henkel Kommanditgesellschaft auf Aktien, Germany have found that 2,4-dioxa-spiro[5.5]undec-8-ens of the following general formula can be used in advantageous manner as perfumes having a wide variety of fragrances.

R_1 R_2 R_3 R_4 R_5

In the above R_1, R_2 and R_3 denote hydrogen or a methyl radical; R_4 represents hydrogen, an alkyl having 1 to 4 carbon atoms or the radicals, $-OCH_3$, $-OC_2H_5$, $-CH{=}CH_2$ or $-CH{=}CH{-}CH_3$; R_5 represents hydrogen of an alkyl having 1 to 4 carbon atoms; and R_4 and R_5 together can be closed to form a cycloaliphatic ring system having the ring members $-(CH_2)_n-$ wherein n represents the numbers 4 to 6.

A simple production of cyclic acetals is described by S.R. Sandler and W. Kawo in *Organic Functional Group Preparations,* Vol. III, Academic Press 1972, pages 41-42, in which the desired spirocyclic acetals and ketals are readily obtained by reacting the diols and aldehydes or ketones with ortho-esters in the presence of small quantities of acid.

The various 2,4-dioxa-spiro[5.5] undec-8-ens specified in the examples were obtained as colorless oils according to the following general working instructions.

A few granules of p-toluene sulfonic acid were added to the mixture comprising 0.1 mol of 3-cyclohexene-1,2-dimethanol, 0.1 mol of aldehyde or ketone and 0.1 mol of triethyl orthoformate. The mixture was agitated for several hours at 20° to 40°C. Ethanol and ethyl formate were subsequently distilled off. The residue was dissolved in ether and the solution obtained was washed with aqueous sodium hydroxide solution and water. After separation, the organic phase was dried over sodium sulfate, and the residue after evaporation was distilled in vacuo. The following were obtained in this manner.

Example 1: *3-Ethyl-2,4-Dioxa-Spiro[5.5] Undec-8-en* – The compound is a colorless oil, boiling point 86°C at 6.0 torrs, n_D^{20} = 1.4773, having a fragrance of roses and catechu.

Example 2: *3,3-Tetramethylene-2,4-Dioxa-Spiro[5.5] Undec-8-en* – The compound is a colorless oil, boiling point 104°C at 1.5 torrs, n_D^{20} = 1.5000, and has a jasmine-like herbal fragrance.

Chemical Designation	Boiling Point °C/Torr	n_D^{20}	Description of Odor
11-methyl-2,4-dioxa-spiro(5,5) undec-8-en	67/1.0	1.4881	Herbal, woody very natural
3,11-dimethyl-2,4-dioxa-spiro(5.5) undec-8-en	60/0.2	1.4804	Green, thujon fragrance, for rosemary-lavanduline compositions
3-ethyl-11-methyl-2,4-dioxa-spiro(5,5)undec-8-en	61/0.1	1.4788	Green, fragrance of buds
3-isopropyl-11-methyl-2,4-dioxa-spiro(5,5)undec-8-en	66/0.08	1.4768	Rose oxide, geranium fragrance
3,3,11-trimethyl-2,4-dioxa-spiro(5,5)undec-8-en	64/0.1	1.4778	Camphoric, woody, fruity, vevdox-fragrance
3,8/9-dimethyl-2,4-dioxa-spiro(5,5) undec-8-en	75/1.5	1.4773	Fruity, green, herbal, anise fragrance
3,9-dimethyl-2,4-dioxa-spiro(5,5) undec-8-en	85/5.0	1.4772	Gardenia, petitgrain fragrance, green
3-ethyl-8/9-methyl-2,4-dioxa-spiro(5,5)undec-8-en	74/0.05	1.4769	Rose fragrance
3-isopropyl-8/9-methyl-2,4-dioxa-spiro(5,5)undec-8-en	74/0.4	1.4740	Rose fragrance, very natural
3,3,8/9-trimethyl-2,4-dioxa-spiro(5,5) undec-8-en	81/1.5	1.4759	Woody, pine fragrance, fresh, sawdust
8/9,11-dimethyl-2,4-dioxa-spiro (5,5)undec-8-en	88/2.4	1.4875	Herbal, woody, rosemary fragrance

(continued)

Chemical Designation	Boiling Point °C/Torr	n_D^{20}	Description of Odor
3-ethyl-8/9,11-dimethyl-2,4-dioxa-spiro(5,5) undec-8-	68/0.05	1.4788	Natural forest soil odor, rose fragrance
3,8,9-trimethyl-2,4-dioxa-spiro (5.5)undec-8-en	58/0.05	1.4822	Herbal, petit-grain fragrance
8,9,11-trimethyl-2,4-dioxa-spiro (5,5)undec-8-en	62/0.05	1.4910	Spicey, thyme fragrance
3-ethyl-8,9,11-trimethyl-2,4-dioxa-spiro(5,5) undec-8-en	75/0.1	1.4826	Woody, green
3-isopropyl-8,9 11-trimethyl-2,4-dioxa-spiro(5,5) undec-8-en	88/0.1	1.4804	Flowery

8/9 designates one substituent in the 8 and 9 position or on a mixture of 8 and 9 positions.

All the compounds given in the above examples have fragrances with excellent clinging properties or persistency which render them suitable for producing a wide variety of perfume compositions. Such compoisitons can be used to perfume a wide variety of products such as cosmetics, washing agents, soaps as well as technical products in concentrations of approximately 0.05 to 2% by weight.

Alkyl-Substituted 1,4-Dioxanes

J. Conrad, U.-A. Schaper and K. Bruns; U.S. Patent 4,124,541; November 7, 1978; assigned to Henkel Kommanditgesellschaft auf Aktien, Germany have synthesized perfumery components which are alkyl-substituted 1,4-dioxanes of the formula

R_2 O R_3

R_1 O R_4

in which R_1 represents an alkyl having 4 to 12 carbon atoms, R_2 represents hydrogen or R_1 may be ring-closed with R_2 to form an optionally alkyl-substituted cycloalkyl radical which has 3 to 12 carbon atoms in the ring and R_3 and R_4 are the same or different and represent hydrogen or an alkyl group having 1 to 4 carbon atoms, particularly a methyl radical. They can be used in advantageous manner as perfumes having an intensive and lasting fragrance.

The production of the 1,4-dioxanes is based on terminal alkyl epoxides or cycloalkyl epoxides on the one hand, and, on the other hand, 1,2-alkylene glycols, production being effected in two stages. In the first stage, the glycol is added to the epoxide in a reaction in the presence of an acid or alkaline catalyst.

In the second stage, the dioxane ring is closed under acid catalysis with p-toluene sulfonic acid with the elimination of water. The water produced is distilled off azeotropically in a known manner by means of an azeotropic agent.

Synthesis of the compounds described in the following examples was carried out as follows.

Production of Ether Diol by Acid Catalysis: 0.5 mol of the particular epoxide, 0.75 mol of the particular 1,2-diol and 1.5 ml of boron trifluoride etherate were heated to 100°C and were agitated for 15 to 20 hours at this temperature. After cooling, the mixture was dissolved in ether, washed neutral with water, dried over sodium sulfate, and recovered by distillation of the solvent in vacuo. The raw yield of ether diol was approximately 90% of theory. The raw ether diols can be purified by distillation in vacuo. Partial decomposition occurred, and a large residue was obtained. The yield was thereby reduced to approximately 40% of theory.

Production of the Ether Diol by Alkaline Catalysis: 1.5 g of sodium was dissolved in 1 mol of the particular diol. 0.25 mol of the particular epoxide was added thereto and the reaction mixture was heated to 180°C for 6 to 8 hours. The surplus diol was distilled off in vacuo and the residue was added to 100 ml of 10% sulfuric acid. The oil which was precipitated was separated, and the aqueous phase was extracted three times with ether. The combined organic phases were washed neutral with water, dried and the product then recovered by distillation of the solvent in vacuo. The raw yield of ether diol was approximately 90% of theory.

Production of the Alkyl-Substituted 1,4-Dioxanes: The raw ether diol, produced as described above, was dissolved in xylene, p-toluene sulfonic acid was added, and the mixture was heated to boiling with azeotropic distillation for several hours to eliminate water.

When the formation of water had been concluded, the reaction mixture was cooled, washed neutral with water, dried over sodium sulfate, the solvent distilled in vacuo, and the product was distilled. The yield was approximately 80% of theory relative to the epoxide used.

All the described alkyl-substituted 1,4-dioxanes were produced by the method described above. They constitute colorless oils or colorless solid products.

Example 1: *Mixture of 2-Butyl-6-Methyl-1,4-Dioxane and 2-Butyl-5-Methyl-1,4-Dioxane –*

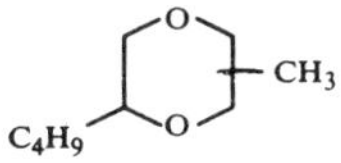

O O CH_3 C_4H_9

BP_{16}80°C. n_D^{20} 1.431. Odor: fresh, bonbon, catechu fragrance.

Example 2: *Mixture of 2-Hexyl-6-Methyl-1,4-Dioxane and 2-Hexyl-5-Methyl-1,4-Dioxane –*

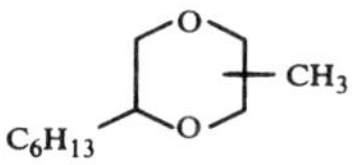

O O CH_3 C_6H_{13}

$BP_{0.2}$54°C. n_D^{20} 1.438. Odor: fatty, helional fragrance, cortex-cyclamal fragrance (alkaline preparation) flowery, fresh, straw-like (acid preparation).

Example 3: *10-Methyl-9,12-Dioxa-Bicyclo-]6.4.0]-Dodecane –*

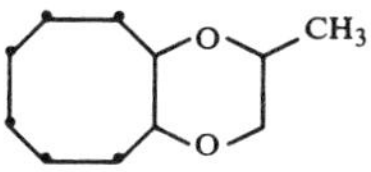

O CH_3 O

$BP_{0.2}$78°C. n_D^{20} 1.4782. Odor: minty, anethole, herbal fragrance.

Example 4: *9,12-Dioxa-Bicyclo[6.4.0]Dodecane –*

$BP_{0.1}$ 60°C. n_D^{20} 1.4814. Odor: minty, potato fragrance.

Example 5: *13,16-Dioxa-Bicyclo[10.4.0] Hexadecane –*

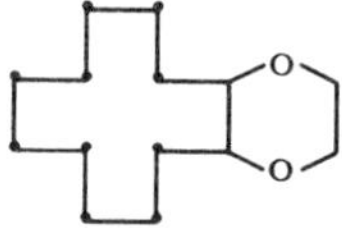

Freezing point 79°C. Odor: earthy, camphoric, woody.

Alkyl-Substituted 1,3-Dioxanes

In similar work *K. Bruns, J. Conrad, P. Meins, H. Möller and H. Schnegelberger; U.S. Patent 4,146,506; March 27, 1979; assigned to Henkel KGaA, Germany* found that 4-isopropyl-5,5-dimethyl-1,3-dioxanes of the following general formula (1) can be used in an advantageous manner as perfumes having a wide variety of fragrances:

(1)

in which R_1 and/or R_2 represent hydrogen or an alkyl or cycloalkyl radical having 1 to 3 carbon atoms.

The dioxanes of the structure specified, which are to be used as perfumes, are produced by conventional methods by acetalization or ketalization of 2,2,4-trimethylpentanediol-(1,3) with aliphatic aldehydes or ketones in the presence of an acidic catalyst.

These dioxanes are valuable perfumes having characteristic fragrances ranging from fruity, herbal to woody. A particular advantage of the perfumes is that they are capable of being combined very satisfactorily.

Examples 1 through 8: The method for the production of representative 4-isopropyl-5,5-dimethyl-1,3-dioxanes will be given by two procedures. The other products are made comparably.

Production of 4-Isopropyl-5,5-Dimethyl-1,3-Dioxane (Example 1) – 146 g (1 mol) of 2,2,4-trimethylpentanediol-(1,3), 33 g (1.1 mols) of paraformaldehyde, 60 ml of ethylene chloride and 0.1 ml of concentrated sulfuric acid were heated in a reactor connected to a water separator until no further water was recovered azeotropically, which took approximately 2 hours. 1.5 g of sodium carbonate was then added to the mixture and the major quantity of the ethylene chloride was distilled off at normal pressure (sump temperature up to 155°C). The residue remaining was subsequently fractionally distilled under vacuum. 150 g of 4-isopropyl-5,5-dimethyl-1,3-dioxane having a BP_{20} = 71°C were obtained, corresponding to a yield of 95% of theory.

Production of 4-Isopropyl-2,2,5,5-Tetramethyl-1,3-Dioxane (Example 5) — 292 g (2 mols) 2,2,4-trimethylpentanediol-(1,3), 108 g (2.02 mols) acetone, 296 g (2 mols) of orthoformic acid triethyl ester and 0.4 g of p-toluene sulfonic acid were agitated for 1 hour at room temperature. Formic acid ester and ethanol were subsequently distilled off at normal pressure (sump temperature up to 130°C). The residue was mixed with ether. Sodium carbonate was added to the solution and the product was washed neutral with water. The solvent was evaporated, and the residue was fractionally distilled under vacuum. 280 g of 4-isopropyl-2,2,5,5-tetramethyl-1,3-dioxane having a BP_{13} of 70°C were obtained, corresponding to a yield of 70% of theory.

The other 4-isopropyl-5,5-dimethyl-1,3-dioxanes, whose odor characteristics and physical data are given in the following table, were obtained in an analogous manner.

Example	Designation	BP, °C	Pressure, torrs	n_D	Odor
1	4-isopropyl-5,5-dimethyl-1,3-dioxane	71	20	1.4390	Herbal, odor of borneol, camphor
2	4-isopropyl-2,5,5-trimethyl-1,3-dioxane	55	5.0	1.4306	Minty, odor of camphor, menthol
3	2-ethyl-4-isopropyl-5,5-dimethyl-1,3-dioxane	32	0.05	1.4325	Sweet, woody, fruity, sassafras, aniseed cedramber fragrance
4	2,4-diisopropyl-5,5-dimethyl-1,3-dioxane	94	20	1.4335	Spicy, woody, fruity
5	4-isopropyl-2,2,5,5-tetramethyl-1,3-dioxane	70	13	1.4305	Fragrance of freshly sawn wood
6	2-ethyl-4-isopropyl-2,5,5-trimethyl-1,3-dioxane	70	4.0	1.4375	Minty, leathery fragrance
7	2,2-diethyl-4-isopropyl-5,5-dimethyl-1,3-dioxane	103	17	1.4420	Nutty, lovage, plum-jam fragrance
8	2-cyclopropyl-4-isopropyl-2,5,5-trimethyl-1,3-dioxane	98	3.0	1.4508	Jasmone fragrance

Substituted Pyridines

The object of the process devised by *S.R. Dolhyj and L.J. Velenyi; U.S. Patent 4,140,690; February 20, 1979; assigned to The Standard Oil Company* is the oxidation of terpenes to provide useful products which can be accomplished with high conversion efficiencies and high selectivities to the desired end products.

This object is accomplished by oxidizing terpenes in the presence of specific oxidation catalysts. In particular, one embodiment contemplates that the oxidation catalysts employed are oxide complexes based on bismuth and molybdenum while in another embodiment the oxidation catalysts employed are oxide complexes based on antimony and uranium or antimony and iron. By contacting the terpene to be oxidized with air or other oxygen-containing gas in the presence of one of these catalysts, it has been found that the objective products can be produced with significantly higher selectivities and conversions than obtainable in prior art processes.

The reaction products produced by the process vary depending upon whether the process is carried out in the gas phase or in the liquid phase. The predominant products of the liquid phase reaction are ketones, aldehydes, alcohols, epoxides and products resulting from isomerization and dehydrogenation of the starting materials. For example, the product obtained upon the liquid phase oxidation of d-limonene contains pulegone, l-p-menthen-9-al, dihydrocarvone, carvone, carveol, dipentene oxide, α-terpineol, carvacrol, α-terpinene, 2,4(8)-p-menthadiene, 1,4(8)-p-menthadiene, cymene (ortho, meta and para), and α-methyl acetophenone.

These products can be recovered by conventional techniques as discussed in U.S. Patent 3,014,047 and are useful as flavoring agents, perfume ingredients and constituents of various pharmaceutical preparations. Also, pulegone can be transformed to menthol (which enjoys wide commercial use in perfumery, flavorings, confectionaries and pharmaceuticals) by hydrogenation of the pulegone in accordance with conventional techniques for hydrogenation of unsaturated ketones.

In a typical experiment, 20 g d-limonene, solvent (benzene) in an amount of 50% with respect to limonene and 2.5 g catalyst (less than 50 mesh) were placed in the round bottom flask and the apparatus was assembled as described above. The reaction took place with constant stirring at 90°C while air was bubbled through the liquid. After a suitable period of time, the reaction was cooled and the catalyst was separated by filtration.

The following examples 1 to 9 were conducted in this manner. The experiments are summarized in the following table. In this table catalyst designations have the following meanings:

Designation	Catalyst Composition
X	82.5% $K_{0.05}Ni_{2.5}Co_{4.5}Fe_3BiP_{0.5}Mo_{12}O_x$ 17.5% SiO_2
V	50% $Cs_{0.05}K_{0.1}Ni_{2.5}Fe_2MnBiCr_{0.5}Mo_{12.6}O_x$ 50% SiO_2

	Example								
	1	2	3	4	5	6	7	8	9
Catalyst	None	X	X	X	X	X	X	X	V
Temp, °C	90	90	144	170	150	95	118	182	140
Time, hr	5½	5½	6¼	2	4½	3	4½	4¾	6½
Gases	air	air	air	air	air	O_2	NH_3, O_2	NH_3, O_2	O_2
Products									
Unreacted hydrocarbon	83.03	67.63	39.06	64.26	49.07	56.61	45.45	89.90	58.95
Isoprene				0.84					0.28

(continued)

	Example 1	2	3	4	5	6	7	8	9
Butylbenzene	0.98	1.87						1.91	
Isomer of limonene			3.02	16.37	8.28	1.62			7.02
Cymene	1.00	2.92	3.99	10.56	1.93	4.81	1.90	0.94	27.96
$C_{10}H_{12}$	2.72	8.72	3.72	3.48	4.89	4.15	17.37	0.57	2.89
Pulegone	1.63	3.87	3.07		4.34	2.32	13.50		0.58
p-Menthane-9-al	0.88	1.49	8.43						
Menthenol	0.50	0.74	2.93	0.34	0.75	2.06	5.70		
Dihydrocarvone	0.38	0.79	9.06	0.86	7.76	8.76	2.54	1.13	0.91
α-Terpineol			4.36	0.46	2.99	2.96			0.44
1-Octene							1.67		
1-Octanol								2.90	
Menthenol				0.31	1.17	0.44			
Carvacrol					0.60	0.70	0.21		
Carvone	1.67	0.80	8.57	0.29	3.47	11.14	3.66	2.67	0.23
α-Methylacetophenone		0.29	2.67	0.23	2.20	3.68	0.41		0.10
Carveol			4.07						
$C_{10}H_{16}O$	1.19	1.43		0.25	1.32		3.30		0.15
$C_{10}H_{14}O$	0.69	1.20	1.02	0.25	1.26		1.34		
$C_{10}H_{12}O$		0.13	0.63	0.23			1.24		
Dipentene oxide			2.53						
Unidentified*	4.85	3.66	2.85	1.28	9.98	0.73	1.71		0.50

*combined

MISCELLANEOUS

Polycyclic Diels-Alder Adducts

W. Hoffmann and K. von Fraunberg; U.S. Patents 4,076,748; February 28, 1978 and 4,143,065; March 6, 1979; both assigned to BASF AG, Germany have developed certain polycyclic Diels-Alder adducts of the formula

(1)

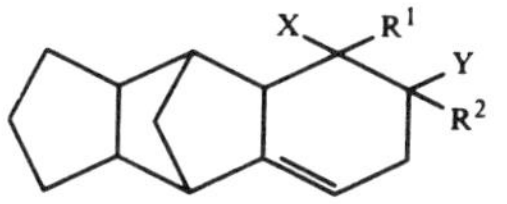

where R^1 and R^2 are $-H$ or CH_3, or R^1 and R^2 together are a further bond between the carbon atoms on which they are present, and X and Y are each $-CH_2OH$, $-CH(OH)CH_3$, $-COOH$, $-COOCH_3$, $-COOC_2H_5$, $-CHO$, $-COCH_3$, $-COC_2H_5$ or $-CN$, or X and Y together are $-COOCO-$, or one of X or Y is $-H$ or $-CH_3$ and the other has one of the above meanings.

These polycyclic Diels-Alder adducts are distinguished by valuable scent characteristics. They have green, fresh, fruity and in some cases woody odors with long-lasting tenacity and/or fixing properties. They may be used as constituents of perfumes and of compositions for perfuming toiletries, e.g., soaps, shampoos and hair lotions, detergents and other consumer products. Furthermore, they may be used for flavoring foodstuffs and tobacco.

They have also developed a process for the manufacture of compounds of the formula (1), wherein the 8-vinyl-tricyclo[5.2.1.0^{2,6}]dec-8-ene of the formula (2)

(2)

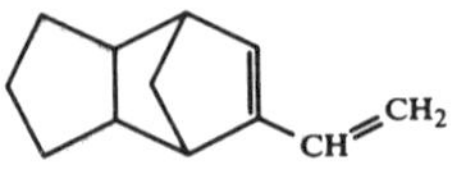

is subjected to a Diels-Alder reaction, by conventional methods, with a dienophile of the general formula (3)

(3)

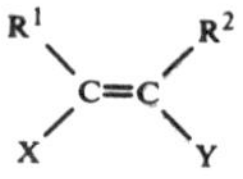

R^1, R^2, X, and Y are as defined above.

The Diels-Alder adducts of formula (1) are obtained by reducing the corresponding oxo compounds by conventional methods.

8-Vinyl-tricyclo[5.2.1.0^{2,6}] dec-8-ene, required as a starting compound, can be manufactured by a simple conventional method, by elimination of water from 8-vinyl-tricyclo[5.2.1.0^{2,6}] decan-8-ol, which, in turn, can be obtained by reacting tricyclo[5.2.1.0^{2,6}]decan-8-one, which is readily available, with a vinyl magnesium halide by the Grignard method.

Suitable compounds of formula (3) include crotonaldehyde, methyl vinyl ketone, acrolein, methacrylonitrile propynal, dimethyl fumarate, maleic anhydride, acetylene dicarboxylic acid and its methyl and ethyl esters, etc.

Examples of the preparation of these compounds of general formula (1) are as follows.

Example 1: 16.0 g (0.10 mol) of 8-vinyl-tricyclo[5.2.1.0^{2,6}] dec-8-ene, 11.2 g (0.2 mol) of acrolein, 10 mg of hydroquinone and 100 ml of toluene are kept at 90°C for 3 hours. The reaction product is washed with three 50 ml portions of water and is then distilled. 19.2 g (89% of theory) of an aldehyde mixture pass over at from 110° to 115°C/0.1 mm Hg. The aldehyde mixture has an n_D^{25} of 1.5332. The scent is waxy, green-pod-like, peachy, fixing.

Gas chromatography shows two compounds in the ratio of 89:11. The ^{13}C spectrum suggests that the main component probably has structure (1) with R^1 and R^2 = –H, X = –CHO and Y = –H.

Example 2: A solution of 21.6 g (0.1 mol) of the aldehyde mixture obtained in Example 1, in 30 ml of ethanol, is added to 4 g of sodium borohydride in 200 ml of ethanol in the course of 60 minutes at 25°C and the reaction mixture is then stirred for 3 hours at 25°C. The ethanol is then distilled off, the residue is taken up in ether and acidified to pH 3 with 10% strength sulfuric acid, and the organic phase is concentrated and distilled. At from 111° to 115°C/0.5 mm Hg, 16.2 g (74% of theory) of an alcohol pass over, which alcohol, according to the analytical data and the nuclear resonance spectrum, probably has the formula (1), where R^1, R^2 and Y are –H and X is $-CH_2OH$. The alcohol has an n_D^{25} of 1.5368. The scent resembles the aldehyde, fainter, but with substantially more persistent tenacity.

Example 3: A solution of 8.0 g (0.05 mol) of 8-vinyl-tricyclo[5.2.1.$0^{2,6}$]dec-8-ene, 10.6 g (0.2 mol) of acrylonitrile, 50 ml of toluene and 10 mg of hydroquinone is heated at 80°C for 12 hours. The reaction mixture is then cooled, washed twice with 30 ml of water, concentrated and distilled. 8.1 g (corresponding to 76% of theory) of an 80:20 mixture of nitriles pass over at from 125° to 128°C/0.2 mm Hg. From the analytical and spectroscopic data, the main product probably has the formula (1), where R^1, R^2 and Y are –H and X is –CN. The main product has an n_D^{25} of 1.5325. The scent is fresh, flowery, slightly fruity (lime), good tenacity.

Example 4: Following the procedure described in Example 3, but using 0.2 mol of crotonaldehyde instead of 0.2 mol of acrylonitrile, 7.2 g (corresponding to 62.5% of theory) of a Diels-Alder adduct of the formula (1), where R^1 and R^2 are –H and one of X or Y is $-CH_3$ and the other is –CHO are obtained; the product boils at from 112° to 115°C/0.05 mm Hg and its refractive index n_D^{25} is 1.5290. The scent is coniferous, woody, suggestive of vetiver oil, good tenacity.

Example 5: Following the procedure described in Example 3, but using 0.2 mol of methacrolein instead of 0.2 mol of acrylonitrile, 8.3 g (corresponding to 72% of theory) of a Diels-Alder adduct of the formula (1), where R^2 is $-CH_3$, Y is –CHO and X and R^1 are –H, or R^1 is $-CH_3$, X is –CHO and R^2 and Y are –H are obtained; the product boils at from 105° to 106°C/0.05 mm Hg and its refractive index n_D^{25} is 1.5292. The scent is pine-like, woody, fresh, slightly fruity.

Example 6: Following the procedure described in Example 3, but using 0.2 mol of methyl vinyl ketone instead of 0.2 mol of acrylonitrile, 10.8 g (corresponding to 94% of theory) of a Diels-Alder adduct of the formula (1), where one of X or Y is $-COCH_3$ and the remaining radicals are –H, are obtained. The product boils at from 115° to 120°C/0.05 mm Hg and its refractive index n_D^{25} is 1.5259. The scent is herbaceous, fruity (raspberries), earthy, good tenacity.

Alkyl Cycloalkyl Carbonates

K. Bruns and P. Meins; U.S. Patent 4,080,309; March 21, 1978; assigned to Henkel KGaA, Germany have found that carbonic acid esters of the general formula

$$R_1-O-\overset{\overset{\displaystyle O}{\|}}{C}-OR_2$$

in which R_1 represents a substituted cyclohexyl radical or a cycloaliphatic radical having 8 to 12 carbon atoms, and R_2 represents a straight or branched chain, saturated or unsaturated aliphatic hydrocarbon radical having 1 to 5 carbon atoms, are perfumes having a very natural and complex scent.

The compounds are produced by reacting cycloalkanols of the general formula R_1-OH with chloroformic acid esters of the general formula $R_2O-COCl$, in which R_1 and R_2 have the aforementioned structure, in anhydrous, inert solvents, such as hexane, benzene, toluene, in the presence of a hydrochloric acid acceptor such as pyridine at a reaction temperature of from 0° to 5°C.

Many perfume esters conforming to the above general formula may be formulated, the most important ones being methyl 1-ethynyl-cyclohexyl carbonate, methyl cis-3,3,5-trimethylcyclohexyl carbonate, methyl trans-3,3,5-trimethylcyclohexyl carbonate, methyl cyclooctyl carbonate, ethyl trans-3,3,5-trimethylcyclohexyl carbonate, ethyl cyclooctyl carbonate, and the methyl and ethyl 2-tert-butyl and 4-tert-butylcyclohexyl carbonates.

The esters are distinguished by particularly intensive and lasting flowery, herbal, fruity and fresh scents.

Example 1: *Methyl Cyclooctyl Carbonate* — 18.9 g of methyl chloroformate were added drop by drop under agitation to a solution of 25.6 g of cyclooctanol and 15.8 g of absolute pyridine in 150 ml of dry benzene under external cooling at 0° to 5°C. After the methyl chloroformate had been added, agitation was continued for 12 hours at room temperature. Then, the benzene phase was drawn off from the precipitated pyridine hydrochloride, and washed with diluted hydrochloric acid, sodium hydroxide solution and water, and dried. After the solvent had been distilled off, the raw ester was distilled in vacuo by means of a Vigreux column. A colorless liquid was obtained which had a herbal, very natural and complex fragrance which is distinguished by a strong and long-clinging flowery jasmine scent.

The characteristic values were as follows: BP, 47°C at 0.01 mm Hg; and the refractive index, n_D^{20} = 1.4580. IR and NMR results confirmed the structure.

Example 2: *Methyl trans-3,3,5-Trimethylcyclohexyl Carbonate* — This substance was produced, analogously to Example 1, from trans-3,3,5-trimethylcyclohexanol and methyl chloroformate. The odor was earthy, fruity, very natural smelling, a fragrance of forest soil or humus. It was a colorless liquid and had a BP of 88°C at 3.2 mm Hg. The refractive index was n_D^{20} = 1.4428.

Example 3: *Methyl 1-Ethynylcyclohexyl Carbonate* — A solution of 36.0 g of 1-ethynylcyclohexanol in 50 ml of benzene was slowly added drop by drop to an agitated suspension, cooled to 0° to 5°C, of 5.5 g of finely distributed sodium in 50 ml of absolute toluene and 250 ml of benzene, and was agitated at room temperature until reaction had been completed. 26.0 g of methyl chloroformate were added under cooling to the sodium salt which had been formed. The mixture was allowed to react for 12 hours at room temperature and was washed several times with water and dried. After distilling off the solvent, the raw ester was fractionated by means of a 20 cm Vigreux column. The product was a colorless liquid having a fruity, herbal, complex odor and a distinctive fragrance of dill. Its BP at 3.5 mm Hg was 47°C; the refractive index, n_D^{20}= 1.4630.

Example 4: *Methyl 4-tert-Butylcyclohexyl Carbonate* — The product was obtained, analogously to Example 1, from 4-tert-butylcyclohexanol and methyl chloroformate. It was a colorless liquid with a fruity, spicy, woody fragrance and a BP of 95°C at 0.8 mm Hg. The refractive index was n_D^{20} = 1.4534.

The compounds given in the above examples have natural, flowery, herbal, fruity, fresh fragrances with excellent clinging properties or persistency which render them suitable for producing a wide variety of perfume compositions. Such compositions can be used to perfume a wide variety of products, such as cosmetics, washing agents, soaps, as well as technical products in concentrations of approximately 0.05 to 2% by weight.

Stereospecific Isomers of Nerolidol and Related Compounds

Y. Fujita, Y. Omura, T. Nishida and K. Itoi; U.S. Patent 4,105,700; August 8, 1978; assigned to Kuraray Company, Ltd., Japan have developed a method for preparing a stereospecific farnesylacetic acid and its esters. Particularly, this is a process for preparing, respectively, Δ^4-cis-Δ^8-trans-, $\Delta^{4,8}$-trans-, Δ^4-trans-Δ^8-cis- and $\Delta^{4,8}$-cis-farnesylacetic acids and esters thereof. More particularly, it is a process for preparing $\Delta^{4,8}$-trans-farnesylacetic acid and its ester. The above stereospecific farnesylacetic acids and esters thereof are represented by the following formula (1):

$$(1) \quad CH_3-\overset{\displaystyle CH_3}{\overset{|}{C}}=CHCH_2CH_2-\underset{\Delta^8}{\underset{\uparrow}{\overset{\displaystyle CH_3}{\overset{|}{C}}}}=CHCH_2CH_2\underset{\Delta^4}{\underset{\uparrow}{\overset{\displaystyle CH_3}{\overset{|}{C}}}}=CHCH_2CH_2CO_2R$$

wherein R represents hydrogen atom, an alkyl, alkenyl, alkynyl, cycloalkyl, cycloalkenyl, aryl, aralkyl, aralkenyl or heterocyclic radical having not more than 20 carbon atoms.

Farnesylacetic acid esters are valuable compounds which are employed as medicines having antiulcer activity and also as perfumes.

The sterospecific geranylacetone used as a starting material is not only used in perfumes, but also very important for preparation of nerolidol and farnesol which are used in perfumes of high grade.

Nerolidol, which is an intermediate compound in the process, has the following formula:

$$CH_3-\overset{\displaystyle CH_3}{\overset{|}{C}}=CHCH_2CH_2\underset{\Delta^6}{\underset{\uparrow}{\overset{\displaystyle CH_3}{\overset{|}{C}}}}=CHCH_2CH_2\underset{\displaystyle OH}{\underset{|}{\overset{\displaystyle CH_3}{\overset{|}{C}}}}-CH=CH_2$$

The stereospecific nerolidol as such has a weak but sweet flower odor and is of high grade as a perfume; it is used for preparation of flower essence oils such as jasmine and violet.

The most preferable method for industrial production involves three steps: (1) the rectification of the cis- and trans- forms of geranylacetone, (2) the conversion of the geranylacetone to cis- or trans-nerolidol and (3) its reaction with an orthoacetic acid ester.

Practically, it has been found that a rectifying column having 20 to 60 plates is necessary for the rectification (1) of the geranylacetone; the reflux ratio is preferably from 5 to 30 at a temperature from 120° to 210°C, under pressure of from 0.1 to 5 mm Hg. An isomerization catalyst suitable for the reaction (1) might be, for instance, a ruthenium, tungsten or organic sulfur compound, plus a small amount of a radical initiator such as azobisisobutyronitrile (AIBN) or benzoyl peroxide (BPO).

Step (2) is carried out by reacting the geranylacetone with (a) a vinyl Grignard reagent or (b) acetylene followed by partial hydrogenation. The rectification of nerolidol is effected in the same way as was geranylacetone. The following examples illustrate the process in more detail.

Example 1: After 6,016 g (purity 95%) of linalool and 3,444 g of diketene were reacted in the presence of 163 g of triethylamine, 240 g of aluminum isopropoxide were added and the mixture was subjected to Carroll rearrangement by heating. Vacuum distillation of the reaction mixture gave 5,200 g of a mixture of cis- and trans-geranylacetone in a ratio of 4:6 having a boiling point of 68° to 74°C (0.4 mm Hg).

Example 2: In a flask provided with a rectifying column having about 40 theoretical plates, were placed 1,000 g of a mixture of cis- and trans-geranylacetone in a ratio of about 4:6 and 2.0 g of acetylacetonatoruthenium [$Ru(AA)_3$] and the mixture was rectified at 175° to 179°C of the column bottom temperature and at 130° to 135°C of the top column temperature at a reduced pressure of 5 to 7 mm Hg under reflux ratio 10 to give 968 g of the distillate, the purity of the cis-isomer of which was 99.4% as determined by gas chromatography.

Example 3: In a 2-liter, 3-necked flask, 1 liter of liquid ammonia was placed and, after addition of 55.2 g of metallic sodium, acetylene gas was passed thereto and continued to pass until the reaction liquid turned from violet to white. 388 g of cis-geranylacetone was added to the ammonia mixture. While the liquid ammonia was under reflux, acetylene gas was bubbled thereto for 4 hours. After removing the ammonia, 110 g of ammonium chloride were added and neutralized. The reaction liquid was poured into water and extracted with ether. The extract was washed with water and dried over sodium sulfate and the solvent was distilled off.

The residue was subjected to distillation under reduced pressure (0.5 mm Hg) in a rectifying column having about 40 theoretical plates (about 30 practical plates) with a reflux ratio of 10:20 and a bottom temperature of approximately 140°C to afford 372 g of cis-dehydronerolidol having a boiling point of 133° to 135°C (5 mm Hg) in a yield of 85%.

Next, in a solution of 320 g of the cis-dehydronerolidol in 1,000 g of n-hexane, 0.15 ml of quinoline and 5.0 g of 0.25% Pd=Lindlar catalyst were added and the mixture was hydrogenated at room temperature under normal pressure. The reaction progress was analyzed with gas chromatography. When the cis-dehydronerolidol of the starting material disappeared, the reaction mixture was filtered out with a glass filter. After the filtrate was distilled, the residue was subjected to high vacuum distillation to give 314 g of cis-nerolidol having a boiling point of 99° to 120°C (0.3 mm Hg).

The trans-geranylacetone was treated in the same way to obtain trans-nerolidol having a boiling point of 107° to 110°C (0.5 mm Hg).

Example 4: In a 2-liter, 3-necked flask was placed a mixture of 648 g of ethyl orthoacetate, 440 g of nerolidol and 22 g of isobutyric acid, which was then heated at 150° to 160°C. Since the reaction was accompanied by a vigorous side reaction to produce ethanol, the ethanol was continuously removed from the reaction system. Progress of the reaction was checked by gas chromatography and the reaction should be discontinued at the point of disappearance of the starting nerolidol. In order to accelerate the reaction, a further amount of isobutyric acid can be added. The reaction gave more than 95% of the conversion ratio based on the amount of nerolidol and more than 98% of the selectivity to ethyl farnesylacetate. The reaction mixture was distilled in vacuo without any posttreatment to give 533 g of the desired product. BP 145° to 152°C (0.4 mm Hg).

The result of gas chromatographic analysis showed that this substance was a mixture of stereoisomers, 15% of $\Delta^{4,8}$-cis-isomer, 45% of Δ^4-cis-Δ^8-trans- and Δ^4-trans-Δ^8-cis-isomers, and 38% of $\Delta^{4,8}$-trans-isomer. Then, the mixture was fractionated through a rectifying column with 40 theoretical plates, which was operated under the conditions of 175° to 185°C bottom temperature and 10 to 20 reflux ratio, to give 150.0 g of ethyl $\Delta^{4,8}$-trans-farnesylacetate, boiling point 130° to 132°C (0.1 mm Hg).

To 362 g of ethyl farnesylacetate obtained as the first distillate on rectification, which was composed of 22.1% of $\Delta^{4,8}$-cis-isomer, 69.2% of Δ^4-cis-Δ^8trans- and Δ^4-trans-Δ^8-cis-isomers, and 8.7% of $\Delta^{4,8}$-trans-isomer, were added 11 g of phenyldisulfide and a small amount of a radical initiator, AIBN, and the mixture was heated for 24 hours at 140°C in a nitrogen atmosphere.

After reaction, the abovementioned ratio was 16.9:50.7:32.4 by gas chromatography. The reaction mixture was roughly distilled in about 0.1 mm Hg to give 356 g of the distillate. Subsequently, this was fractionated through a rectifying column with 40 theoretical plates to give 237.8 g of the first distillate and 84.2 g of the second distillate, ethyl $\Delta^{4,8}$-trans-farnesylacetate.

β-Phellandrene

L.M. Hirschy, B.J. Kane and S.G. Traynor; U.S. Patent 4,136,126; January 23, 1979; assigned to SCM Corporation have devised a process for the preparation of β-phellandrene (1-methylene-4-isopropyl-2-cyclohexene), which is widely used in perfumes and artificial essential oils because of its peppery, minty, refreshing, and slightly citrusy odor. It is of importance as an intermediate in various synthesis schemes, such as in the preparation of 1-menthol. However, it has been difficult to synthesize β-phellandrene in good yields, such as from p-menthadienes, because of its ease in isomerizing under most reaction conditions to its β-terpinene isomers.

This process provides a method for synthesizing β-phellandrene in good chemical and optical yields and comprises heating a p-menth-1-ene-7-sulfonate salt at a temperature of about 150° to 350°C under nonacidic conditions until the β-phellandrene product is formed.

The following examples show in detail how the process can be practiced. All temperatures are in degrees centigrade and all percentages are weight percentages, unless otherwise expressly noted.

Example 1: Into a 1-liter distillation vessel was charged 3 g of (+)p-menth-1-ene-7-sulfonic acid sodium salt. The sodium sulfonate salt was heated to 260°C under reduced pressure (22 mm Hg) for about 15 minutes. After the initial removal of water from the vessel by distillation, a yellow liquid was distilled from the vessel and collected in the distillation flask. This yellow liquid was extracted with ether and the extract dried over magnesium sulfate. GLC analysis of the extract (0.7 g) showed the ether (89%) and one major component which was identified by GC, MS and IR as (–)β-phellandrene (9.4%). Minor amounts of α-terpinene and α-phellandrene were noted in the extract also.

Example 2: The procedure of Example 1 was repeated with 50 g of the sodium sulfonate salt to which was added 22 g of sodium carbonate. Evaporation of the ethereal extract yielded a yellow oil which contained ether (25.4%) and (–)β-phellandrene (60.2%). The theory yield of product, thus, was 52%. The (–)β-phellandrene then was purified by distillation. $[\alpha]_D = -11.01°$.

Example 3: The procedure of Example 1 was repeated with 3 g of the sodium sulfonate salt and 20 g of NaOH. The yellow oil collected (1.3 g) analyzed by GLC to be mostly (about 85%) the desired (–)β-phellandrene product.

Unsaturated Nitriles

R.S. DeSimone; U.S. Patent 4,156,690; May 29, 1979; assigned to Polak's Frutal Works, Inc. has found a method for preparing unsaturated conjugated nitriles by a relatively simple and inexpensive method. The nitriles are alpha-, beta-dialkyl-substituted conjugated aliphatic nitriles having the following general structural formula:

$$R''-\underset{\displaystyle R'}{C}=\underset{\displaystyle R}{C}-C\equiv N$$

where R is an aliphatic hydrocarbon radical of about 1 to 15 carbon atoms having no unsaturation in conjugated relationship with either the nitrile unsaturation or the alpha, beta-olefinic unsaturation, and R' and R'' are the same or different aliphatic hydrocarbon radicals meeting the description of R; or R' and R'', taken together, form a cycloaliphatic radical having about 6 to 15 carbon atoms. Exemplary, but by no means all inclusive, of alpha-, beta-dialkyl conjugated nitriles within the scope of the above formula are, e.g., 2,3-dimethyl-2-nonene nitrile, 2,3-dimethyl-2-octene nitrile, 2-ethyl-3-methyl-2-octene nitrile, 2,3,7-trimethyl-2,6-octadiene nitrile,

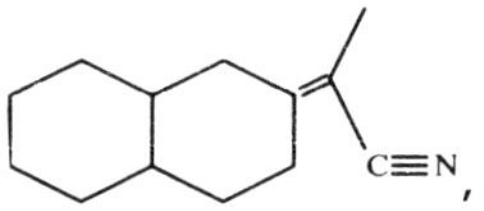

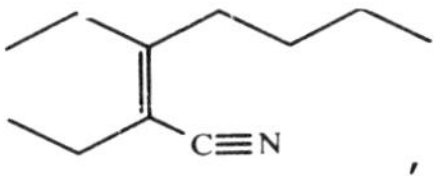

C≡N, C≡N,

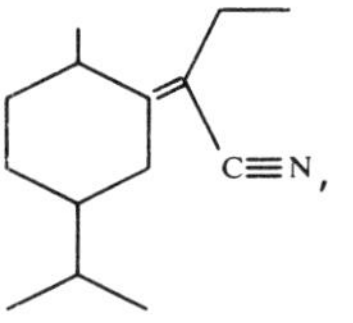

C≡N, and $CH_3-(CH_2)_9-C-C\equiv N$.

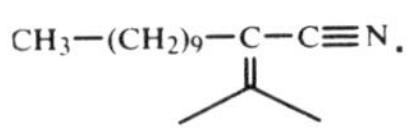

These α,β-dialkyl conjugated aliphatic nitriles have been found to be useful as perfumery ingredients exhibiting a variety of commercially useful fragrance notes. Some members of the class have a jasmine-like fragrance, others have other floral notes, while still others exhibit a woody, cinnamic fragrance. They are useful alone, mixed with each other, or, more normally, mixed with other perfumery compounds in about 0.1 to 30% concentration to form a perfume, when mixed with a suitable carrier.

In accordance with this process, a ketone is reacted with higher alkyl nitriles, in the presence of a basic catalyst, in an aldol-type condensation to give a class of α,β-dialkyl conjugated aliphatic nitriles (trialkyl-substituted acrylonitriles) having carbon skeletons not readily available through classical synthetic routes.

The reaction can be described as follows:

$$R'R''C{=}O + R{-}CH_2{-}C{\equiv}N \xrightarrow[-H_2O]{\text{Base}} R'R''C{=}C(R){-}C{\equiv}N$$

where R, R' and R'' are as defined above.

The bases employed are selected from the group of alkali or alkali earth hydroxides, metal alcoholates and quaternary ammonium hydroxides, either used alone or in combination. When calcium hydroxide is used, a phase-transfer agent, such as crown ether or quaternary ammonium chloride or hydroxide, must also be employed.

The nitrile α-carbanion attacks the carbonyl carbon to form an intermediate aldol which dehydrates to form the trialkyl-substituted acrylonitrile.

Example 1: *3,7-Dimethyl-2-Ethyl-2,6-Octadiene Nitrile* — Methyl heptenone (6-methyl-hepten-5-one-2), 25.2 g was combined with 138 g of butyronitrile and 12.2 g of KOH was then added. The mixture was heated to reflux with stirring at 118°C for a total of 15 hours under N_2. Crude butyronitrile was recovered by distillation at between 85° and 141°C pot temperature and 81° to 89°C head temperature using a short path still head at <300 mm Hg. The pot contents were then cooled to 100°C, 100 ml of H_2O was added, and the mixture was stirred one hour.

Hydrochloric acid, 10%, 100 cc was then added with stirring along with 100 ml of benzene. Another 200 ml of 4% HCl were added with stirring and the oil phase separated. The aqueous phase was extracted with one 15 ml portion of toluene and then with 10 ml portions five times. The combined oil phases were distilled at 208 to 213 mm Hg, 68° to 167°C pot temperature and 55° to 118°C head temperature to remove solvent on an 18-inch Vigreaux column.

The product nitrile distilled at 118° to 135°C head temperature and 142° to 205°C pot temperature at 17 mm Hg. The fraction containing the product nitrile was taken up in hexane and washed with water three times to remove crystals of butyronitrile. The washed nitrile was redistilled to give the product as a mixture of cis-trans isomers in 16% conversion yield. The product was interpreted by some perfumers to have a jasmine or immortelle-type fragrance.

Example 2: *2-(4-Methyl Cyclohexylidene) Heptanonitrile* — Into a 25 ml flask, equipped with magnetic stirrer, heating mantle, thermometer, condenser and calcium sulfate drying tube, was charged 1.12 g of 4-methyl cyclohexanone (0.01 M), 11.1 g of heptanonitrile, and 0.66 g of 85% KOH. The mixture was heated to reflux with stirring at about 83°C for a total of 32 hours. After cooling, 30 ml of hexane was added and the mixture washed seven times with 30 ml of H_2O, each washing being back-extracted twice with 30 ml portions of hexane. The combined organic phase and hexane extracts were dried over anhydrous sodium sulfate and the solvent removed by rotary evaporation at about 15 mm Hg to give 6.3 g of crude oil.

Gas chromatographic analysis (6' x ¼" stainless steel column packed with 20% SE30 on Chromosorb W, programmed at 135° to 220°C, 4°/min, He flow ~60 ml/min) showed two product peaks at RF 19.0 and 20.2 minutes, which were 6.9% in ratio to heptanonitrile. An IR spectrum of the product peaks, trapped as a group from the gas chromatograph, showed a nitrile band at 4.5 microns and an olefinic band at 6.25 microns. The product was evaluated by perfumers as having a jasminic nitrile-type odor with a strong, nerol oxide note.

Examples 3 through 20: Following the general procedure as described in the previous examples, the ketones set forth in the table below were also reacted, under the conditions set forth, i.e., 10 M nitrile reacted with 1 M ketone in the presence of 1 M KOH at 90° to 100°C for 19 hours with the usual workup.

Ex.	Ketone	Nitrile	Conversion (%)	Yield (%)	Odor Type
3	Methyl isobutyl ketone	Butyronitrile	~8	–	–
4	2-Pentanone	Butyronitrile	~8	–	Lovage
5	2-Octanone	Butyronitrile	~2	5.3	Anise, lovage, heliotropine
6	2-Pentanone	Valeronitrile	2.5	–	–
7	3-Heptanone	Butyronitrile	~4	~9.5	Lovage, butyric
8	2-Hexanone	Propionitrile	32.8	54	Waxy, terpenic, cumin
9	β-Decalone	Propionitrile	66.2	75.5	Sweet, rosy, coumarinic
10	4-Methyl cyclohexanone	Butyronitrile	52.8	58	Resinous, myrrh, celery, tobacco
11	4-Methyl cyclohexanone	Valeronitrile	42.9	50.1	Similar to Ex. 10, plus soapy, jasmone
12	4-t-Butyl cyclohexanone	Propionitrile	77.5	78	Bready, coconut, musty
13	3-Heptanone	Propionitrile	19	20	Strong, floral, almond, jasmone, lactonic
14	Cyclohexanone	Propionitrile	29.6	37.4	Strong, cinnamic, almond
15	6,9-Dimethyl-5-decen-2-one	Propionitrile	48.8	51.8	Fatty, lemony, floral, green water
16	4-Methyl cyclohexanone	Propionitrile	75	79.4	Fruity, rosy, ylang benzoate
17	2-Dodecanone	Propionitrile	49	51.7	Watery, aldehydic
18	Dihydro-α-ionone	Propionitrile	41.6	41.6	Mixture of α, β, and γ isomers, woody, fatty, hairy, indolic
19	Carvomenthanone	Propionitrile	25.6	63.8	Woody
20	Cyclododecanone	Propionitrile	14.6	83.3	Earthy, musky, musty, woody

PERFUME INTERMEDIATES AND EXTENDERS

PERFUME INTERMEDIATES

Certain Norsesquiterpene Derivatives

P. Maupetit and P.J. Teisseire; U.S. Patent 4,083,812; April 11, 1978; assigned to SA Roure Bertrand Dupont, France have found that certain ketoalcohol tricyclic norsesquiterpene derivatives having the formulas:

(1)

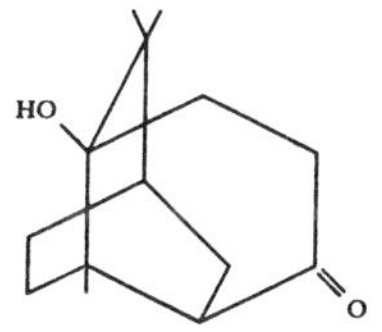

(1a)

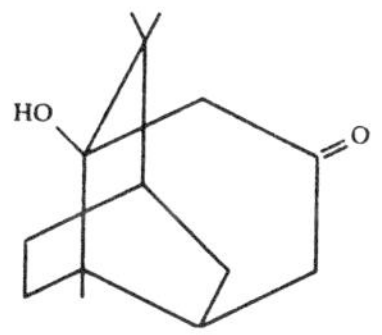

are useful as odorants, as intermediates for the manufacture of odorants, and as fixatives. They produce camphorous, musty, and woody scents.

Compound 1 is prepared by subjecting the unsaturated tricyclic alcohol, norpatchoulenol of Formula 2 (which can be isolated from natural patchouli oil by known methods), to epoxidation for the preparation of an epoxy alcohol of the following Formula 3; reducing an epoxy alcohol of Formula 3 for the preparation of a glycol of Formula 4; and oxidizing the glycol of Formula 4 for the preparation of Formula 1.

Compound 1a is prepared by subjecting norpatchoulenol to hydroboration and oxidation to obtain a glycol and then oxidizing the glycol.

The reaction scheme for making compound 1 may be illustrated and described as follows:

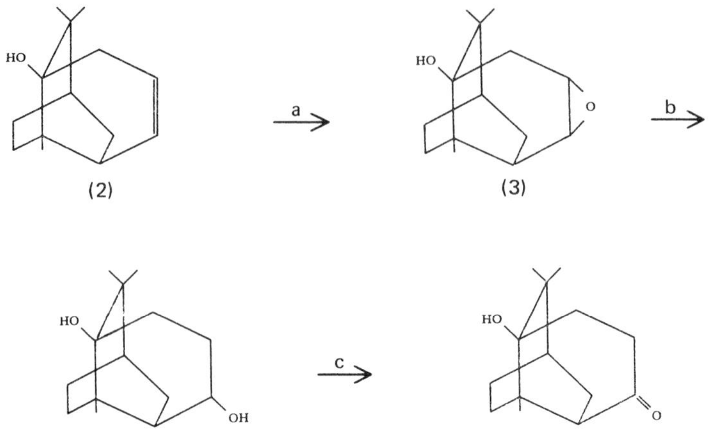

(2) (3)

(4) (1)

Example: (a) 1 g (4.8 mmol) of norpatchoulenol, dissolved in 50 ml of methylene chloride, and 1 g of dry sodium acetate are added to a 500 ml flask. The thus obtained suspension is stirred vigorously, cooled and then mixed with 15 ml of 35% peracetic acid. The mixture is then left for 48 hr at room temperature, until the norpatchoulenol has practically disappeared. After the addition of 300 ml of water, the reaction mass is extracted with methylene chloride. Then the organic extracts are washed with 9% sodium bicarbonate solution, 10% sodium sulfite solution and finally to neutrality with water.

The solvent is then distilled off. There are thus obtained 1.05 g of crystallized, crude epoxyalcohol of the intermediate of Formula 3, which can be obtained analytically pure (90% yield) by chromatography on silica gel and vacuum sublimation.

(b) 100 ml of anhydrous petroleum ether and 10 ml of diisobutyl aluminum hydride are added to a 500 ml flask provided with stirrer, reflux condenser and dropping funnel. The reaction medium is kept under a dry nitrogen atmosphere. Then 0.82 g (3.7 mmol) of the epoxyalcohol of Formula 3 obtained according to part (a) of this example, dissolved in 30 ml of dry petroleum ether, are added at ambient temperature to the hydride solution. After this addition, the reaction mixture is held under reflux for 3 hr, then cooled to approximately 0°C, slowly mixed with 20 ml of absolute ethyl alcohol and finally with 250 ml of saturated sodium chloride solution. The reaction mixture is extracted with petroleum ether, whereupon the organic extracts are washed to neutrality with water.

Distillation of the solution produces 0.85 g of the glycol intermediate of Formula 4. By chromatography on silica gel, there are obtained 0.80 g of white crystallized Formula 4 product (ca 82% yield).

(c) 0.64 g (2.86 mmol) of the glycol of Formula 4, obtained according to part (b) of this example, are dissolved in 100 ml of methylene chloride. After the addition of 12 g of CrO_3/pyridine complex, the mixture is stirred for 5 hr at

20° to 25°C, then filtered and the filtrate taken up in ethyl ether. The solution is washed with 10% hydrochloric acid to eliminate the pyridine, then with 9% bicarbonate solution, and finally until neutral with water. After distillation of the solvent, there are obtained 0.60 g of red, crystallized ketoalcohol of Formula 1, which is purified by chromatography over silica gel. Yield 0.53 g (ca 85%).

3-Cyanomethylcyclopentanone

K. Kondo, D. Tunemoto, K. Sugimoto and Y. Takahatake; U.S. Patent 4,100,184; July 11, 1978; assigned to Sagami Chemical Research Center, Japan have developed a process for producing 3-cyanomethylcyclopentanone derivatives of the following formula:

(6) [structure: 2-R^1-3-(cyanomethyl)cyclopentanone; labels O, R^1, CN]

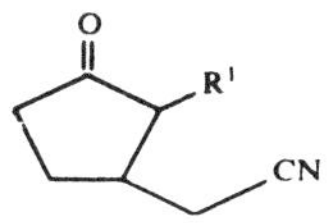

in which R^1 represents hydrogen or an alkyl, alkenyl or alkynyl group. These cyanomethylcyclopentanone derivatives are useful precursors for producing jasmonoids which are the fragrant components in jasmine, e.g., methyl jasmonate which is the cis isomer of a compound having the following formula:

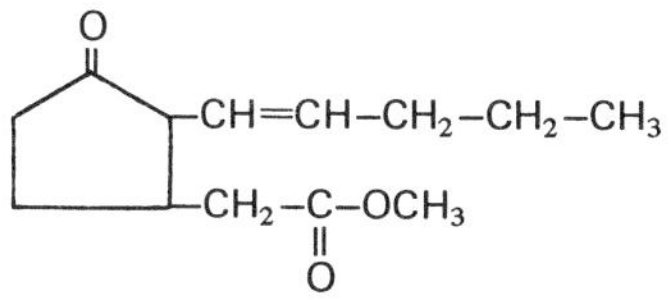

[structure: cyclopentanone ring bearing $-CH{=}CH-CH_2-CH_2-CH_3$ and $-CH_2-\overset{O}{\overset{\|}{C}}-OCH_3$]

The process, which utilizes commonly available reagents, is summarized in the following reaction sequence in which R^1 represents hydrogen or an alkyl, alkenyl or alkynyl group, and R^3 represents an alkoxy group.

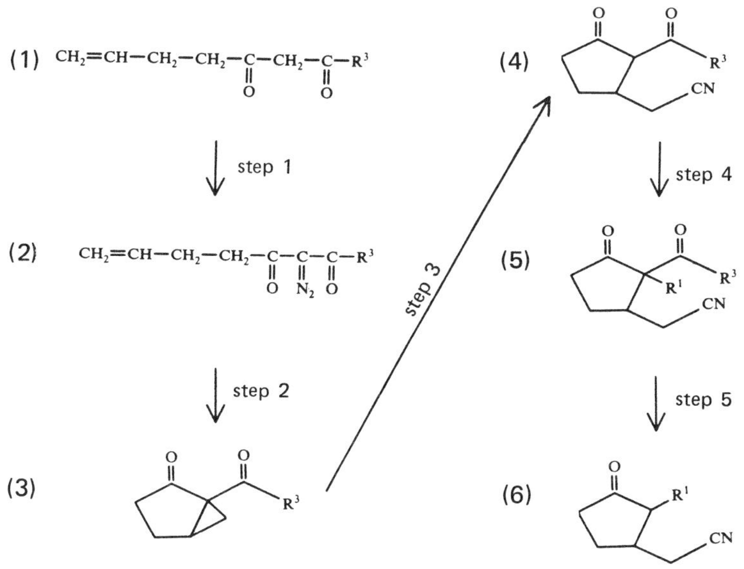

(1) $CH_2{=}CH-CH_2-CH_2-\overset{\|}{\underset{O}{C}}-CH_2-\overset{\|}{\underset{O}{C}}-R^3$

↓ step 1

(2) $CH_2{=}CH-CH_2-CH_2-\underset{O}{\overset{\|}{C}}-\underset{N_2}{\overset{\|}{C}}-\underset{O}{\overset{\|}{C}}-R^3$

↓ step 2

(3) [structure: bicyclic cyclopropane-fused cyclopentanone bearing $-C(=O)R^3$; labels O, O, R^3]

→ step 3

(4) [structure: cyclopentanone with $-C(=O)R^3$ and CH_2CN; labels O, O, R^3, CN]

↓ step 4

(5) [structure: cyclopentanone with R^1, $-C(=O)R^3$ and CH_2CN; labels O, O, R^1, R^3, CN]

↓ step 5

(6) [structure: 2-R^1-3-(cyanomethyl)cyclopentanone; labels O, R^1, CN]

Step 1 involves a reaction between compound (1) and an azide such as p-toluene sulfonyl azide under alkaline conditions.

Step 2 involves subjecting the α-diazo-β-dicarbonyl derivative (2) to conditions promoting carbene or carbenoid formation. This can be attained either by use of a metal or metal salt catalyst, e.g., copper sulfate, or by photoirradiation, e.g., with a mercury vapor lamp.

Step 3 involves treating the bicyclo[3.1.0]hexan-2-one derivative (3) with a cyanating reagent such as potassium cyanide, hydrogen cyanide or acetone cyanohydrin, under alkaline conditions.

Step 4 involves selective alkylation at the 1-position of the 5-cyanomethyl-2-oxocyclopentanecarboxylic acid esters (4) by reaction with an alkyl, alkenyl or alkynyl halide in the presence of a base such as an alkali metal carbonate or hydroxide.

Step 5 involves eliminating the alkoxy carbonyl groups at position 2 of the 3-cyanomethylcyclopentanone derivative (5) by hydrolyzing it and then heating the hydrolyzed product or by heating it in the presence of an alkali metal salt.

The conversion of the 3-cyanomethylcyclopentanone derivatives (6) to the corresponding carboxylic acid esters, which are jasmonates, is straightforward.

Chain Terpene Alcohols

A. Murata, S. Tsuchiya, H. Suzuki, H. Ikeda; U.S. Patent 4,107,219; August 15, 1978; assigned to Nissan Chemical Industries Ltd., Japan have developed a process for producing chain terpene alcohols having the formula (1) or (2) by a hydrogenation of N,N-dialkyl hydroxylamine having the formula (3) or (4) with hydrogen in the presence of a catalyst. More particularly, it relates to a process for producing chain terpene alcohols such as geraniol, nerol, citronellol, hydroxygeraniol, hydroxynerol or hydroxycitronellol.

(1) R^1 R^2 OH

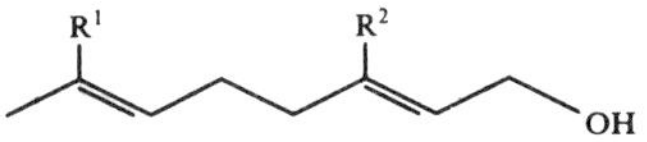

(2) R^1 OH R^2 OH

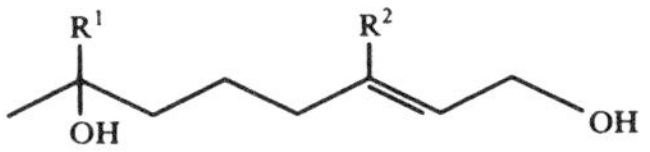

(3) R^1 R^2 ONR^3_2

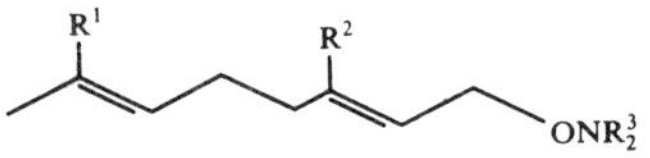

(4) R^1 OH R^2 ONR^3_2

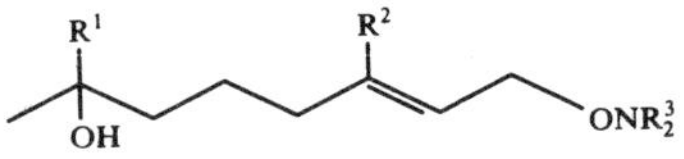

R^1 and R^2 are the same or different and respectively represent a hydrogen atom or a lower alkyl group and R^3 represents an alkyl group.

Many of the compounds have usefulness as perfume ingredients or intermediaries for producing perfume ingredients.

The catalysts used in the process can be iron family elements, copper family elements, oxides thereof and copper chromite. The catalysts are suitable for the hydrogenation cleavage of the –O–N bond without hydrogenating the remaining unsaturated bonds. Suitable catalysts include Raney catalysts of aluminum alloy containing Ni, Co, Fe or Cu and Urushibara nickel catalysts having similar characteristics to Raney nickel, the reduced Ni, Co, Fe or Cu catalysts obtained by reducing a metal oxide prepared by a precipitation method from a water-soluble metal salt with hydrogen, etc.

Example 1: *Preparation of 3,7-Dimethyl-2,6-Octadienyldiethylamine N,N-diethylnerylamine* – In a glass reactor (100 cc) purged with nitrogen, 0.25 mol of isoprene, 0.05 mol of diethylamine, 0.005 mol of n-butyllithium and 15 g of benzene were charged and the mixture was stirred at 65°C for 8 hr to react them.

After the reaction, a small amount of ethanol was added to cease the reaction. The reaction products were measured by gas chromatography.

As the result, the yield of N,N-diethylnerylamine was 32.3% (based on isoprene) or 80.7% (based on diethylamine).

The conversion of isoprene was 38.5% and the selectivity of isoprene to N,N-diethylnerylamine was 83.8%.

Example 2: The process of Example 1, using 9.9 g of myrcene, 3.6 g of diethylamine and 0.1 g of metallic sodium, was carried out at 40°C for 3 hr.

After the reaction, the reaction mixture was distilled to obtain 7.7 g of N,N-diethylgeranylamine (boiling point: 85° to 87°C/2 mm Hg).

Unsaturated Ketones

T. Ohnishi, Y. Fujita, F. Wada, T. Nishida, Y. Omura, F. Mori, T. Hosogai and S. Aihara; U.S. Patent 4,122,119; October 24, 1978; assigned to Kuraray Co., Ltd., KK Japan have developed a process for preparation of α-, β-, γ-, and δ-unsaturated ketones from easily obtainable and inexpensive starting materials. The process comprises the step of subjecting a substituted propargyl alcohol of formula (2):

$$\begin{array}{c} CH_3 \\ | \\ R-CH-C=CH_2 \\ | \\ CH_3-C-C\equiv CH \\ | \\ OH \end{array} \qquad (2)$$

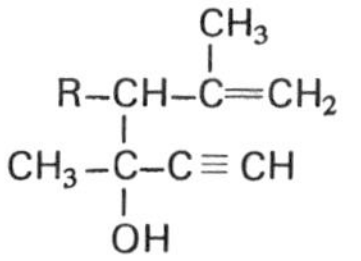

wherein R represents the group

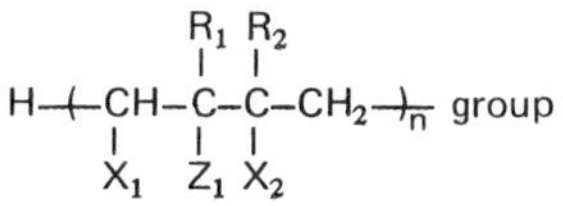

$$H\!\left(\!\begin{array}{c} R_1\;\;R_2 \\ |\;\;\;\;| \\ CH-C-C-CH_2 \\ |\;\;\;\;\;\;\;|\;\;\;\;| \\ X_1\;\;\;\;Z_1\;X_2 \end{array}\!\right)_n \text{group}$$

wherein one of the substituents X_1 and X_2 is hydrogen and the other and Z_1 together form a bond or both X_1 and X_2 are hydrogen and Z_1 represents hydrogen, hydroxyl or lower alkoxy, R_1 represents a hydrogen or lower alkyl, R_2 represents hydrogen or lower alkyl and n represents 1 or 2 whereby if n is 2 the substituents X_1, X_2, Z_1, R_1 and R_2 within the 2 units are alike or different from each other to a thermal treatment sufficient to rearrange its structure in order to obtain a rearranged ketone product containing structurally isomeric unsaturated ketones of formula (1).

(1)

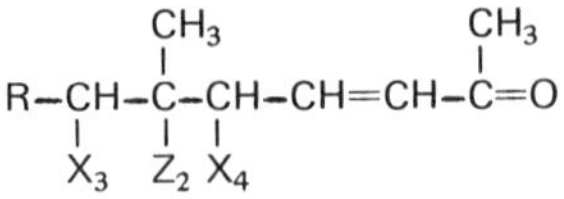

$$R\text{-}CH(X_3)\text{-}C(CH_3)(Z_2)\text{-}CH(X_4)\text{-}CH{=}CH\text{-}C(CH_3){=}O$$

wherein one of the substituents X_3 and X_4 is hydrogen and the other and Z_2 together form a bond and R is as defined above.

The unsaturated ketones of formula (1) are useful not only as perfumes, but also as intermediates for the production of other perfumery compounds such as ionone and irone.

A preferable embodiment of this method for producing unsaturated ketones comprises a step of producing a mixture of unsaturated ketones of formula (1-1-a) and of formula (1-2-a) or a mixture of unsaturated ketones of formula (1-1-b) and of formula (1-2-b) through the rearrangement of the substituted propargyl alcohol of formula (2-a) or formula (2-b) and a step of isomerizing the unsaturated ketone of formula (1-2-a) or (1-2-b) to the unsaturated ketone of formula (1-1-a) or (1-1-b) in the presence of a catalyst. The two steps are performed in independent reaction systems or in a single reaction system as illustrated.

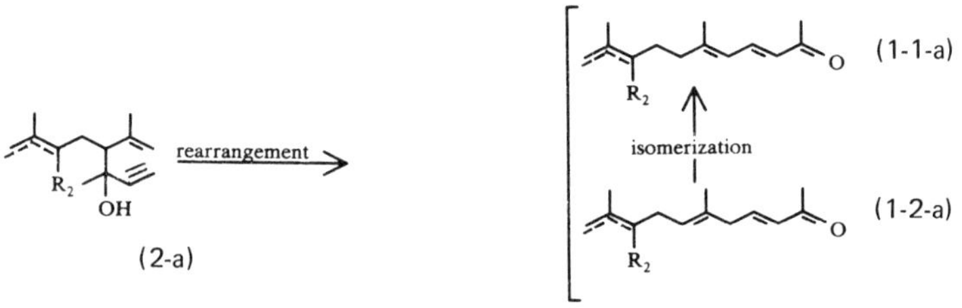

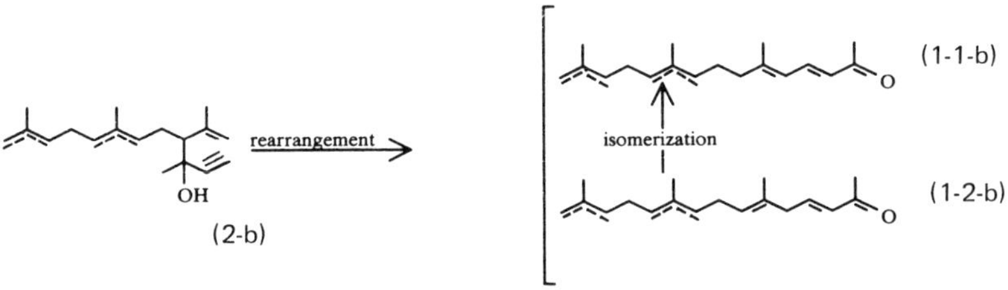

R_2 OH (2-a) rearrangement → (1-1-a) R_2 isomerization (1-2-a) R_2

OH (2-b) rearrangement → (1-1-b) isomerization (1-2-b)

An example of a particularly desirable unsaturated ketone produced by the above method is 6,10-dimethyl-3,5,9-undecatrien-2-one:

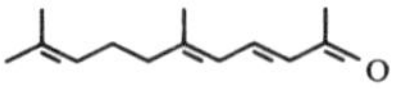

which is obtained from 4-isopropenyl-3,7-dimethyl-1-octyn-6-en-3-ol. This unsaturated ketone may be cyclized in the presence of an acid catalyst to ionone which is of value as a perfumery product or as an intermediate for the production of vitamin A and colorants. A second example is 6,9,10-trimethyl-3,5,9-undecatrien-2-one

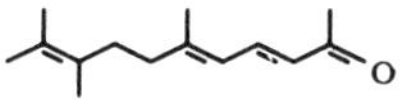

which is obtained from 4-isopropenyl-3,6,7-trimethyl-1-octyn-6-en-3-ol. This unsaturated ketone is of value as an intermediate for the production of irone.

Isochroman Precursors

Certain isochromans have been described which have an outstanding musk fragrance. *M. A. Sprecker; U.S. Patent 4,123,469; October 31, 1978; assigned to International Flavors and Fragrances Inc.* has reported a single economic and efficient process for producing monocycloalkylated aromatic compounds which are precursors for such materials as isochromans or cycloalkylated aromatic nitriles. The process involves carrying out the disproportionation reaction of an aromatic compound with a bicycloalkylated aromatic compound in the presence of an aluminum chloride catalyst whereby when 1 mol of the aromatic compound is reacted with 1 mol of the bicycloalkylated aromatic compound, 2 mols of the monocycloalkylated aromatic compound are produced as follows:

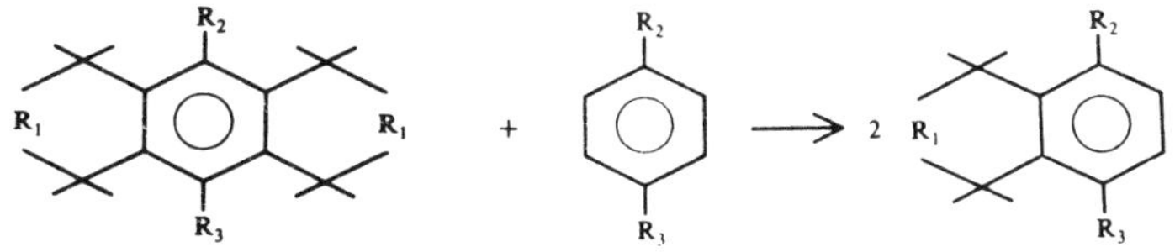

wherein R_1 is selected from the group consisting of the moieties:

and R_2, R_3 and R_4 are the same or different and each represents hydrogen, methyl or ethyl.

The reaction may be either carried out (1) in situ after the original monocycloalkylation reaction is carried out and the bicycloalkylated product is formed together with the desired monocycloalkylated product, or (2) it may be carried out on the neat bicycloalkylated aromatic compound. In both cases, a specified range of quantities of aluminum chloride catalyst and a specified range of ratios of aromatic compound:bicycloalkylated compound should be used.

An in situ reaction is carried out using a large molar excess of aromatic compound (which is recoverable) in the presence of an aluminum chloride catalyst (which is critical to the process). The mol ratio of aromatic compound:diol is in the range of from about 15:1 up to 50:1 with a preferred ratio of from 20:1 up to 35:1.

If the mol ratio is below about 15:1 then solubility problems exist. The mol ratio range of diol compound:aluminum chloride is from about 2:1 up to about 5:1 with a preferred mol ratio of 2.1:1.

In the case where the reaction is carried out by means of reacting the aromatic compound with the neat bicycloalkylated aromatic compound, a large molar excess of aromatic compound (which is recoverable) is used and the reaction is carried out in the presence of aluminum chloride at a temperature in the range of from about 40°C up to reflux temperature (e.g., 80°C) and for a period of time of from about 1 hr up to 10 hr.

The mol ratio of bicycloalkylated aromatic compound: aluminum chloride catalyst may vary from about 1:100 to about 1:0.25 with a preferred ratio of bicycloalkylated aromatic compound:aluminum chloride catalyst being about 1:2. The mol ratio of aromatic compound:bicycloalkylated aromatic compound in this reaction may vary from about 2:1 to about 40:1 with a preferred mol ratio range of from about 10:1 to 15:1.

The resulting monocycloalkylated aromatic compound can be further reacted as is or it can be purified and then further reacted with, for example, acylation reagents such as acetyl chloride or acetic anhydride or with the following reagents, in sequence, in order to add a C≡N moiety to the aryl ring:

(1) a chloromethylation reagent such as p-formaldehyde and hydrochloric acid thereby forming a chloromethyl substituent on the ring;

(2) nitropropane in order to transform the chloromethyl group into a carboxaldehyde group;

(3) hydroxylamine in order to transform the carboxaldehyde group into a alkoxime group and

(4) acetic anhydride in order to transform the aldoxime group into a C≡N group.

Example: *Preparation of 1,1,4,4-Tetramethyltetrahydronaphthalene –*

OH
OH
+
$AlCl_3$

Into a 12-liter reaction flask equipped with stirrer, thermometer, reflux condenser, heating mantle, cooling bath and Gooch tubing is placed: (1) 596.5 g (3.9 mols) of 2,5-dimethylhexane-2,5-diol; (2) 7,200 g (91 mol) of benzene.

The mixture is heated to 45°C in order to dissolve all of the dimethylhexanediol. The reaction mass is then cooled to 25°C and 1,093 g (8.19 mols) of aluminum trichloride is added while maintaining the reaction mass temperature at 21° to 25°C, the addition taking place over a period of 1 hr. The reaction mass is then heated up to 40°C over a period of 30 min and maintained at 40°C, with stirring, for a period of 3 hr. The reaction has substantially gone to completion after operation at 40°C for a period of 1 hr. The reaction mass is then poured into a 22 liter flask containing 500 ml HCl (concentrated), 500 ml water and 3 liters of crushed ice. The resulting mass is then stirred for a period of 30 min and the water layer is separated from the organic layer. The organic layer is then washed with 2 liters of water followed by 1 liter of water, followed by 50 ml 50% aqueous sodium hydroxide. 20 ml acetic acid is then added to the mass in order to bring the pH to 6-7.

The benzene is then stripped off at atmospheric pressure and the resulting oil is transferred to a small distillation flask and rushed over at 92° to 110°C at a vacuum of 7 to 10 mm Hg yielding 625 g of tetrahydrotetramethylnaphthalene (75% yield). NMR, IR and Mass Spectral analyses yield the information that the resulting product has the structure of the product shown in the reaction equation given above.

PERFUME EXTENDERS AND DILUENTS

Alpha-Methylstyrene Dimers

Compounded perfumery compositions contain a number of ingredients which may be of natural or synthetic origin. The ingredients are blended by the perfumer to create the desired odor effect. Such essential oils which contain high percentages of hydrocarbon constituents such as patchouli oil (an essential oil derived from *Pogostemon patchouli)* have, for example, warm aromatic spicy odors. When the perfumer wishes to include this type of note, for example, in a perfumery composition of an oriental type, he will use patchouli oil.

However, such natural oils as oil of patchouli are expensive essential oils and are of limited availability. Even more extreme examples are natural sandalwood oil and natural vetiver oil. Although attempts have been made to simulate the odor of patchouli oil, sandalwood oil, and vetiver oil by use of blends of synthetic perfumery chemicals, the creation of such oils having identical aromas with reference to the natural oils has not been achieved.

J.B. Hall, W.J. Wiegers, I.D. Hill, R.M. Novak and F.L. Schmitt; U.S. Patent 4,142,998; March 6, 1979; assigned to International Flavors & Fragrances Inc. have found that certain compounds can be used as extender materials for essential oils. These compounds are miscible with the essential oils and do not appreciably alter their aromas in quality or strength. They may be used in a proportion of from 1 to 30% of the total weight of essential oil extender composition.

These extenders are dimerization products of α-methylstyrene, or a methyl or other C_{2-4} lower alkyl homologue thereof, or a hydrogenated derivative thereof, or a mixture of these, having one of the generic structures:

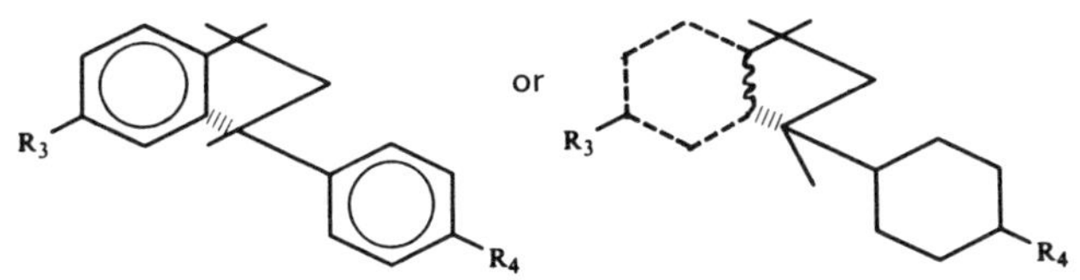

R_3 ... R_4 or R_3 ... R_4

wherein R_3 and R_4 are the same or different and each represents hydrogen or methyl or other C_{2-4} lower alkyl, and wherein the dashed lines and wavy lines represent carbon-carbon single bonds or carbon-carbon double bonds, with the proviso that when there is one double bond present only the wavy line is a double bond, and when there is more than one double bond present, the ring containing the dashed lines and the wavy line is a benzene ring, and wherein the line |||||||| represents either a carbon-carbon single bond or no bond.

These extended perfumery materials will find use as constituents of compounded perfumery compositions in which a number of perfumery materials of natural and/or synthetic origin will be blended together to produce a particular desired odor effect. Such compositions may then be used in space sprays or can be blended in soap, detergent or deodorant compositions, including bath salts, shampoos, toilet waters, face creams, talcum powders, body lotions, sun cream preparations and shave lotions and creams. The perfumery compositions can also be used to perfume substrates such as fibers, fabrics and paper products.

Example 1: *Preparation of Alpha-Methylstyrene Dimerization Product –*

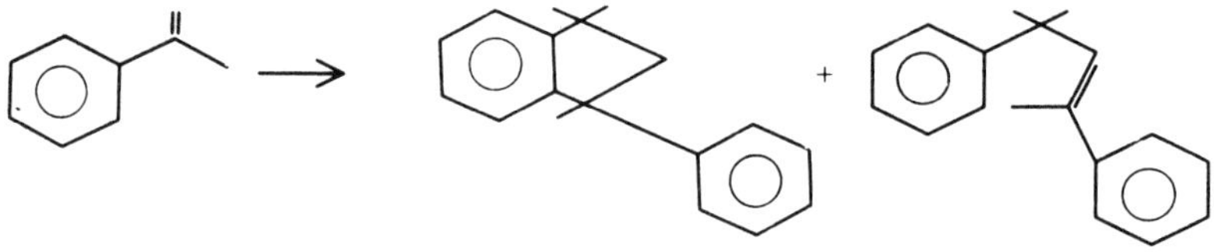

(minor) (major: cis and trans isomer mixture)

Into a 1 liter reaction flask equipped with thermometer, addition funnel, heating mantle, reflux condenser, stirrer, Y-adapter and distillation head is added 100 g of cyclohexane followed by 5 g of p-toluene sulfonic acid. The resulting mixture is heated to 50°C and over a 1 hr period, 500 g of alpha-methylstyrene is added to the reaction flask. The reaction mass is then heated to 100°C and maintained at that temperature for a period of 4 hr. 529.3 g of crude product is then recovered which is then mixed with 15 g of Primol and 0.2 g of Ionox. The resulting mixture is fractionally distilled.

Example 2: Patchouli oil (80 parts) obtained from the Seychelle Islands is blended with the alpha-methylstyrene dimerization product produced according to Example 1 (20 parts). The alpha-methylstyrene dimerization product is found to act as an extender for the patchouli oil in that the characteristic odor effect of the latter is substantially not modified.

Example 3: A patchouli oil extender base is prepared by blending the following ingredients:

Ingredients	Parts
Dimerization product of alpha-methylstyrene (produced according to Example 1)	38
Galaxolide	27

(continued)

Ingredients	Parts
Isolongifolene oxidate	20
Omega-hydroxymethyl longifolene	10
Cedrol	3
Sandalwood oil (East Indian)	2

This mixture (46 parts) is then blended with natural patchouli oil (Seychelles) (60 parts) to provide a satisfactory extended patchouli oil.

Terpene Dimers

W.J. Wiegers, J.B. Hall, I.D. Hill, R.M. Novak, F.L. Schmitt, B.D. Mookherjee, C.-K. Shu and W.L. Schreiber; U.S. Patent 4,165,301; August 21, 1979; assigned to International Flavors & Fragrances Inc. have found that dimerization products of (1) monocyclic terpenes containing two carbon-carbon double bonds, (2) bicyclic terpenes containing one carbon-carbon double bond and (3) a monocyclic terpene containing two carbon-carbon double bonds and a bicyclic terpene containing one carbon-carbon double bond or mixtures of same or hydrogenation products thereof or mixtures of the hydrogenation products and the dimerization products may be used as diluents or extenders of various perfumery materials without appreciable loss of the characteristic odor effect of such perfumery materials.

These dimerization products are produced by dimerizing such compounds as camphene; alpha- or beta-pinene; d-limonene; alpha- or beta-phellandrene; alpha-, beta-, or gamma-terpinene; delta-3-carene; terpinolene; or by dimerizing mixtures of two or more of such compounds, such as the mixture of C_{10} terpenes commonly known as sulfate turpentine or by dimerizing a C_{10} terpene and a dehydrogenated terpene (e.g., cymene) in the presence of acid catalysts, such as sulfuric acid and hydrofluoric acid or in the presence of acid clay catalysts, such as Japanese Acid Clay or Fuller's earth or acidic cation exchange resin catalysts.

Turpentines containing the 10-carbon terpene ingredients which may be dimerized by this process are derived from a number of species of pine trees, including the Engelmann, Douglas, Torrey, Linnaeus, Virgina pines, etc.

Example 1: *Preparation of Dimers of Camphene* – Into a 2 liter reaction flask equipped with stirrer, thermometer, addition funnel and reflux condenser with Bidwell Trap are placed: 336 g of hexahydropentamethylindane and 32 g of Filtrol 25. The mixture is heated with stirring to 155°C. Over a period of 2.25 hr, while maintaining the reaction mass at 155°C, 547 g of camphene is added thereto. The reaction mass is then stirred for 7½ hr at 155° to 158°C and the progress of dimerization is monitored on GLC apparatus (Conditions: 5% SE 30 column, 10' x ¼", programmed at 80° to 240°C at 8°C per minute). GLC analysis shows very little change after 2 hr. The reaction mass is then filtered. The filter cake is washed with 200 g of hexahydropentamethylindane.

The weight of filtrate is 1,056 g. The resultant filtrate is distilled in the presence of Primol (30 g) and Ionox (1 g) through an 18" Vigreux column equipped with reflux head. Fourteen fractions were obtained of which fractions 8 through 12 were practically pure dimers.

Example 2: A patchouli oil extender base is prepared by blending the following ingredients:

Ingredients	Parts
Camphene dimer produced according to Example 1	38
Guaioxide	27
Isolongifolene oxidate	20
Omega-hydroxymethyl longifolene	10
Cedrol	3
Sandalwood oil (East Indian)	2

This mixture (46 parts) is then blended with natural patchouli oil (Seychelles) (60 parts) to provide a satisfactory extended patchouli oil.

Example 3: *Preparation of Dimerization Products from Sulfate Turpentine* – Into a 500 ml reaction flask equipped with thermometer, stirrer, condenser and addition funnel are placed 5 g Primol and 2 g of 20% phosphoric acid on silica catalyst (Chemtron Corporation). The reaction mass is heated to 150°C and 50 g of sulfate turpentine is added dropwise over a period of 2 hr with stirring. The reaction mass is then heated for another 2 hr at 150°C.

The reaction mass is then cooled and filtered and the resulting dimerization product is a mixture of compounds containing unreacted terpene monomers. The unreacted terpene monomers are distilled and again dimerized using a borontrifluoride etherate catalyst. The resulting dimerization products are then combined and distilled and used in the following examples.

Example 4: Vetiver oil (70 parts) obtained from Haiti is blended with the dimerization product produced according to Example 3. The thus formed dimerization product is found to act as an extender for the vetiver oil in that the characteristic odor effect of the latter is substantially not modified.

Example 5: Sandalwood oil (75 parts) obtained from Indonesia is blended with the dimerization product produced according to Example 3 (25 parts). The dimerization product thus produced is found to act as an extender for the sandalwood oil and the characteristic odor effect of the latter is substantially not modified.

Terpene Dimers plus Alpha-Methylstyrene Dimers

J.B. Hall, W.J. Wiegers, I.D. Hill, R.M. Novak and F.L. Schmitt; U.S. Patent 4,170,576; October 9, 1979; assigned to International Flavors & Fragrances Inc. describe the use of the dimers described in the last patent (A) and mixtures of these dimerization products with (B) alpha-methylstyrene dimerization products, dimerization products of methyl or other C_{2-4} lower alkyl homologues thereof and hydrogenated derivatives thereof as diluents or extenders for various perfumery materials.

Dimerization products (B) are produced by dimerizing alpha-methylstyrene or a C_{2-4} homologue having, for example, the structure:

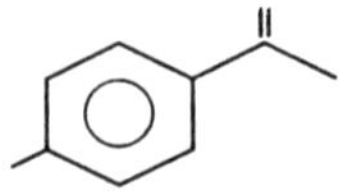

in the presence of Lewis acid catalysts, Bronstedt acid catalysts such as sulfuric acid or in the presence of acid clay catalysts such as Japanese acid clay or Fuller's earth or cation exchange resin catalysts. The useful dimerization products of alpha-methylstyrene have the structures:

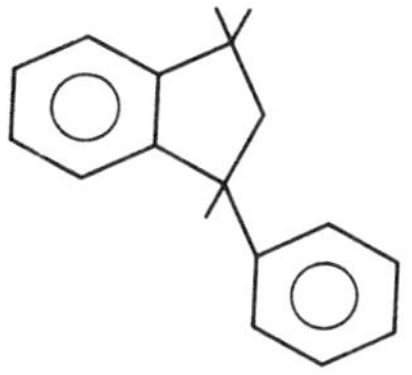

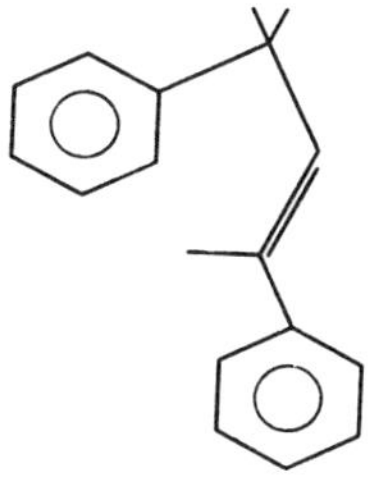

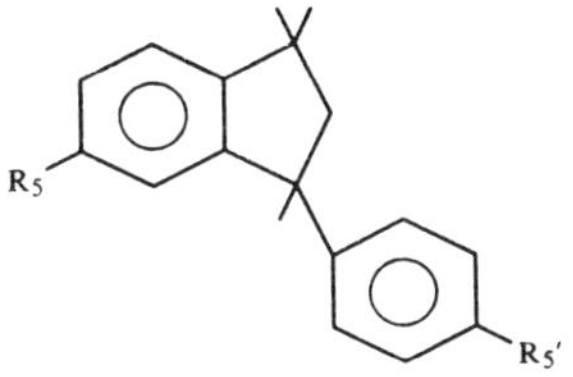

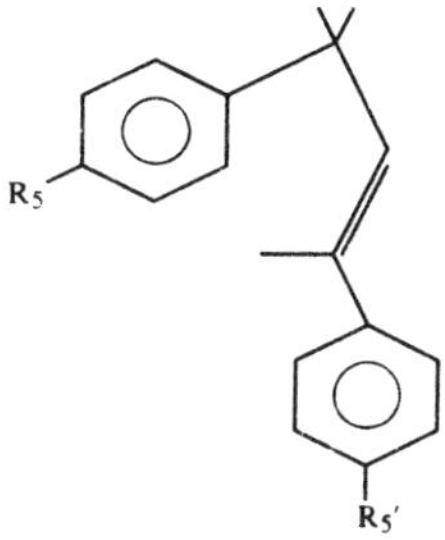

where one of R_5 or R_5' is methyl or other C_{2-4} lower alkyl and the other of R_5 or R_5' is hydrogen or each of R_5 and R_5' are the same or different C_{1-4} lower alkyl, e.g., methyl.

Hydrogenation products of these compounds can be represented by the generic structures:

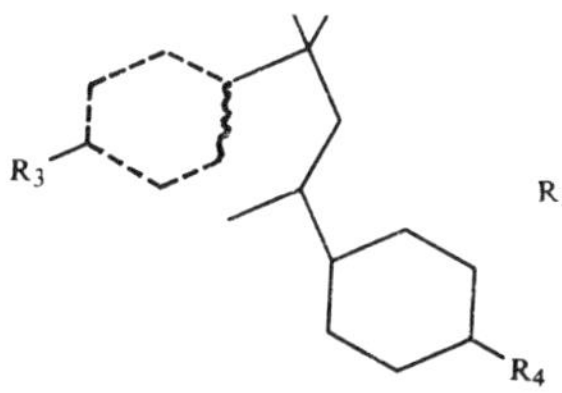

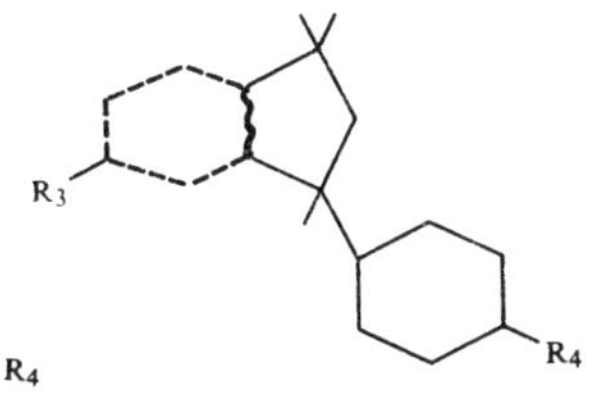

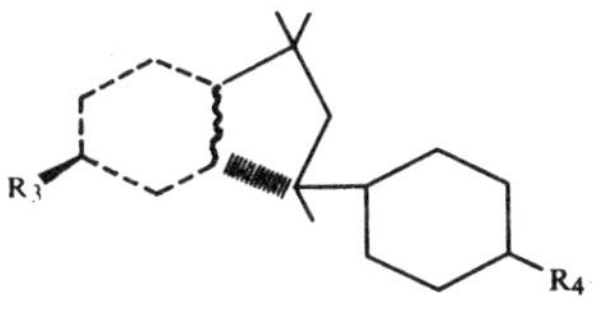

wherein R_3 and R_4 are the same or different and represent hydrogen or methyl or other C_{2-4} lower alkyl; wherein the dashed lines and wavy line represent carbon-carbon single bonds or carbon-carbon double bonds with the proviso that when there is one double bond present, only the wavy line is a double bond and when there is more than one double bond present, the ring containing the dashed lines and the wavy line is a benzene ring and where the line |||| represents either a carbon-carbon single bond or no bond.

In the case of the hydrogenation product, when R_3 and/or R_4 are lower alkyl, for example, methyl, the methyl groups may be in a cis or trans relationship to one another and with respect to the cyclohexyl moieties.

A significant property of the described mixtures of dimerization products and hydrogenated dimerization products is that they have a broad range of solubilities for various types of perfumery materials including complete solubility for certain alcohols, esters, pyrans, aldehydes, ketones, cyclic ethers, cyclic amines, nitriles and natural oils.

Example 1: *Preparation of Alpha-Methylstyrene Dimerization Product –*

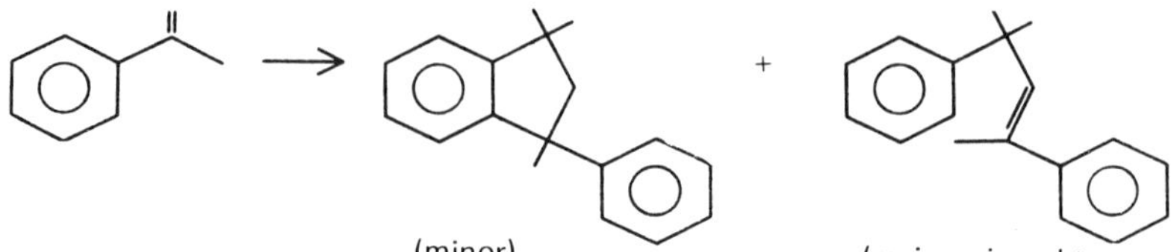

(minor)

(major: cis and trans isomer mixture)

Into a 1 liter reaction flask equipped with thermometer, addition funnel, heating mantle, reflux condenser, stirrer, Y-adapter and distillation head is added 100 g of cyclohexane followed by 5 g of p-toluene sulfonic acid. The resulting mixture is heated to 50°C and over a 1 hr period, 500 g of alpha-methylstyrene is added to the reaction flask. The reaction mass is then heated to 100°C and maintained at that temperature for a period of 4 hr. 529.3 g of crude product is then recovered which is then mixed with 15 g of Primol and 0.2 g of Ionox. The resulting mixture is fractionally distilled through a Y adapter distillation column.

Example 2: Patchouli oil (85 parts) obtained from the Seychelle Islands is blended with 10 parts by weight of the camphene dimer produced according to Example 1 of the previous patent, and 5 parts of the alpha-methylstyrene dimer of this patent. The camphene dimer/alpha-methylstyrene dimer mixture is found to act as an extender for the patchouli oil in that the characteristic odor effect of the latter is substantially not modified.

Example 3: The extended patchouli oil prepared according to Example 2 is successfully incorporated into a compounded composition of the Chypre type by blending the following ingredients:

Ingredients	Parts
Cinnamic aldehyde	1
Ethyl methylphenylglycidate	1
Methylnonylacetaldehyde	2
Oakmoss (absolute)	20
Sandalwood oil (East Indian)	20
Vetiveryl acetate	20
Ylang oil No. 1	20
Benzoin resin (Sumatra)	30
Alpha ionone (100%)	30
Clove stem oil (Zanzibar)	36
Bergamot oil	40
Hydroxycitronellal	40
Isoeugenol	40
Extended patchouli oil (Example 2)	40
Coumarin	50

(continued)

Ingredients	Parts
Musk ketone	50
Amyl salicylate	60
Cedarwood oil (American)	60
Citronellol	60
Benzyl acetate	80
Phenylethyl alcohol	150
Terpinyl acetate	150

PRODUCT APPLICATIONS

GELLED PERFUME

Polymeric Carbonate Gelling Agent

J. Teng, M.C. Stubits, A. Minton and J.H. Baker; U.S. Patent 4,067,824; Jan. 10, 1978; assigned to Anheuser-Busch, Incorporated have developed a gelled perfume comprising a conventional perfume base and a polymeric carbohydrate gelling agent. This perfume possesses aesthetically pleasing clarity and enduring fragrance. The perfume can be conveniently carried and easily dispensed.

About 0.5 to 3.5% by weight gelling agent is used in the composition. The gelling agent is prepared by dissolving it in denatured ethyl alcohol (SD-40). Other solvents can be used in place of ethyl alcohol. The alcohol makes up about 45 to 70% by weight of the final composition.

About 0.1 to 0.3% by weight of a preservative agent is used in the composition. About 15 to 35% by weight of an emollient is also used in the composition. About 10 to 18% by weight perfume oil is used in the composition. Conventional perfumes may also be used to provide the perfume oil.

The gelling agent that is employed can be used with both alcoholic base fragrances and oil base fragrances. The perfumes in the alcoholic base category should contain at least 12 oz of perfume oil per gallon of alcohol, and the alcohol shall contain no more than 15% water (by volume). A preferred perfume contains 20 oz of perfume oil per gallon of alcohol, and the alcohol contains 5% (by volume) of water. There are no such limitations for perfumes in the oil base category.

The gelling agents used are polymeric carbohydrate derivatives selected from the group consisting of hydroxypropyl cellulose esters and hydroxypropyl starch esters and mixtures of these esters. These esters can have a degree of substitution (DS) of acetyl groups of about 1.2 to 3 and a degree of molar substitution (MS) of hydroxypropyl groups of about 2 to about 8. Suitable esters are hydroxypropyl cellulose acetate, hydroxypropyl starch acetate, hydroxypropyl cellulose laurate, hydroxypropyl starch laurate and methyl hydroxypropyl cellulose acetate.

The important physical properties of the gelling agent are summarized in the table. The usual form is that of a white powder of 20 to 40 mesh. Depending upon the equipment, the particle size can be varied. The material is basically not hygroscopic and therefore has low moisture content. The gelling agents are highly substituted derivatives and are generally inert to enzymatic activity.

Physical Properties of Gelling Agent

Color	White
Physical form	Soft powder, 20 to 40 mesh
Moisture	0.5%
Ash	1.0%
Specific gravity	1.017
Glass transition temperature	85°C
Melting range	190° to 210°C
Char point	240°C

This gelling agent does not support microbial growth and is inert to proteolytic amylolytic degradation.

Example 1: 0.025 g of hydroxypropyl cellulose acetate was dispersed in 2.5 ml of Chantilly perfume (Houbigant), and the dispersion was agitated for approximately 3 minutes on a Vortex Mixer. The resulting clear gel was smooth and highly viscous.

Example 2: 6 ml of perfume oil, Perry F-73-295, were blended in denatured alcohol (100 ml), SDA 40. To the blend were added 1.8 g of methyl hydroxypropyl cellulose acetate. The resulting solution was mixed until the gel was formed.

Phosphates or Carbonates as Perfume Stabilizers

A gelatinous fragrance-imparting composition comprises, as a gel substrate, a water-soluble macromolecular compound such as carrageenan, agar, an alginate, polyethylene oxide, a polyacrylate or the like, and a perfume is dispersed and retained in the gel substrate. When water is evaporated from the composition, the perfume is evaporated into the open air together with the evaporated water, to produce an aromatizing effect.

In general, such gelatinous fragrance-imparting compositions comprise 1.0 to 5.0% by weight of gel substrate, 0.1 to 20.0% by weight of a perfume and 0.1 to 20% by weight of a water-soluble surface active agent, with the balance being essentially water. When a perfume is emulsified and dispersed in such an aqueous gel, the stability of some perfumes is very bad and a change of the fragrance or a generation of a bad smell often occurs during long-time storage. Simple perfumes of the ester and the aldehyde types, such as linalyl acetate, benzyl acetate and cyclamen aldehyde, possess particularly low stabilities in such compositions.

A. Sakurai and M. Fujita; U.S. Patent 4,137,196; January 30, 1979; assigned to Kao Soap Co., Ltd., Japan have discovered that the stability of a perfume incorporated in such aqueous gel compositions can be improved when a specific phosphate or carbonate is incorporated in the aqueous gel having a perfume emulsified and/or dispersed in it, and when the pH of the gel is adjusted to 6.0 to 10.0 by addition of an appropriate pH adjusting agent.

More specifically, they provide a stable, gelatinous, fragrance-imparting composition comprising a gel substrate, a perfume, at least one first additive selected from the group consisting of phosphates having the formula:

$$(1) \qquad (M)_m(H)_{3-m}PO_4 \cdot xH_2O$$

wherein M is sodium, potassium or ammonium, m is an integer of from 1 to 3, and x is 0 or an integer of from 1 to 12, and carbonates having the formula:

$$(2) \qquad (M)_n(H)_{2-n}CO_3 \cdot xH_2O$$

wherein M and x are the same as defined above and n is an integer of 1 or 2, and at least one pH adjusting agent selected from phosphates and carbonates having the above formulas (1) and (2), with the proviso that the pH adjusting agent is different from the first additive, and wherein the pH of the composition is in the range of from 6.0 to 10.0.

As simple perfumes which are especially effective in the process, there can be mentioned p-cresyl methyl ether, isoeugenol, octyl aldehyde, alkyl aldehydes having 12 to 26 carbon atoms in the alkyl group, benzaldehyde, hyacinth aldehyde, cinnamic aldehyde, jasmine aldehyde, lily aldehyde, 1-carvone, jasmone, linalyl acetate, linalyl formate, methyl benzoate, civet, isopropylquinoline, p-methylquinoline, dimethylbenzyl carbinol, ethyl acetate, pineapple oil, melon natural, coffee oil base, methyl citronellal and acetaldehyde.

Example 1: In 84.8 parts of deionized water there were incorporated 0.15 part of monopotassium dihydrogen phosphate and 0.15 part of disodium monohydrogen phosphate, and the mixture was heated at 90°C to dissolve the salts. Then, 2.0 parts of agar as the gel substrate, 3.0 parts of sodium dodecylbenzenesulfonate, 3.0 parts of a perfume (green compounded perfume or ester aldehyde simple perfume) and 5.0 parts of ethanol were added to the above solution under agitation, and the mixture was poured into a vessel and cooled to room temperature to solidify and form a gelatinous aromatizing agent.

After storage at 40°C for 20 days, an organoleptic test was carried out. It was found that the fragrance was not significantly changed.

Example 2: A basic composition comprising 2.0 parts of an alginic acid derivative (AG-1), 2.0 parts of a surface active agent (polyoxyethylene alkyl phenyl ether) and 3.0 parts of a perfume (floral compounded perfume) was incorporated into 93.0 parts of deionized water to form a comparative gelatinous aromatizing agent.

Separately, 0.6 part of a sodium dihydrogen phosphate dihydrate and 0.5 part of disodium monohydrogen phosphate dodecahydrate were added to the above basic composition, and the mixture was added to 91.9 parts of deionized water to form an aromatizing agent according to this process. After these samples had been stored at 40°C for 20 days, an organoleptic test was carried out. The fragrance was not significantly changed and good results were obtained.

AIR FRESHENERS

Substituted Pyran Deodorizing Agents

J.G. Kaufman; U.S. Patent 4,107,289; August 15, 1978; assigned to Naarden International Holland, B.V., Netherlands has found that malodors may be destroyed by undergoing a chemical reaction with substituted pyran compounds in a liquid or vaporous state, the vapors being released from substrates at normal conditions of temperature and pressure whereby the malodor will be destroyed.

By utilizing substituted pyran compounds such as a substituted dihydropyran molecule, it is possible to destroy the malodors by the formation of addition products across the double bond of the heterocyclic ring. In addition, when utilizing substituted dihydropyrans which possess a side chain containing an active double bond which is active by nature of its being alpha,beta to carbonyl groups, the activity of the compound to destroy the malodor will be significantly enhanced, thus in effect providing what may be termed as a double-barrelled action. This activity is particularly useful against malodors in which the offensive odor is due to the presence of sulfur-containing compounds such as mercaptans, the substituted pyran being especially effective on acidic sulfur compounds.

It is therefore an object of this process to provide substituted pyran compounds which may be utilized for the elimination of malodors.

A further object of this process is to provide a method for the deodorization of malodors which may be found in living quarters, offices or other enclosed spaces by contacting the offensive malodors with the vapors of substituted pyran compounds which may form one component of a deodorizing composition of matter.

One embodiment is found in the use of a compound having the formula:

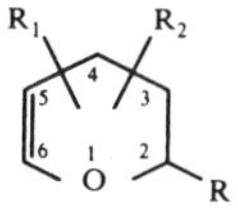

wherein R_1 and R_2 are each selected from the group consisting of hydrogen, lower alkyl (1 to 6 carbons) and cycloalkyl and R represents an organic radical; as a chemically active deodorizing agent.

A specific embodiment is found in the use of 3,4-dihydro-2H-pyran-2-carbinyl acetate or its 2,5-dimethyl homologue in admixture with a gelatinous substrate to chemically destroy malodors.

The substituted pyrans may be utilized to chemically destroy malodors in various ways. For example, one method of utilizing the pyrans is to composite the pyran in a gelatinous substance to form a deodorizing compound, the active ingredient thereof being the substituted pyran. The gelatinous substances which may be used as the substrate for the pyrans may be prepared by compositing a gelling agent with deionized water and adding thereto inert liquids plus other inert solids. The gelling agents which may be employed may include fatty acid soaps containing from about 10 to about 22 carbon atoms in length.

If so desired, the gel may include an aroma composite consisting of a mixture of aroma materials which may be present in any proportion necessary to give the composite a pleasant fragrance, the aroma materials including alcohols, aldehydes, ketones, etc., in such proportions so as to provide a perfume composition. This perfume composition may be present in a range of from about 1% or less to about 10% or more by weight of the finished composition of matter.

The amount of substituted pyran compound which may be incorporated into the gel may be varied within a wide range and will depend upon the particular application to which the gel is to be used, or upon the particular type of perfume which is incorporated into the gel. Generally speaking, the amount of substituted pyran, and particularly substituted dihydropyrans, which will possess the side chain containing a double bond which is active for purposes of destroying malodors by nature of its being alpha,beta to a carbonyl group, may range from about 0.1% to about 5% or more of the finished composition of matter.

Another method of utilizing the dihydropyrans is to incorporate these dihydropyrans in containers such as aerosol dispensers whereby the active ingredient may be sprayed into a space which is permeated with the malodor.

In addition to the use of dihydropyrans in gelatinous substances and spray dispensers, these compounds may also be utilized as a chemically active ingredient for the destruction of malodors by incorporating the dihydropyrans in a cream base or lotion.

These compounds may find further use in powders, such as talcum powders, in other cosmetic preparations such as lipsticks, hair sprays or lacquers, in soaps, shampoos and other detergents for personal use, in natural or synthetic sponges or fibers etc. These compounds can be added to products which on account of their composition develop malodors (e.g., cold wave preparations), or to products that may develop malodors during their use such as certain feminine hygiene products, kitty litter and the like.

Example: A deodorizing gel containing the substituted dihydropyran as the chemically active ingredient thereof is prepared by adding 2% by weight of agar-agar and 0.2% by weight of a powdered silica absorbent (Dowacil 200) to 91.5% by weight of deionized water. The mixture is mechanically agitated and heated to boiling. After reaching the boiling point, heating is discontinued and after the mixture has cooled to a temperature below 70°C, 1.3% by weight of ethylene glycol, 1.5% by weight of an anionic ethoxysulfate solubilizer (Neocol 2040) and 0.5% by weight of a surface active agent (Triton X-100) is added with agitation. After thoroughly admixing the solution, 0.5% by weight of 2,5-dimethyl-3,4-dihydro-2H-pyran-2-carboxaldehyde and 2.5% by weight of a perfume composition are then added to the mixture. The mixture is again thoroughly agitated and poured into a mold where, after reaching room temperature, the mixture will solidify and become a gelatinous substance.

To test the efficacy of the deodorizing gel a test is effected by spraying alpha-furfuryl mercaptan for 4 seconds from an aerosol dispenser into a controlled smelling booth, the odor of the mercaptan simulating an obnoxiously strong coffee odor. Following this, the gelatinous substance containing the 2,5-dimethyl-3,4-dihydro-2H-pyran-2-carboxaldehyde is then placed in the booth. An odor evaluation by a panel of professional fragrance evaluators disclosed the fact that the deodorizing gel containing the substituted pyran effectively eliminated the obnoxious mercaptan odor from the booth.

Perfumed Gels of Hydroxypropyl Cellulose

Many solid air freshener compositions have been based on carrageenan. However, the problem of syneresis, separation of moisture from the gel, which occurs with carrageenan plus its increasing cost makes it desirable that another matrix for a gel be utilized.

B.M. Mitzner; U.S. Patent 4,128,507; December 5, 1978; assigned to Polak's Frutal Works, Inc., has found that stable, heat and syneresis resistant shaped bodies suitable for use as air fresheners can be prepared using, as a matrix, a hydroxypropyl cellulose gel. More specifically, the solid, self-supporting perfumed gel suitable for use as an air freshener comprises about 3 to 10% by weight of a solid phase consisting essentially of hydroxypropyl cellulose and about 90 to 97% by weight of a liquid phase consisting essentially of about 35 to 90% by weight of a linear polyol plasticizer, about 10 to 55% of a perfume oil, about 0 to 10% of a solvent for hydroxypropyl cellulose selected from the class consisting of water and C_{1-3} aliphatic alcohols.

Hydroxypropyl cellulose is a water-soluble, organosoluble, thermoplastic cellulose derivative (Klucel). It is prepared by reacting alkali cellulose with propylene oxide in an inert solvent whereby the hydroxypropyl moiety attaches to the cellulose molecule through one of the hydroxyl groups present thereon, forming an ether.

Three sites for potential substitution are available on each cellulose molecule and, additionally, the hydroxypropyl group, once attached, offers another site for reaction so that, theoretically, any number of hydroxypropyl molecules can be affixed to a single anhydroglucose unit. Normally, however, commercially available hydroxypropyl cellulose will have a substitution level between about 2 and 10 hydroxypropyl molecules per anhydroglucose unit and preferably about 3 to 4. (The substitution level is referred to as the MS or molecular substitution).

It is available in low, medium, and high viscosity grades, indicating low, medium, and high molecular weight cellulose. Any of these viscosity grades can be used in the products of this process. For further information concerning preparation and properties of hydroxypropyl cellulose, reference is made to U.S. Patent 3,278,521.

The linear polyols which can be employed in this perfumed gel are exemplified by propylene glycol, trimethylene glycol, trimethylol propane, glycerin, sorbitol, and lanolin. These materials are weak solvents for hydroxypropyl cellulose and are frequently employed as plasticizers for hydroxypropyl cellulose in other applications, although in lower amounts relative to the amount of hydroxypropyl cellulose.

Gel matrices are easily prepared by simply mixing the ingredients and kneading the mixture until the hydroxypropyl cellulose particles are evenly distributed throughout the liquid (plasticizer-perfume oil-solvent) phase and are partially swollen by the liquid. While the mixture is still fluid, it is placed in a mold and allowed to stand at room temperature until the hydroxypropyl cellulose is thoroughly solvated and the mass is sufficiently rigid to be self-supporting.

In the following examples parts and percentages are by weight unless otherwise indicated.

Example 1: Ten parts of hydroxypropyl cellulose (Klucel G) was dispersed in a mixture of 40 parts propylene glycol and 50 parts jasmin perfume oil. This mixture was mixed at 25°C until the hydroxypropyl cellulose was completely softened and swollen, at which point the mass was a very high viscosity fluid. This was placed in a cylindrical mold for about one hour to harden.

When removed from the mold, the hardened body was found to possess the odor of the perfume unchanged from the original. The odor was given off by the body for a period of at least 8 months.

Example 2: The procedure of Example 1 was repeated using 5 parts of high molecular weight hydroxypropyl cellulose in 95 parts of a fluid. The fluid mixture contained 45 parts of rose oil, 40 parts propylene glycol and 10 parts methanol.

The mixture was agitated well at 25°C until a high viscosity fluid was obtained. At this point, the mass was poured into a 3 x 4 inch nonwoven polymer bag lined with perforated polypropylene film and the bag was heat sealed. The perfume odor, substantially unchanged from the original, was given off over a period of at least 6 months.

Containing Color Change Perfume Systems

D.R. Munden; U.S. Patent 4,128,508; December 5, 1978; assigned to International Octrooimaatschappij "Octropa" BV, Netherlands has provided a color change perfume system for use in an air freshening device to give visual indication of the time when it becomes ineffective.

The indicator may be a standard pH indicator and the reactive component a suitable volatile acid or base. In this case the volatility of the acid or base is selected to be similar to the volatility of the perfume composition, so that when the acid or base has volatilized and the pH indicator changed color, the user is informed that the effective life of the perfume has, or is about to cease.

In a preferred embodiment, the reactive component comprises an alkali, such as sodium hydroxide, which reacts chemically with carbon dioxide which is absorbed from the air. In this case, no odor problem arises from the reactive component.

The following examples further illustrate the process. All parts are expressed by weight.

Example 1: A balanced color change perfume discharge system having a blue to yellow color change was prepared from the following components.

Components	Percent
Deionized water, up to	100.00

(continued)

Components	Percent
Perfume*	10.00
Alcohol	20.00
Carbitol	20.00
Formaldehyde (40% aqueous)	0.20
Triton X100 (surfactant)	10.00
Bromothymol blue	0.10
Morpholine (amine)	1.00

*Fruity pine type, details given below.

The alcohols are used to solubilize the perfume in the aqueous components and modify its evaporation rate, and the morpholine is the reactive component which volatilizes into the atmosphere changing the pH of the system.

Fruity Pine Perfume Components	Percent
Benzyl acetate	8.00
Bornyl acetate	69.00
Citronellol standard	4.00
Dimethyl benzyl carbinol acetate	0.50
Linalol	4.00
Methyl ionone	6.00
Phenyl ethyl acetate	3.50
Terpineol	4.00
Undecalactone 10%	1.00

Using this level of morpholine, the color took four days at an average temperature of about 27°C to change completely and the perfume had lost its effectiveness over the same period when applied to a porous ceramic disc.

Example 2: A green to yellow color change air freshener system was prepared from the following components. In this example sodium hydroxide is the reactive component and is changed by contact with CO_2 from the atmosphere to cause a drop in pH in the system.

Components	Percent
Perfume*	75.00
Carbitol, up to	100.00
Bromothymol blue	0.025
Oil soluble dye (FD&C yellow No. 11)**	0.10
Sodium hydroxide (as 20% solution)	0.10

*Pine needle green sweet type, details given below.
**1971 Color Index 47000.

Pine Needle Green Sweet Perfume Components	Percent
Aldehyde C_9 10%	0.50
Aldehyde C_{10} 10%	1.00
Methyl nonyl aldehyde 10%	3.00
Anisic aldehyde	5.00
Bergamot synthetic	8.00
Bornyl acetate	50.00
Citronellol	2.00
Coumarin	5.00

(continued)

Pine Needle Green Sweet Perfume Components	Percent
Geranyl acetate	3.00
Lavandin	2.00
Lixetone	2.50
Musk ambrette	2.00
p-tert-Butyl cyclohexyl acetate	3.00
Terpinoline	5.00
Versalide	2.00
Galbanum	0.50
Orange oil sweet	0.80
Rosemary	2.00
Linalyl acetate	1.70
Elemi gum	1.00

This air freshener system applied to a compressed cellulose board 2 mm thick had an effective life of approximately 5 weeks at an average temperature of 22°C.

PERFUMED DETERGENTS

With Water-Insoluble Microcapsules

The process devised by *D.K. Brain and M.T. Cummins; U.S. Patent 4,145,184; March 20, 1979; assigned to The Procter & Gamble Company* is one for preparing a detergent which contains a perfuming agent which remains intact on a fabric throughout the laundering operation. Manipulation of the fabric laundering then causes release of perfume from the perfuming agent. The laundry detergent composition comprises:

(a) From 2 to 95% of a surfactant selected from the group consisting of anionic, nonionic, ampholytic and zwitterionic surfactants; and

(b) An effective amount of a perfuming agent comprising a perfume encapsulated in a water-insoluble, friable microcapsule.

When fabrics are washed in the above detergent composition, at least a portion of the microcapsules become entrained in the fabric. Upon drying and manipulating the fabric at least a portion of the microcapsules are ruptured to release the perfume.

The microcapsules utilized comprise a core of perfume material, usually liquid, and a thin polymeric shell surrounding the core. The microcapsules can vary in size from 5 to about 300 μ and generally have a shell thickness of between about 0.1 to 50 μ.

Perfume microcapsules were prepared using the process of U.S. Patent 3,516,941. The perfumes used were of the type which is conventional in detergent compositions and the capsules (0.3 wt %) were then mixed into an unperfumed granular laundry detergent composition containing 21% of anionic surfactant (linear C_{12} alkylbenzene sulfonate), 25% of sodium tripolyphosphate, 12% of sodium silicate (SiO_2/Na_2O ratio 2.0) and 16% of sodium sulfate.

Microcapsules of varying sizes were employed and it was found that capsule sizes varying from 14 to 200 μ produced the best results.

Capsules having an average particle size of 32 μ and a perfume loading of 75% were employed in the following example which is illustrative of the process.

Example: Spray-dried laundry detergent compositions having the following formula are useful in this process. In each case, the perfume microcapsules are admixed after the basic detergent granule is spray-dried.

Ingredients	. .Composition (wt %). . A	B
Sodium alkylbenzene sulfonate	7	12
Tallow alcohol sulfate	6	—
C_{14} alcohol ethoxylate sulfate	6	8
Sodium tripolyphosphate	24	25
Sodium silicate solids	5	12
Sodium carbonate	6	—
Sodium sulfate	17	33
Zeolite	18	—
Perfume microcapsules	0.6	0.6
Moisture and miscellaneous, up to	100	100

With Cyanomethyl-Substituted Tetrahydropyrans

A. Prevedello, M. Brunelli and E. Platone; U.S. Patent 4,150,037; April 17, 1979; assigned to Anic, SpA, Italy have achieved a method for cyclization of gamma-delta or delta-epsilon unsaturated alcohols which gives certain tetrahydrofuran and tetrahydropyran derivatives which are useful for perfuming detergents by reacting the alcohol with a cationic ion-exchange resin at temperatures between 20° and 150°C.

More particularly, this process relates to a method for cyclization of 3,7-dimethyl-3-hydroxy-6-octenenitrile and to the products thus obtained, namely 2,6,6-trimethyl-2-cyanomethyl-tetrahydropyran and 2-methyl-2-cyanomethyl-5-isopropyl tetrahydrofuran.

The tetrahydropyran and tetrahydrofuran derivatives are mainly used as perfuming agents in the formulation of detergents, on account of their considerable stability in alkaline surroundings.

It has been found that it is possible to direct the cyclization of the gamma-delta or delta-epsilon unsaturated alcohols towards the formation of either the tetrahydropyran or the tetrahydrofuran derivative or a mixture of them, as a function of the amount of resin used, or, more exactly, of the ratio of the cationic-type ion-exchange resin to the unsaturated alcohol and/or of temperature.

The compound 2,6,6-trimethyl-2-cyanomethyl-tetrahydropyran, has the following structural formula:

(1)

CH_2
H_2C CH_2
H_3C CH_3
C C
H_3C O CH_2CN

This compound, in addition to possessing very pleasing odoriferous properties as well as a good stability to alkalies, has a nitrile group which can be converted, according to known methods, into an ester, an amide, an amine, an aldehyde and other compounds.

Example 1: 40 g of 3,7-dimethyl-3-hydroxy-6-octenenitrile are charged into a 1,000 ml flask equipped with a mechanical stirrer and a bubble condenser. There are subsequently added 400 ml toluene and 8 g of Amberlite 15(H) (a coarse-lattice cationic resin formed by a styrene-divinylbenzene copolymer which contains 4.6 meq of sulfonic groups/g of resin). The mixture is kept boiling and vigorously stirred during 3 hours. Upon cooling the resin is filtered off, washed with toluene and the solvent is then evaporated off.

On the raw reaction product, the conversion rating, the yield and the selectivity are as follows: conversion, 98%; selectivity, 92%; yield, 90%. Both gas chromatography and mass spectrometry confirm that the tetrahydropyran derivative shown in formula (1) is produced.

Example 2: 70 g of 3,7-dimethyl-3-hydroxy-6-octenenitrile (98% purity) are charged in a 2 liter flask equipped with a mechanical stirrer, a bubble condenser, a thermometer and screwed-in plug having a silicone diaphragm to effect occasional samplings. Subsequently, 700 ml of toluene and 70 g of Amberlite 15(H) are added. The mixture is refluxed while maintaining it stirred during 14 hours. Then the mixture is allowed to cool, filtered and the resin is washed with toluene and the solvent is distilled off.

There are obtained 75 g of a tetrahydrofuran product (it still contains some toluene) on which conversion, yield and selectivity are determined by means of gas chromatography. Conversion was 99% and selectivity and yield were each 70%.

By stopping the reaction within a shorter time, mixtures of various compositions of tetrahydrofuran and tetrahydropyran derivatives are obtained.

CONTROLLED RELEASE PERFUMES

Multiple Emulsions

J. Versteeg; U.S. Patent 4,083,798; April 11, 1978; assigned to Exxon Research & Engineering Co., discusses certain multiple ernulsions, methods for their preparation and their use for the purposes of slow or controlled release of active ingredients such as fragrances. These multiple emulsions are prepared by dispersing an aqueous phase into an oil to form a first emulsion. The first emulsion is then dispersed as droplets in a second aqueous phase in the presence of a water soluble protein and a gelling polysaccharide. These multiple emulsions are characterized as stable to layering and coalescence of the droplets of the first emulsion while retaining pourability.

In the internal phase of the droplets of the first emulsion, active ingredients such as fragrances may be retained and can be released either over a long period of time by permeation due to the solubility of the active ingredients in the external phase of the first emulsion or can be retained indefinitely until the first emulsion is broken to release instantaneously the active ingredients.

In producing the multiple emulsion, a first emulsion is prepared which is a water-in-oil emulsion, that is, the internal phase of the emulsion is an aqueous solution while the external phase of the emulsion is immiscible with water. The internal phase of this emulsion contains an active fragrance ingredient. Normally, the active ingredient is either insoluble in the oil phase of the emulsion or soluble to the extent of less than 0.5 weight percent at 25°C. The first emulsion is formed by methods known in the art, for example, an aqueous solution is mixed with an aqueous immiscible solution, i.e., an oil, under conditions of agitation and preferably in the presence of surfactants.

The ratio of volume of the aqueous phase to the volume of the oil phase may vary from 1/10 to 1/1, preferably from 1/3 to 1/2. Agitation is provided at sufficient shear rates to disperse the aqueous phase into droplets and form stable emulsions. Suitable surfactants are disclosed in U.S. Patent 3,779,907.

The first emulsion is then dispersed as droplets in a second aqueous solution. This second aqueous solution, which may be designated as the continuous phase of the compositions, contains a water solubilized protein and a gelling polysaccharide, such as agar, pectin, starch, which can be dissolved in hot water and form gelled solutions when cooled.

The system's storability of fragrance compositions provides the opportunity for making products which slowly release their active ingredient and can also be shipped, distributed and dispersed from containers by pouring or coarse spraying. They are therefore useful fragrances applied to the skin or other surfaces, air freshener and other products where slow release is desirable.

Lactone-Modified Hydrophilic Polyurethane Resins

The process developed by *F.E. Gould; U.S. Patent 4,156,067; May 22, 1979; assigned to Tyndale Plains - Hunter Ltd.,* is one for preparing polyurethane polymers characterized by a molecular weight above 6,000 and having lactone groups and hydroxyl groups in the polymer backbone. They are prepared by reacting a mixture of polyols, a polyfunctional lactone and a polyfunctional isocyanate proportioned so as to provide the desired polymer properties. The product is soluble in alkaline solutions and may be used for many purposes.

The water absorptivity of the polyurethane lactone polymers is above 10%, preferably in the range of about 20% to 60%, and these polymers may range in their physical properties from rigid solids to completely gel-like high water absorptive polymers. The polymers can provide a leachable substrate wherein the leaching agent may be water, gases, alcohols, esters and body fluids, e.g., animal or human.

Example 1: A diethylene glycol solution of polyethylene glycol is prepared by heating 109.2 parts (0.075 mol) of polyethylene glycol having a molecular weight of 1,450 in 17.4 parts (0.164 mol) of diethylene glycol with stirring. The solution is cooled to below 60°C and to it is added a solution of delta gluconolactone prepared by dissolving 11.6 parts (0.065 mol) of delta gluconolactone in 46.4 parts of dimethyl sulfoxide. Eighty and eight tenths parts (0.316 mol) of methylenebiscyclohexyl-4,4'-isocyanate (Hylene W) is added to the mixture with stirring. One half part by weight of an organic tin catalyst solution, dibutyltin dilaurate (T_{12}, Metal and Thermite Co.) is added to the reaction mixture with stirring at a temperature below 45°C to avoid undue temperature rise caused by the heat of reaction. After stirring for 20 minutes the temperature increases to 80°C. The reaction mixture is then transferred to a tray and placed in an oven at 90°C for 1 hour to complete the reaction. This polymer, in the wet state, is soft, pliant and flexible.

Example 2: A polyurethane polyether resin is prepared by the method described in Example 1 above substituting for the polyethylene glycol a block copolymer having a molecular weight of 7,500 obtained by adding poly(oxyethylene) groups to a poly(oxypropylene) chain having a molecular weight of 2,250.

	Parts
Block copolymer (MW 7,500)	1,157.4
Diethylene glycol	32.75
Delta gluconolactone*	116
Hylene W	808
Dibutyltin dilaurate	5

*As a 20% solution in dimethyl sulfoxide.

Example 3: A hair set composition is prepared by dissolving the polyurethane polyether resin of Example 2 above in 95% ethanol. Sufficient 1 N sodium hydroxide is added to effect solution and the pH of the final composition adjusted to 7.0. The resulting solution may be applied to the hair in the form of a spray and provides superior holding with a soft feel.

Example 4: To the hair set composition of Example 3 is added 2 wt % of a perfume essence. When this composition is applied to the hair the solvent rapidly evaporates but the perfume is bound to the polyurethane polymer and is slowly released over a period of 24 hours.

MISCELLANEOUS

Roll-On Perfume

The preparation of roll-on perfume compositions has presented many problems to those skilled in the art. To be an effective roll-on perfume, the composition must be viscous, clear and stable. Moreover, the composition must not contain harmful or irritable skin ingredients. Also, since most perfume oils are water-insoluble, an alcohol or other solvent must be incorporated into the composition in order to obtain homogeneous compositions and also to assist in the volatility of the composition. However, the presence of the alcohol or solvent often results in excessive drying of the skin. Accordingly, there is a need in the art to minimize the presence of the alcohol or solvent in perfume compositions and it is to this end that the process is directed.

I.R. Schmolka; U.S. Patent 4,089,814; May 16, 1978; assigned to BASF Wyandotte Corporation has developed clear, viscous, stable, roll-on perfume compositions having a significantly reduced alcoholic content, prepared by employing certain polyoxyethylene polyoxypropylene block copolymers. These block copolymers serve four functions, namely: providing thickening to the system, solubilizing the water-insoluble essential oils, reducing the alcohol content of the compositions, and providing lubricity to the composition.

These roll-on perfume compositions comprise, based on 100 parts by weight of total composition: (a) from 5 parts to 15 parts of an essential oil; (b) from 25 parts to 40 parts of an alcohol; and (c) from 20 parts to 40 parts of polyoxyethylene-polyoxypropylene block copolymer represented by the formula:

$$HO(C_2H_4O)_b(C_3H_6O)_a(C_2H_4O)_bH$$

wherein a is an integer such that the hydrophobe represented by (C_3H_6O) has a molecular weight of from 3,250 to 4,000 and b is an integer such that the hydrophile portion represented by (C_2H_4O) constitutes from about 50 to 90, and most preferably from 70 to 90, weight percent of the copolymer and (d) from 5 parts to 50 parts of water.

Illustrative block copolymers of the formula above, which may be employed in the preparation of the compositions of the process are presented in the table.

Copolymer	Molecular Weight of Hydrophobe (average)	Weight Percent of Hydrophile (average)	Approximate Total Molecular Weight of Copolymer
A	3,250	50	6,550
B	3,250	60	8,125
C	3,250	80	16,250
D	4,000	60	10,000
E	4,000	70	13,500
F	4,000	80	20,000
G	4,000	90	40,000

The compositions are prepared by adding a block copolymer to a water-alcohol solution at a temperature of from about 5°C to 25°C and thereafter adding the essential oil thereto. If the temperature of the solution is below room temperature, the resulting solution is allowed to warm to room temperature whereby the viscous, clear solutions of the process are obtained.

The following example illustrates the process. All parts are by weight unless otherwise indicated.

Example: The following compositions were prepared by adding the copolymer to the water and alcohol maintained at a temperature of from 5° to 10°C. Stirring is continued until a homogeneous solution was obtained at which time the essential oil was added and the solution was allowed to warm to room temperature. A viscous, clear composition resulted.

Composition A

Ingredients	Parts
Copolymer E	28
Ethanol	28
Essential oil B	10
Water	34

(continued)

Composition B

Ingredients	Parts
Copolymer E	21
Ethanol	33
Essential oil C	10
Water	36

Composition C

Ingredients	Parts
Copolymer B	33
Isopropanol	26
Essential oil B	10
Water	31

Composition D

Ingredients	Parts
Copolymer F	25
tert-Butanol	35
Essential oil D	5
Water	35

All compositions were viscous, clear compositions which may be employed as roll-on perfume compositions. In the above compositions, the essential oils employed are as follows: essential oil B–floral composition containing jasmin, neroli, sandalwood, patchouli, vetivert, and rose oils; essential oil C–chypre oil containing oakmoss, rose, sandalwood, jasmin, ylang and vetivert oil; and essential oil D–rose oil.

Polymer-Petroleum Wax Fragrance-Emitting Article

J.W. Newland; U.S. Patent 4,110,261; August 29, 1978; assigned to W & F Mfg. Co. Inc., has developed an improved thermoplastic material which has desirable physical properties rendering it suitable for formation into fragrance-emitting decorative articles such as figurines for placement in a room to mask unpleasant odors or to supply a pleasant odor. For generic identification, the process composition is identified as a polymer-petroleum wax composition.

The petroleum wax ingredient of the composition is present in principal amount by weight. Petroleum waxes suitable for use as ingredients in the composition would include, for example, refined paraffin waxes, semirefined paraffin waxes, and microcrystalline waxes.

The polymer ingredient is present in an amount from 3 to 25% by weight of the composition. This polymer ingredient must be one, or a mixture of one or more, of the polymers selected from the group consisting of ethylene vinyl acetate, ethylene vinyl acetate acid terpolymer, ethylene ethyl acrylate, ethylene isobutyl acrylate, polyethylene and polypropylene.

It is essential that each of the aforementioned six operative members of the polymer group have a weight average molecular weight of below about 10,000. This low molecular weight characteristic minimizes blending problems of the polymer with other ingredients, and also keeps the viscosity of the composition low during formation of the article as by the simple and inexpensive method of flush molding, which is the preferred technique.

The third essential part of the composition is a perfume fragrance ingredient, which is present in an amount to make up substantially the balance of 100% by weight of the composition. The perfume fragrances employed in the practice of the process are those known to persons skilled in this art, and are available commercially.

Example:

	Percent
Refined paraffin wax (MP 143° to 145°F)	74.959
Elvax 250 (ethylene vinyl acetate)	10.0
Perfume fragrance	15.0
BHT (butylated hydroxytoluene)	0.001
UV531 (ultraviolet stabilizer)	0.010
UV5411 (ultraviolet stabilizer)	0.010
TiO_2 (titanium dioxide)	0.020

To blend this formulation, the Elvax 250, which has a weight average molecular weight below 10,000, is added to the paraffin wax along with the BHT and heated to a temperature of 240° to 250°F and put into solution using a Cowles dissolver. As soon as the mixture is homogenous the temperature is decreased to 180°F and the perfume fragrance, which was previously mixed with the TiO_2, UV521 and UV5411, is added to the homogenized polymer-paraffin wax blend and the temperature is maintained at approximately 170°F.

The BHT acts as an antioxidant. The UV531 and UV5411 are ultraviolet light stabilizers and do not function in any other capacity. The TiO_2 is a whitener and filler and functions only in that capacity. Other colorants may be substituted as desired. While the composition has been described as prepared from two separately prepared mixtures, this is merely preferred and not critical. The TiO_2, UV531 and UV5411 may be combined with Elvax 250 and paraffin wax prior to blending, as desired.

The resulting composition is then utilized to mold a figurine according to a preferred flush molding technique described below. According to this technique, a mold of the described configuration having a filling opening is prepared by immersion in a solution of water and surfactant, an emulsion of silicone oil in water such as Dow Corning 36, at 105°F and then blown free of water droplets.

The polymer-paraffin wax composition of the above example at 170° to 180°F is then poured through the mold opening into the mold cavity which is filled. The fill begins to harden from its outside surface, in contact with the wall of the mold cavity, progressively inwardly toward the center of the cavity, to form a shell surrounding a still fluid body. The fill is left to dwell for a time period depending upon the weight of the shell or shell thickness desired to be formed within the mold. The mold is then inverted and the unsolidified material within the solidified shell is poured out. The mold is then returned to an upright position with the mold opening at the top, and the excess of solidified fill at such opening is cut off.

Preferably the hollow shell so molded is then filled with core wax that has been aerated and is at a temperature approximately 5°F below its melting point.

Fabric Conditioning Product Containing Deodorant Perfume

The products described by *D.C. Hooper, G.A. Johnson and D. Peter; U.S. Patent 4,134,838; January 16, 1979; assigned to Lever Brothers Company* are fabric conditioning compositions which are capable of softening and deodorizing fabric. More particularly, the compositions are those which additionally impart to the fabric of a garment the property of suppressing body malodor when the garment is subsequently worn adjacent the skin.

It has been discovered that certain combinations of perfume materials, hereinafter referred to as deodorant perfumes, when incorporated in the fabric of a garment together with a fabric softening material are capable of imparting to the fabric of that garment the property of reducing body malodor when the garment is subsequently worn in contact with the skin.

The effectiveness of deodorant perfumes when employed in this way can be measured by a test based on that devised by Whitehouse & Carter as published in *The Proceedings of the Scientific Section of the Toilet Goods Association,* No. 48, (1967), pp 31-37, under the title "Evaluation of Deodorant Toilet Bars."

For a treated shirt to be effective in reducing body malodor, it is necessary for the odor reduction value of the shirt as measured by this test to be at least 0.50.

Preferably, the odor reduction value is at least 0.70, more preferably at least 1.00 and most preferably at least 1.20. It is clear that the deodorant effect demonstrated in this way is not solely that of odor masking, since in many instances there is no detectable smell of perfume on the treated fabric after a few hours.

The fabric conditioning composition comprises compounds generally classified as fabric softeners that are employed during the washing, rinsing or drying cycle of the home laundering operation. Such fabric softeners are inorganic clays or water-soluble or water-dispersible organic, waxy materials having a preferred melting (or softening) point between about 25°C and 150°C. Softener materials of this type are also fabric substantive in the sense that they are readily deposited onto the surfaces of fabrics treated therewith. Many fabric softeners of this type also impart some degree of static control to the fabrics being treated therewith.

The essential materials required for making deodorant perfumes that are operative in these formulations are those that depress the partial vapor pressure of morpholine by at least 10% more than that required by Raoult's Law, as determined by a test (described in the patent) which is designated "The Morpholine Test."

Deodorant perfumes can be incorporated in fabric conditioner compositions according to the process, at a concentration of from about 0.01 to 10%, preferably from 0.05 to 3% and most preferably from 0.1 to 1% by weight.

Example 1: This example illustrates the treatment of shirts with a liquid fabric conditioning composition which is added during the rinse cycle of a laundry process.

Forty polyester cotton shirts were washed, line dried and then rewashed according to the standard procedure for liquid fabric conditioning treatment during the rinse cycle of a laundry process. Half of the shirts were treated as test shirts and the remainder as control shirts.

The formulation of the fabric conditioning composition employed was as follows:

	Weight Percent
Arquad 2HT	3
Ammonyx 4080*	3
Deodorant perfume (Formulation B6)	0.2
Formaldehyde	0.0075
Sodium chloride	0.02
Coloring matter	0.0025
Deionized water, up to	100

*Dialkyl (C_{16-18}) imidazoline methosulfate.

The formulation of the deodorant perfume was as follows:

Deodorant Perfume Formulation B6	Parts
Benzyl propionate	4.0
Bergamot oil	15.0
o-tert-Butylcyclohexyl acetate	2.0
p-tert-α-Methylhydrocinnamic aldehyde	15.0
Clove leaf oil	10.0
Diethyl phthalate	9.25
Dimethylbenzylcarbinyl acetate	5.0
Ionyl acetate	10.0
Isobutyl benzoate	5.0
LRG-201	1.25
3a-Methyl-dodecahydro-6,6,9a-trimethyl-naphtho-2-(2,1-b)furan	0.5
Neroli oil	3.0
Petitgrain oil	10.0
Phenylethyl alcohol	10.0
Total	100

The test and control shirts were assessed according to the standard test procedure as described hereinbefore.

The results of the modified Whitehouse and Carter test were as follows. The average scores were 2.42 for the control composition and 1.36 for the test composition.

The reduction in odor intensity of the test composition was the difference between these two scores, which was 1.06. This was well in excess of 0.50 which defines the lower limit of reduction of odor intensity (odor reduction value) of compositions of the process.

Example 2: This example illustrates the treatment of shirts with a substrate impregnated fabric conditioning composition which is added to a damp laundry load immediately prior to tumble drying.

Forty polyester cotton shirts were washed, line dried and rewashed and tumble dried according to the procedure described hereinbefore for substrate impregnated fabric conditioning composition treated during the tumble drying sequence of a laundry process. Half of the shirts were treated as test shirts and the remainder as control shirts.

The formulation of the fabric conditioning composition employed to impregnate the standard substrate tissue (of the type described hereinbefore with reference to the standard and test procedure) were as follows:

	Weight Percent
Di(hardened tallow) dimethyl ammonium chloride	75
Condensation product of a secondary linear (C_{11-15}) alcohol with 12 mols of ethylene oxide	25

The formulation of the deodorant perfume (B6) was that described in Example 2.

The test and control shirts were assessed according to the standard test procedure as described hereinbefore.

The results of the modified Whitehouse and Carter test were as follows. The average scores were 2.48 for the control composition and 1.74 for the test com-

The reduction in odor intensity of the test composition was the difference between these two scores which was 0.74. This was well in excess of 0.50 which defines the lower limit of reduction of odor intensity (odor reduction value) of compositions of the process.

MEAT FLAVORINGS

SULFUR-CONTAINING COMPOUNDS

Heterocyclics

G.A.M. van den Ouweland and H.G. Peer; U.S. Patents 4,080,367; March 21, 1978 and 4,134,901; January 16, 1979; both assigned to Lever Brothers Company describe food flavoring substances which can impart to food a savory flavor of roast or fried meat. These compounds have the following formulas, in which Y is either a sulfur or oxygen atom and R^1 and R^2 are hydrogen, or an alkyl or hydroxy alkyl group having 1 to 9 carbon atoms.

(1)

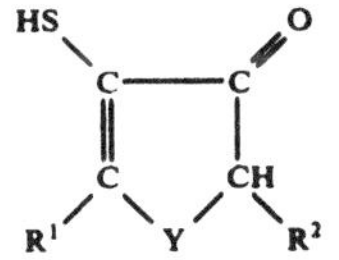

HS O
C—C
C CH
R^1 Y R^2

Examples of compounds of this class are: 4-mercapto-5-methyl-2,3-dihydrothiophene-3-one and 4-mercapto-5-ethyl-2,3-dihydrofuran-3-one.

(2)

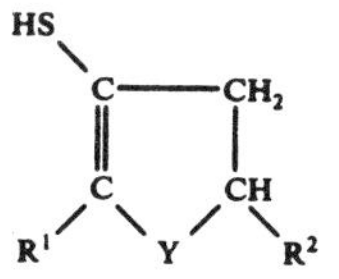

HS
C—CH_2
C CH
R^1 Y R^2

Examples of compounds of this class are: 3-mercapto-5-methyl-4,5-dihydrofuran and 3-mercapto-2,5-dimethyl-4,5-dihydrothiophene.

(3)

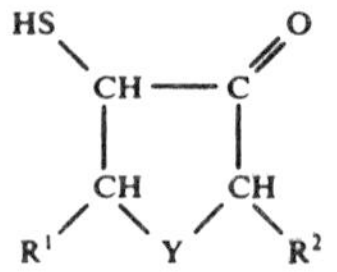

HS
CH — C=O
| |
CH CH
R^1 Y R^2

Examples of compounds of this class are: 4-mercapto-5-methyl-tetrahydrothiophene-3-one and 4-mercapto-2,5-dimethyl-tetrahydrofuran-3-one.

(4)

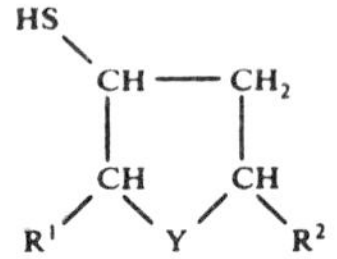

HS
CH — CH_2
| |
CH CH
R^1 Y R^2

Examples of compounds of this class are: 3-mercapto-2-methyl-tetrahydrothiophene (cis and trans) and 3-mercapto-5-ethyl-tetrahydrofuran (cis and trans).

(5)

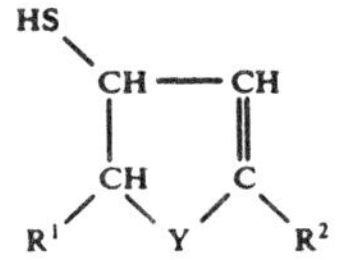

HS
CH — CH
| ||
CH C
R^1 Y R^2

Examples of compounds of this class are: 3-mercapto-2-methyl-2,3-dihydrothiophene and 3-mercapto-2,5-dimethyl-2,3-dihydrofuran.

The flavoring characteristics of compounds satisfying the above five general formulae and their tautomers were found to be particularly interesting in the case where R^1 and R^2 represent a hydrogen atom, a methyl group or a hydroxymethyl group.

Compounds like structure (1), except that the 4-position has a hydroxyl rather than a mercapto group, which may be used as starting furanone compounds to be reacted with hydrogen sulfide are, for example, 4-hydroxy-5-methyl-2,3-dihydrofuran-3-one, 4-hydroxy-2,5-dimethyl-2,3-dihydrofuran-3-one and 4-hydroxy-2-methyl-5-ethyl-2,3-dihydrofuran-3-one. Preferred examples of the pyrones which may be reacted with hydrogen sulfide are: 3-hydroxy-2-methyl-4-pyrone (maltol) and 3-hydroxy-2-ethyl-4-pyrone.

As an illustration of suitable quantities of the flavoring substances that may be added to specified types of foodstuff, as little as 1 to 10 ppm w/w is sufficient to impart a pleasant roast meat flavor to soups which are bland or otherwise lightly flavored. On the other hand, when incorporating a similar roast meat flavor to already flavored foodstuffs such as those based on vegetable protein, it may be necessary to incorporate larger amounts, for example, from 600 to 8,000 ppm w/w of the flavoring substance in order to obtain a desirable flavor.

Example 1: *Preparation of 4-Hydroxy-5-Methyl-2,3-Dihydrothiophene-3-one* – 140 g of commercially available 1-butyn-3-ol (BP 107°C at atmospheric pressure) were treated in an aqueous solution with 200 g of a 30% formaldehyde solution in the presence of 10 g CuCl and refluxed for 50 hr. The resulting 156 g (70%) of 2-pentyn-1,4-diol (BP 115°C at 2.5 mm Hg) were isolated by evaporating off the water and distilling the residue.

50 g (0.5 mol) of 2-pentyn-1,4-diol were dissolved in 250 ml of dry pyridine. The solution was stirred and cooled to -10°C in an ice-salt mixture. With stirring, a cold solution of 286 g (1.5 mols) of p-toluene sulfonyl chloride in 550 ml of dry dichloromethane was added dropwise, under exclusion of atmospheric moisture, from the dropping funnel, in such a manner that the temperature did not exceed -5°C. After completion of the addition (about 1.5 hr) stirring at 0°C was continued for 5 hr, and water (30 ml) was added in portions at intervals of 5 min, with stirring and cooling, so that the temperature did not rise above 5°C.

The solution was then poured into 1,000 ml of cold water. The mixture was extracted three times with dichloromethane; the combined extracts were successively washed with portions of ice-cold dilute sulfuric acid, water, sodium hydrogen carbonate solution and water. The dichloromethane solution was then dried with anhydrous sodium sulfate and evaporated to dryness, affording a syrup which crystallizes on standing. It was recrystallized from ethanol; yield 125 g = 61%; MP 80° to 80.5°C.

A solution of the ditosyl compound (98 g = 0.24 mol) and potassium thioacetate (60 g = 0.527 mol) in dimethylformamide (1.5 liters) was stirred for 45 min at 40°C under nitrogen, then concentrated under reduced pressure and diluted with water (1 liter). The mixture was extracted five times with dichloromethane, the combined extracts were washed with water, dried with anhydrous sodium sulfate and evaporated to dryness. The residue was distilled through a short path column, affording 43.8 g = 84.5% of the dithioacetate; BP 129° to 130°C at 1.6 mm Hg; n_D^{20} = 1.5440. 20 g of the 1,4-dithioacetoxy-2-pentyn were then dissolved in 1,000 ml of an alcohol-water mixture (90:10 by volume) and the solution was cooled to -25°C. A solution of 32 g potassium permanganate and 48 g magnesium sulfate heptahydrate in 700 ml of water was slowly added in 2 hr while maintaining the temperature at -20° to -25°C.

The reaction mixture was stirred for another 2 hr at the same temperature, and 600 g of ice were then added. The reaction mixture was then extracted with cold chloroform. The light yellow colored organic solution yielded after drying and evaporation of the solvent 13.5 g = 59% of a yellow oil (pentane-2,3-dione-1,4-dithioacetate).

10 g of the yellow oil thus obtained (pentane-2,3-dione-1,4-dithioacetate) were dissolved in 1,500 ml of 0.5 N aqueous hydrochloric acid and stirred for 1.5 hr at 95°C. After cooling the reaction mixture was extracted five times with chloroform, the combined extracts were washed with water, dried with anhydrous sodium sulfate and evaporated to dryness, affording a syrup which crystallized on standing. After recrystallization, from dichloromethane, white crystals of 4-hydroxy-5-methyl-2,3-dihydrothiophene-3-one were obtained; MP 152° to 153°C; yield = 40%.

Example 2: A beef-flavored composition was prepared by adding 250 ml of water to a mixture of 5.7 g of 4-hydroxy-5-methyl-2,3-dihydrofuran-3-one and 25.0 g of cysteine and heating the mixture at about 100°C for 2½ hr. The resulting mixture was cooled and quantities of between 0.2 and 2.0 ml of the reaction mixture were sprayed over 100 g portions of dehydrated textured vegetable protein containing no meat. An excellent roast meat flavor was thereby imparted to this material as assessed by eleven out of a total panel of twelve expert tasters.

Dextrin-maltose was added to a portion of the flavored mixture which resulted from the reaction described above in an amount which provided a composition containing about 70 pbw of dextrin-maltose to each part of the substance calculated on a solid basis. The composition was freeze-dried and a beef-flavored product was obtained.

Example 3: A mixture of 1.5 g of 4-hydroxy-5-methyl-2,3-dihydrofuran-3-one and 1.5 g of cysteine in 30 ml of water was heated at about 100°C for 2½ hr. To the resulting solution was added 33 g of maltodextrin. The solution thus obtained was carefully freeze-dried. The powder obtained was used as a good beef flavor in soup or gravy.

Example 4: 6.3 g of 3-hydroxy-2-methyl-4-pyrone, 10.5 g of sodium sulfide ($Na_2S{\cdot}9H_2O$) and 100 ml of water were heated together in a round-bottomed flask at 100°C for 2½ hr.

To the reaction mixture was added 117 g of maltodextrin. The resulting solution was spray-dried immediately. The powder thus obtained proved to have a good beef flavor.

Example 5: A mixture of 5 g 4-hydroxy-5-methyl-2,3-dihydrothiophene-3-one (cf Example 1), 0.5 g of hydrogen sulfide and 50 ml of water were heated in an autoclave for 4 hr at 100°C and was subsequently allowed to cool. A product with a roasted meat flavor was obtained which was diluted to a volume of 1 liter, forming a liquid meat flavor.

Example 6: A basic composition for a dry beef soup was obtained by mixing the following ingredients:

Ingredient	Grams
Onion powder	0.5
Spice mix	0.5
Fat	4
Dried soup vegetables	1
Monosodium glutamate	2
Modified potato starch	3
Noodles	20
Salt	8

One liter of water was added to the mixture and the whole was boiled for 5 min. The soup so obtained was divided in two portions of 500 ml. In the first portion 150 mg of maltodextrin was dissolved and in the second portion 150 mg of the flavor powder prepared according to Example 3.

Both soups were assessed in a paired comparison test by a panel consisting of 8 persons. The soup containing the flavor powder had a characteristic beef flavor and was preferred by 7 out of the 8 testers.

Furyl Alkyl Disulfides

Reproduction of sweet meat, roasted meat, liver flavors and aromas and hydrolyzed vegetable-protein-like flavors and aromas has been the subject of the long and continuing search by those engaged in the production of foodstuffs, e.g., luncheon meats such as liverwurst sausages. The severe shortage of foods, especially protein foods, in many parts of the world has given rise to the need for utilizing nonmeat sources of proteins and making such proteins as palatable and as meat-like as possible. Hence, materials which will closely simulate or exactly reproduce the flavor and aroma of roasted meat products (e.g., roast-beef-like) and liver products are required.

Moreover, there are a great many meat-containing or meat-based foods presently distributed in a preserved form. Examples include condensed soups, dry-soup mixes, dry meat, freeze-dried or lyophilized meats, packaged gravies and the like. While these products contain meat or meat extracts, the fragrance, taste and other organoleptic factors are very often impaired by the processing operation and it is desirable to supplement or enhance the flavors of these preserved foods with versatile materials which have sweet meat, roasted meat and/or liver taste and aroma nuances.

W.J. Evers, H.H. Heinsohn, Jr. and M.H. Vock; U.S. Patent 4,098,910; July 4, 1978; assigned to International Flavors & Fragrances Inc. have provided certain 3-furyl alkyl disulfides for altering the organoleptic properties of foodstuffs. These compounds have the formula:

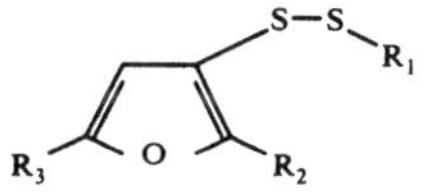

wherein R_1 is C_{1-7} lower alkyl or C_{5-6} cycloalkyl; R_2 and R_3 are each selected from the group consisting of hydrogen and methyl, at least one of R_2 or R_3 being methyl provided that when R_1 is methyl, each of R_2 and R_3 is methyl. The furyl alkyl disulfides may be produced by several processes. One method, particularly suitable when the alkyl substituents contain three or more carbon atoms, involves reaction of a 2-alkyl-3-furanthiol with an approximately equimolar amount of an alkyl sulfenyl chloride. Preferably the alkyl sulfenyl chloride is previously formed in situ by reaction of sulfuryl chloride and corresponding dialkyl disulfide. This reaction sequence is illustrated by the following:

$$R_1SSR_1 + SO_2Cl_2 \longrightarrow 2R_1SCl$$

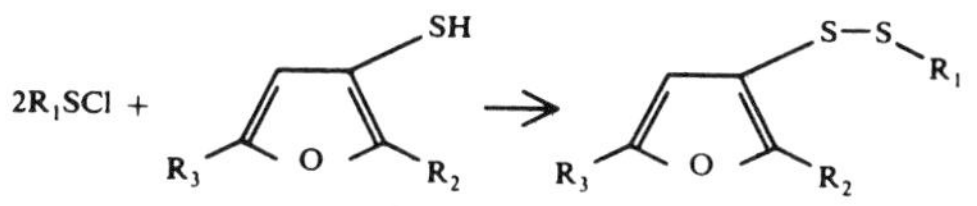

Thus, a thiol such as 2-methyl-3-furanthiol can be reacted with an equimolar amount of n-pentanesulfenyl chloride, $C_5H_{11}SCl$, to produce n-pentyl (2-methyl-3-furyl) disulfide. This reaction also can be carried out in a solvent such as diethyl ether, cyclohexane, hexane, carbon tetrachloride, benzene and the like. The reaction temperature is preferably from -60° up to 0°C at atmospheric pressure. The following compounds are examples produced using the above process which have useful organoleptic properties giving rise to their use as foodstuff flavors as set forth in an illustrative manner in the following table. These compounds are usually combined with other flavoring adjuvants as well known to those skilled in the art.

3-Furyl Alkyl Disulfide Compounds	Structure	Flavor Properties
ethyl(2-methyl-3-furyl) disulfide		Sweet, roasted, liver-like aroma and sweet, pot roast beef flavor with nutty, liver, bloody and metallic nuances.
isoamyl(2-methyl-3-furyl) disulfide		Meaty-liver, chicken fat, roasted, aroma with metallic, rubbery, sweet and nutty nuances and a meaty, rubbery, roasted flavor with sweet and "baked goods" notes.
cyclohexyl(2-methyl-3-furyl) disulfide		Meaty, liver and sweet aroma with a rubbery nuance and a sweet, liver, meaty flavor with rubbery and nutty nuances.
isoamyl(2,5-dimethyl-3-furyl) disulfide		Sulfury, rubber, liver aroma and sweet, rubbery, sulfury flavor with skunky, liver and bitter nuances.

Example 1: The following ingredients are refluxed for 4 hr.

Ingredient	Parts by Weight
L-cysteine hydrochloride	0.9
Carbohydrate-free vegetable protein hydrolysate	30.9
Thiamine hydrochloride	0.9
Water	67.30

The resulting mixture is then aged for three days and an aliquot portion is withdrawn and dried. Based on the weight of the dry solid obtained, sufficient gum arabic is added to the batch to provide a composition containing 1 pbw of gum arabic. The composition is then spray-dried.

Ethyl(2-methyl-3-furyl) disulfide is added to the spray-dried material at the rate of 4 ppm. The resulting material has an excellent beef-liver flavor.

Example 2: A beef-liver gravy is made by formulating a composition in the amounts indicated.

Ingredient	Parts by Weight
Cornstarch	10.50
The final product of Ex. 1	3.00

(continued)

Ingredient	Parts by Weight
Caramel color	0.30
Garlic powder	0.05
White pepper	0.05
Salt	1.90
Monosodium glutamate	0.20

To one unit of gravy flavor concentrate, eight ounces of water is added and the mixture is stirred thoroughly to disperse the ingredients, brought to a boil, simmered for 1 min, and served. This meatless gravy exhibits an excellent beef-liver flavor.

Example 3: 10 ppm cyclohexyl(2-methyl-3-furyl) disulfide is added to beef broth prepared from a commercial dried mixture and 250 ml hot water. The cyclohexyl(2-methyl-3-furyl) disulfide increases the sweet meaty and liver-like character and imparts a pleasant nutty note. The resultant beef broth has an improved more blended meaty flavor than does the unflavored beef broth.

Thioethers

Thioether compounds useful as flavoring agents to impart meat-like flavors to foods are described by *S. van den Bosch, D.K. Kettenes, K.B. de Roos, G. Sipma and J. Stoffelsma; U.S. Patent 4,119,737; October 10, 1978; assigned to P.F.W. Beheer BV, Netherlands.* The thioethers are generically represented by the following general structural formula:

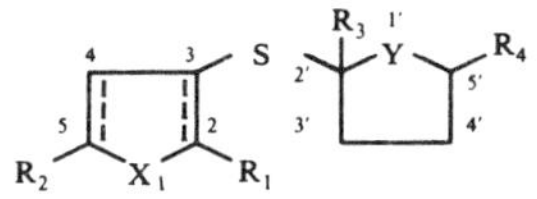

wherein the X-containing ring is a member selected from the class consisting of: the cyclopentane ring, the $\Delta^{2,3}$-cyclopentene ring, the tetrahydrofuran ring, the tetrahydrothiophene ring, the 4,5-dihydrofuran ring, the 4,5-dihydrothiophene ring, the furan ring, and the thiophene ring. Y is selected from the class consisting of oxygen and sulfur, and R_1, R_2, R_3 and R_4 are selected from the class consisting of hydrogen and 1 to 3 carbon alkyl groups. The compounds of the process are subclassified hereinafter by the structural formulae:

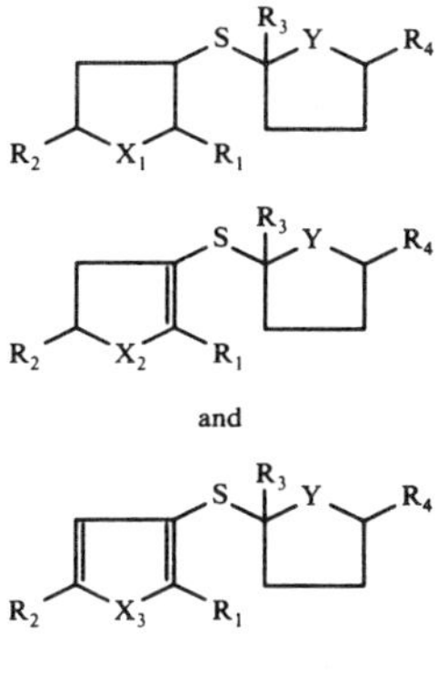

and

wherein X_1 and X_2 represent a $-CH_2-$ group, oxygen or sulfur, X_3 represents oxygen or sulfur, and Y, R_1, R_2, R_3 and R_4 are as specified above.

The compounds can be prepared by known methods, e.g., by addition of a mercaptan to a cyclic vinyl ether or vinyl thioether. As an example of such a method, the compounds, exemplified by a compound of Formula 1 may be prepared according to the following reaction scheme:

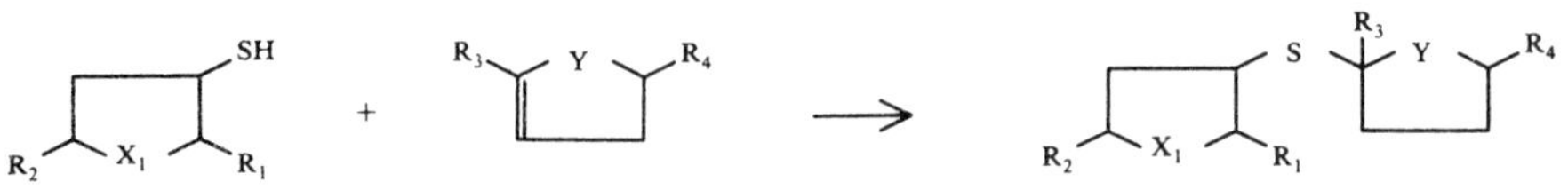

(1)

wherein X_1, Y, R_1, R_2, R_3 and R_4 possess the aforedescribed meaning.

The addition reaction can be conducted in the presence of a catalytic amount of acid and with or without a solvent. A variety of solvents can be used, hydrocarbons such as pentane, ethers such as diethyl ether, tetrahydrofuran, etc. As acids can be used p-toluenesulfonic acid, thionyl chloride, gaseous hydrogen chloride, etc. The addition can be effected at room temperature or at slightly elevated temperatures, e.g., at the reflux temperature of the solvent.

A few representative compounds include: (1a) 2-methyl-3-(2'-methyl-2'-tetrahydrofurylthio)tetrahydrofuran; (1b) 2-methyl-3-(2',5'-dimethyl-2'-tetrahydrofurylthio)tetrahydrofuran; (1c) 2-methyl-3-(2'-methyl-2'-tetrahydrofurylthio)-tetrahydrothiophene; and (1d) 2-methyl-3-(2',5'-dimethyl-2'-tetrahydrofurylthio) tetrahydrothiophene.

Example 1: An instant beef gravy powder was prepared from the following ingredients:

Ingredient	Weight Percent
Spray-dried vegetable oil base powder	20.0
Spice extract powder	0.2
Spray-dried tomato powder	7.0
Sodium chloride	12.0
Monosodium glutamate	8.0
Hydrolyzed vegetable protein	6.0
Caramel powder	1.5
Granulated sugar	3.0
Modified potato starch	25.0
Autolyzed yeast powder	5.0
Carboxymethylcellulose	0.8
Nonfat dry milk solids	11.5
Total dry matter	100.0

60 g of this mixture were dissolved in 940 g of boiling water, thus giving 1 kg of gravy. The gravy so obtained was divided into two portions of 500 g each. To one of the portions 0.25 g of a 0.3% solution of Compound (1b) in ethanol was added and the mixture was well stirred. The level of the compound in the gravy can be expressed as 1.5 ppm. Both gravies were compared by a panel of

qualified testers. The gravy containing the Compound (1b) was definitely preferred by the majority of the testers, since it was found to possess a well-recognizable beef note which was absent in the gravy without the compound.

Example 2: 1½ kg of gravy was prepared according to the method described in Example 1. The gravy was divided into three portions of 500 g each. To one of the portions 0.10 g of a 0.3% solution of Compound (1c) in ethanol was added. This was called Mixture A. To a second portion of the gravy 0.10 g of a 0.3% solution of Compound (1d) in ethanol was added. This mixture was called Mixture B. Both Mixtures A and B were well stirred. Mixtures A and B were panel tested against the control which was the gravy without the additions described above. The panel showed a clear preference for Mixtures A and B over the control. Mixture A was described as having a cooked meat note, whereas Mixture B was described as possessing a more general beef note.

Substituted Furyl Sulfides

S. van den Bosch; U.S. Patent 4,120,985; October 17, 1978; assigned to P.F.W. Beheer BV, Netherlands uses as flavoring agents sulfur-containing compounds quite similar to those in the previous patent and prepared in the same way, by addition of a mercaptan to a cyclic vinyl ether or thioether. The compounds are represented by the structural formulae (I) and (II):

(I)

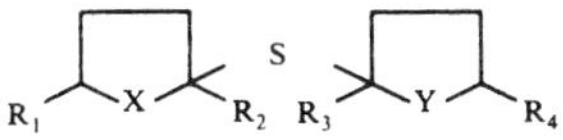

R_1 X R_2 S R_3 Y R_4

(II)

R_1 X S R_3 Y R_4

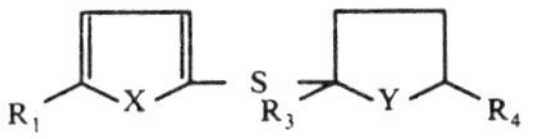

wherein X and Y are selected from the group consisting of sulfur and oxygen; R_1, R_2, R_3 and R_4 are selected from the group consisting of hydrogen and alkyl radicals having up to three carbon atoms. The particularly preferred materials for imparting desirable flavor and fragrance notes include the compounds in which R_1, R_2, R_3 and R_4 are hydrogen or methyl radicals.

Specific representatives of these compounds included within the above structural formulae (I) and (II) are: bis(tetrahydrofuryl-2)sulfide, bis(2-methyltetrahydrofuryl-2)sulfide, bis(2,5-dimethyltetrahydrofuryl-2)sulfide, bis(2-methyltetrahydrothienyl-2)sulfide, 2-methyltetrahydrofuryl-2-(2'methyltetrahydrothienyl-2')sulfide, thienyl-2-(2'-methyltetrahydrofuryl-2')sulfide, thienyl-2-(2'methyltetrahydrothienyl-2')sulfide, 5-methylfuryl-2-(2'-methyltetrahydrofuryl-2')sulfide, and 5-methylfuryl-2-(2',5'-dimethyltetrahydrofuryl-2')sulfide.

Example 1: A gravy was prepared by mixing the following ingredients.

Ingredient	Grams
Whey powder	12.5
Fat flakes (edible)	20
Sodium chloride	17.5
Monosodium glutamate	5
Hydrolyzed vegetable protein	7.5
Cornstarch	30
Caramel color	5.5
Locust bean gum	2

40 g of this mixture were dissolved in 960 g of boiling water to obtain 1 kg of the gravy. The gravy was well stirred and simmered for 5 min. The gravy was divided into two portions.

To one portion of the gravy thienyl-2-(2'-methyltetrahydrothienyl-2')sulfide was added at a level of 0.03 ppm. The obtained gravy was tested by the panel against a control, which was the gravy without the compound. The gravy containing the compound was unanimously preferred by the panel because of its pronounced meaty, soupy taste.

Example 2: 1 kg of gravy was prepared according to the method described in Example 1. The gravy was divided into two portions. To one portion of the gravy bis(tetrahydrofuryl-2)sulfide was added at a level of 2 ppm. The obtained gravy was compared by the panel with a control, which was the gravy without the compound of the process. The gravy containing the compound was unanimously preferred by the panel, because of its meaty body, onion-like taste.

Example 3: 1 kg of gravy was prepared according to the method described in Example 1. The gravy was divided into two portions. To one portion of the gravy 5-methylfuryl-2-(2'-methyltetrahydrofuryl-2')sulfide was added at a level of 1 ppm. The obtained gravy was compared by the panel with a control, which was the gravy without the compound of the process. The gravy containing the compound was unanimously preferred by the panel, because of its pronounced roast meaty taste.

Alpha-Substituted Alkylidene Methionals

D.A. Withycombe, A. Hruza, M.H. Vock, C. Giacino, B.D. Mookherjee, A.O. Pittet, and W.L. Schreiber; U.S. Patent 4,156,029; May 22, 1979; assigned to International Flavors & Fragrances Inc. describe mixtures of cis and trans isomers of 2-[(methylthio)methyl]-3-phenyl-2-propenal having the structures:

(I) 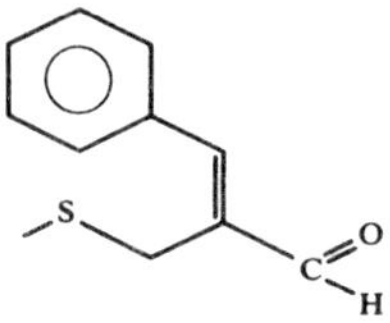 and (II)

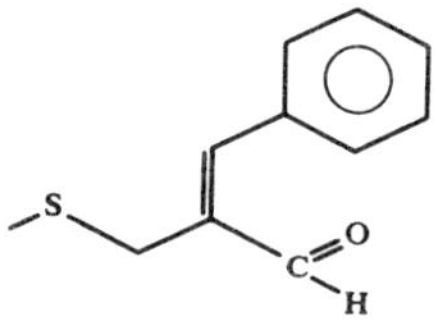

S C O H S C O H

These compounds augment and enhance the aroma and taste of meat extracts, meat, and hydrolyzed vegetable protein. They are obtained by reacting benzaldehyde with methional in the presence of a base such as NaOH, $Ba(OH)_2$, KOH, etc. The preferred concentration of base is from 0.3 to 0.8 molar and the reaction temperature varies between -10° and 50°C. The cis (Compound I) and trans (Compound II) forms are produced in a ratio of from 70 to 95% cis to 30 to 5% trans.

Substances suitable for use as coingredients or flavoring adjuvants are well known in the art for such use, being extensively described in the relevant literature. Apart from the requirements that any such materials be organoleptically compatible with the alpha-substituted alkylidene methionals, nonreactive with the alpha-substituted alkylidene methionals and ingestibly acceptable and thus nontoxic or otherwise nondeleterious, nothing particularly critical resides in the selection thereof. Accordingly, such materials which may in general be characterized as flavoring adjuvants or vehicles comprise broadly stabilizers, thickeners, surface active agents, conditioners, other flavorants and flavor intensifiers.

Example 1: *Preparation of a Mixture of 87% Cis- and 13% Trans-2-[(Methylthio)Methyl]-3-Phenyl-2-Propenal* – Four mols of methional and 4½ mols of benzaldehyde are placed in a 2 liter reaction flask fitted with a 250 ml addition funnel, Friedrichs condenser, Y-adapter, thermometer, and mechanical stirrer and cooled to 0°C at which time dropwise addition of 25% NaOH solution is begun. The reaction mass temperature is maintained at 0°C; the total addition time is 1.25 hr. The reaction is then permitted to warm to 20°C and is stirred for an additional 3 hr.

The reaction mass is transferred to a 4 liter separatory funnel and 500 ml of methylene chloride are added. An emulsion is formed which is treated with 300 ml of 10% H_2SO_4, carefully separated, and a second 300 ml aliquot of 10% H_2SO_4 is added thereby permitting the emulsion to be extracted. The aqueous acid layers are combined and extracted with 500 ml of methylene chloride. The methylene chloride layers are combined and washed with 300 ml of 10% H_2SO_4, 3 x 300 ml of saturated $NaHCO_3$, and 1,000 ml of NaCl brine.

The last wash is permitted to stand overnight at which time it is drained and extracted a second time with 1,000 ml of NaCl brine. The product is dried over sodium sulfate and the methylene chloride is stripped on a rotary evaporator. The residue is subjected to rush-over distillation using a 2" splash column and Fraction 4 (88%) is again subjected to rush-over distillation to obtain a yellow liquid in Fractions 2 to 5 boiling at 151°C at 3.8 mm Hg. The cis-trans isomer mixture was determined to be 87:13 by NMR analysis.

Example 2: The 87:13 cis-trans isomer mixture of 2-[(methylthio)methyl]-3-phenyl-2-propenal prepared according to Example 1 is dissolved in propylene glycol to provide a 0.1% solution. This solution in the amount of 0.9 g is added to 7.3 g of a soup base consisting of the following.

Ingredient	Parts by Weight
Fine ground sodium chloride	35.5
Hydrolyzed vegetable protein	27.5
Sucrose	11.0
Beef fat	5.5
Sethness caramel color (powder B & C)	2.7

The resulting mixture has a hydrolyzed vegetable protein-like meaty, beef broth-like aroma and flavor with an excellent mouthfeel, the effect of which is enhanced as if monosodium glutamate (18 g) is added thereto even though the monosodium glutamate is absent from the mixture.

Example 3: 330 g of gelatin is dissolved at 40°C in 8,250 g of deionized water to form a gelatin solution. 600 cc of 5% acetic acid is then added. 330 g of spray-dried gum arabic is dissolved at room temperature in 8,250 g of deionized water to form a gum arabic solution.

The gum arabic solution is placed in a 30 liter vessel and 2.5 liters of the gelatin solution is added. The temperature of the mixture is adjusted from 37° to 40°C.

Through a tube beneath the surface of the gum arabic solution, 4,000 g of 0.1 wt % solution of the 87:13 cis-trans isomer mixture of 2-[(methylthio)-methyl]-3-phenyl-2-propenal prepared according to Example 1 in propylene glycol is added over a period of approximately 30 min.

The mixture is agitated at 37° to 40°C until an average droplet size of 25 μ is obtained. The remaining gelatin solution (6 liters) is then added. The pH of the solution is then adjusted to 4.5 to 4.6 with a 10% sodium hydroxide solution. After the 25 μ droplet size is achieved, the temperature is allowed to drop to 25°C over a period of approximately 25 hr while maintaining the pH at 4.5 to 4.6.

The capsule slurry is then cooled to 5°C with stirring and is maintained at 5°C with stirring for at least 2.5 hr. The slurry is then sprayed-dried. The capsules thus formed are filtered and mixed with the soup base of Example 2 in the weight ratio of 1:6. 20 g of the resulting capsule soup base mixture is then added to 30 oz of boiling water thereby creating a soup having an excellent meaty flavor.

The resulting mixture has an excellent meaty, beef broth-like aroma and flavor with an excellent mouthfeel and an enhanced effect as when monosodium glutamate is added thereto, even though the monosodium glutamate is absent from the mixture.

Meat Aroma Precursor

C.A. Bernhardt and M.J. Mohlenkamp, Jr.; U.S. Patent 4,161,550; July 17, 1979; assigned to The Procter & Gamble Company have developed a composition which generates a meat-like aroma which comprises the following.

(a) A hydrogen sulfide precursor selected from the group consisting of cysteine, edible salts of cysteine, glutathione, edible salts of glutathione, edible protein-containing sulfhydryl and cystine; preferably cysteine;

(b) A dimethyl sulfide precursor selected from the group consisting of edible salts and derivatives of S-methyl methionine, S-methyl-4-methyl thiobutyric acid, S-methyl-methyl cysteine, S-methyl-3-methyl thiopropionic acid, and S-methyl-methyl thioacetic acid; preferably S-methyl methionine; and

(c) An edible proteinaceous material, preferably soy protein.

When cooked, the precursor composition generates an aroma surprisingly similar to that of cooked beef. It is contemplated that the precursor compositon will be used in combination with other foodstuff materials and flavorants which will also generate a meat-like flavor and/or modify and enhance the aroma of the precursor composition when cooked by the consumer. It is one of the advantages of the aroma precursor composition that a meat-like aroma is generated therefrom under normal home-cooking conditions.

As used herein, dimethyl sulfide precursors include compositions which generate, physically or chemically, dimethyl sulfide. Dimethyl sulfide encapsulated in hardstock, e.g., fatty acid material having a melting point of about 60°C, is an acceptable dimethyl sulfide precursor. As used herein, derivatives of S-methyl methionine and S-methyl-methyl cysteine include N-alkyl amides and alkyl esters. As used herein, derivatives of S-methyl-4-methyl thiobutyric, S-methyl-thiopropionic and S-methyl thioacetic acids are defined as the alkyl or aryl esters or amides thereof. Suitable salts are the chloride, bromide and iodide, and S-methyl methionine chloride is the most preferred salt.

The cysteine reactant of the process must be edible and may be in either salt or free-base form. D-cysteine, L-cysteine, DL-cysteine, cysteine hydrochloride or other edible salts of cysteine may be used. It is also contemplated that precursors of cysteine may also be employed, particularly, edible substituted thiazolidines. Edible substituted thiazolidines selected from the group consisting of reducing sugar thiazolidines, aldehyde thiazolidines, and ketone thiazolidines are stable precursors of cysteine and are preferred. Edible protein-containing sulfhydryls include egg white and other proteins which contain sulfhydryl groups.

Proteinaceous materials suitable for use in the aroma precursor composition include edible proteins such as those derived from vegetable protein sources, e.g., soybeans, sunflower seeds, safflower seeds, corn, peanuts, wheat, peas, cottonseeds, coconut, rapeseed, sesame seed, leaf proteins and single-cell proteins, such as yeast. Since meat analog products are generally made from proteinaceous material, the protein of the meat analog itself can serve as the proteinaceous substance required in this formulation. Animal protein sources can also be used. They include milk, egg, poultry, meat and fish. It has been found that the protein can be heat-coagulable or not heat-coagulable and in denatured and undenatured forms.

Example: An aroma precursor composition was prepared by first preparing a dimethyl sulfide precursor by adding 0.0468 g of S-methyl methionine to 52 g of bland, dry (4% moisture) soy protein granules. These were mixed in 138 ml of water and allowed to hydrate and were then freeze-dried. About 1.41 parts of the freeze-dried dimethyl sulfide precursor granules were mixed with 51.2 parts bland soy protein, 23.0 parts of Fri-al (mixed animal-vegetable fat) and 0.18 part of L-cysteine hydrochloride monohydrate.

This precursor composition was cooked by boiling 145 ml of water and mixing the composition therein. The mixture was removed from the heat, stirred and covered for 5 min and evaluated. Expert sniffers described the head-space aroma as beef-like.

FROM YEAST

Balancing Inorganic Ions

Yeast extracts prepared from baker's yeast, Japanese sake yeast and beer yeast which are easily obtainable in large quantities at low price have been used as components of seasonings. However, these extracts have unfavorable odors peculiar to yeast and are lacking in the thickness and body which can be obtained with meat extracts, and they are inferior to the meat extracts in their taste quality.

It has been discovered that the quality of yeast extracts are generally greatly affected by the balance of inorganic ions, particularly cations, included in the final yeast extracts produced, and also that an interaction between the inorganic ions and 5'-nucleotides is quite important for attaining the thickness and body which is desirable in beef or meat extracts.

H. Eguchi; U.S. Patent 4,066,793; January 3, 1978; assigned to Ajinomoto Co., Inc., Japan have prepared high-quality seasonings which are characterized as having less of the unfavorable odors peculiar to yeast itself, and by having a thickness or body in taste which resembles that of beef extract or shin extract prepared by concentrating extract of boiled shin. These characteristics are apparent from tasting a soup containing 2% of yeast extract and 0.5 to 0.8% NaCl at 60° to 70°C.

The process for preparing the seasonings comprises the steps of decomposing suspended yeast cells by autolysis in the presence of potassium ions at a pH of 5 to 7 and a temperature of 30° to 60°C, and thereafter performing the following steps in any convenient order. Add from 5 to 20 wt % sodium chloride, so that the K^+/Na^+ ratio is more than 0.5. Separate the resulting clear extract from the insoluble residue; heat at 90° to 100°C for 10 to 30 min, and add 0.5 to 4 wt % of a 5'-nucleotide.

The favorable taste of the seasoning compositions is substantially realized by a mutual interaction between 5'-nucleotide, inorganic ions and amino acids from the yeast extract. It is also very important to control a balance of inorganic ions and 5'-nucleotide.

While the precise quantitative balance between sodium ions and potassium ions cannot be exactly determined, it appears desirable that the amount of potassium ions which may be provided as inorganic or organic compounds is larger than 0.5 times, and preferably between 1 and 5 times, the amount of sodium ions in the final seasoning compositions. The inorganic or organic compounds which provide potassium ions are preferably added at an early stage in the hydrolysis.

Example: Baker's yeast (*Saccharomyces cerevisiae* CBS 1172) was cultured at 30°C with aerobic shaking in a medium containing 3% cane molasses calculated as sugar, 0.1% KH_2PO_4, 0.1% corn steep liquor, 0.5% $(NH_4)_2SO_4$ and 0.05% $MgSO_4{\cdot}7H_2O$. After 24 hr cultivation, yeast cells were harvested by centrifugation and washed with water to obtain yeast cells in a creamy state (yeast cream) with a water content of 42%.

Water was added to 2.24 kg of the yeast cream to prepare 10 kg water suspension of yeast cells. Into the suspension, 49.5 g KH_2PO_4, 169 g K_2HPO_4 and 218.4 g d-potassium bitartrate was dissolved completely and thereafter its pH value was adjusted to 6.5.

The suspension was mixed with 50 ml ethyl acetate and mixed well by shaking, and allowed to stand for 20 hr at 52°C for autolysis of the yeast cells. The solution after being autolyzed was adjusted to pH 5.5 with hydrochloric acid, and 1 kg NaCl was dissolved in the solution. From the solution treated as mentioned above, clear soluble fraction was obtained by centrifugation. The solution used to wash the residue was added to the clear solution.

After being heated at 98°C for 30 min, 2% of 5'-sodium inosinate was added to the clear solution. Then by concentrating the fraction under reduced pressure, 2.72 kg of modified yeast extract of which water content is 50% were obtained (Sample A).

Potassium ion concentration and sodium ion concentration were determined by conventional method, and were found to be 5.7 and 4.2% respectively; therefore K^+/Na^+ was 1.36.

The same kinds of modified yeast extracts, Samples B, C, D and E were similarly prepared except for variations in the amount of KH_2PO_4, K_2HPO_4 and d-potassium bitartrate. The amounts used were such that the ratio, K^+/Na^+, was 0.34 for Sample B, 0.50 for Sample C, 0.75 for Sample D and 2.00 for Sample E. Sample F contained no potassium salts.

The yield of modified yeast extracts based on the yeast cells employed in Samples A through F were all about 50% relative to the weight on a dry basis. As a control, shin extract was prepared by the following process. 5.5 kg shin which were cut to about 2 cm cubes were added to 18 liters of water and boiled down to ⅓ of the original volume with a low fire during which process the lye and oily materials rising to the surface were removed. The extract prepared was concentrated under reduced pressure to obtain 500 g of shin extract with a water content of 50%.

The Samples A through F were subjected to organoleptic testing by a panel of 20 members who had been specially trained for this kind of test. The panel test was performed by the following procedure. Solutions of sodium chloride of 0.6% concentration were added to Samples A through F and to the shin extract which was used as a control in an amount of 2% concentration and panel members were asked to give marks about degree of thickness and body of each sample solution in contrast to those of shin extract marked as 10.

The average marks were 9.2 for the Samples A and E, 9.0 for the Sample D, 8.5 for the Sample C, 6.0 for the Sample B whose K^+/Na^+ ratio was under 0.5, and 5.0 for the Sample F to which potassium salt was not added at all. These results apparently show the importance of maintaining a K^+/Na^+ ratio of more than 0.5.

The panel was also asked to comment on the degree of unfavorable strange odors peculiar to yeast in each sample solution prepared as mentioned above. The strange odors in Samples A through E were all weaker than that of Sample F to which potassium salt was not added at all, and it was found that the strange odor decreased with increasing K^+/Na^+ ratio.

All panel members indicated that Samples A through E had more favorable taste than Sample F, and that the favorable taste increased with increasing K^+/Na^+ ratio.

Bouillon Base

R.J. Gasser and L.B. Huster; U.S. Patent 4,073,961; February 14, 1978; assigned to Societe d'Assistance Technique pour Produits Nestle, SA, Switzerland have devised a method of preparing a starting material for the preparation of bouillons from a yeast protein degradation product, which comprises diluting a yeast autolysate with at least the same quantity by weight of water, precipitating substantially insoluble salts, especially alkaline earth metal phosphates, by heating the diluted autolysate at a pH-value of from 7 to 8.5, separating solid fractions from the treated autolysate by heating so that only a solution is left, treating the residual solution thus obtained by steam distillation, concentrating the treated solution by evaporation to a dry matter content of from 50 to 80 wt %, mixing 0 to 8% of fats, 0 to 5% of a monosaccharide and 0 to 2% of a nucleotide with the concentrated solution obtained, reacting the mixture obtained by heating at a pH of from 5.8 to 6.8 and drying the mixture.

The product obtained by this process is clearly distinguished from known products by its totally neutral flavor and by the complete absence of any after-taste indicative of the origin of the starting material used for its production. It may be used, preferably after grinding, as a starting material for and as the main constituent of edible compositions, such as bouillons in cube or powder form. The taste or specific note of these bouillons may be imparted by incorporating in the composition a small percentage of vegetable or meat extract. In the following example, the percentages are by weight.

Example: 33% of yeast extract are diluted in 66% of water in a double-jacketed stirrer-equipped boiler. The dry matter content of the solution is adjusted to 26%. The pH value of the solution is adjusted to 7. The solution is then heated to a temperature of 95°C and kept at that temperature for 10 min. The solution is then cooled to ambient temperature and the solids which have precipitated are separated from it by centrifuging. The residual solution is heated to 97°C.

The solution thus preheated is passed downwards through a vertical column filled with glass elements in countercurrent to steam ascending through the column at the same throughput by weight. The purified solution obtained is concentrated in a vacuum evaporator to a dry matter content of 60%. The pH value of the concentrated solution is adjusted to 6.2 by the addition of hydrochloric acid.

The concentrated solution is heated to 95°C in a double-jacketed stirrer-equipped tank, and 2% of prime juice, 0.5% of glucose and 0.5% of inosine monophosphate are added to it while stirring. The mixture is kept at 95°C for 20 min, after which it is concentrated in an evaporator to a dry matter content of 82%.

The paste obtained is applied to two metal plates which are introduced into a vacuum drying cabinet where they are left for 6 hr at 65°C, followed by rapid cooling to ambient temperature. The dry mass obtained is immediately ground. A powder with a completely neutral taste is obtained.

Yeast Autolysate and Whey Solids Added to Hydrolyzed Vegetable Protein

C.R. Corbett; U.S. Patent 4,165,391; August 21, 1979; assigned to Stauffer Chemical Company has provided an improved flavor agent for providing meaty flavor and increased flavor intensity to foods such as gravies, sauces, soups, cheese spreads, dairy-based snack spreads or dip and seasoning mixes. The product comprises 50 to 85% hydrolyzed vegetable protein, 10 to 25% yeast autolysate and 5 to 25% on a dry solids basis of soluble modified whey solids selected from the group of delactosed whey solids, the second fraction obtained by passing a liquid mixture of cheese whey solids through a bed of molecular sieve resin, and the permeate and the delactosed permeate resulting from the ultrafiltration of whey, all percents being by weight.

The second fraction powder in the following example is available from Stauffer Chemical Company under the trademark ENR-EX, and the yeast autolysate is NIV, also from Stauffer. The HVP paste is a corn-based hydrolysate, Vi-zate 268, from A.E. Staley.

Example; Three flavoring agent paste blends were prepared by blending the following ingredients with a fork in a small bowl.

	. . . . . . Example (% by wt)		
	1-A	1-B	1-C
HVP paste	75	80.03	77.41
Yeast autolysate paste	12.5	13.34	12.90
Second fraction powder	12.5	6.63	9.69

The three blends were evaluated organoleptically in two different commercial brown gravy formulations against a control of one of the better products of industry. The formulations utilized are as follows.

	Control	Process
Brown Gravy Formulation 1	. . . (parts by weight) . . .	
HVP*	4.50	4.50
Flavoring agent Ex. 1-A, 1-B or 1-C	–	10.32
Flavoring agent (control)**	10.32	–
Wheat flour	25.50	25.50
Waxy maize starch	23.00	23.00
Salt, granulated	5.50	5.50
Sugar, granulated	4.50	4.50
Monosodium glutamate	4.25	4.25
Onion powder, toasted	4.50	4.50
Maltodextrin	10.90	10.90
Caramel color powder (beef red)	1.65	1.65
White pepper, ground	0.20	0.20
Hydrogenated vegetable oil	6.40	6.40

*Maggi HPP Type RF-B powder (Nestle Co., Inc.) with partially hydrogenated vegetable oil.

**Maggi HPP Type 4BE paste (Nestle Co., Inc.) with partially hydrogenated vegetable oil added.

The brown gravy was prepared by combining 28 g of the formulation with 236.6 ml of water, heating until boiling and simmering for 2 min.

	Control	Process
Brown Gravy Formulation 2	. . . (parts by weight) . . .	
HVP*	8.80	8.80
Flavoring agent Ex. 1-A, 1-B or 1-C	–	14.48
Flavoring agent (control)**	14.48	–
Maltodextrin	16.00	16.00
Salt, granulated	6.00	6.00
Caramel color powder	1.00	1.00
Pepper, white	0.10	0.10
Beet root powder	0.30	0.30

(continued)

Brown Gravy Formulation 2	Control	Process
	. . . (parts by weight) . . .	
Nonfat milk solids, agglomerated	7.00	7.00
Cornstarch	45.50	45.50
Onion powder, toasted	2.30	2.30

*Maggi HPP Type 245 powder (Nestle Co., Inc.) with partially hydrogenated vegetable oil added.

**Maggi HPP Type 4BE paste (Nestle Co., Inc.) with partially hydrogenated vegetable oil added.

The brown gravy was prepared by combining 22 g of the formulation with 236.6 ml of water, heating to a boil while stirring constantly and simmering for 1 min under reduced heat.

These compositions were tested by a small informal taste test panel by tasting just the heated gravy itself. Five out of five panelists preferred the gravies made using the flavoring agent of Example 1-A over the control. More enhanced flavor was noticed by four of the panelists. Gravies made using the flavoring agents of Examples 1-B and 1-C were judged to have slight flavor difference when compared to the control. The gravy prepared using the flavoring agent of Example 1-C was closest to the control in intensity and type of flavor. Even though the quality of HVP used in preparing this flavoring agent was significantly lower than that used in the control as evidenced by its harsh flavor, generally dark color and low cost, equivalent results were obtained.

FROM ANIMAL PRODUCTS

For Specifically Reproducible Meat Flavors

L.C. Chhuy and E.A. Day; U.S. Patent 4,081,565; March 28, 1978; assigned to International Flavors & Fragrances Inc. provide specifically reproducible beef, veal, pork, lamb and poultry flavors, without the use of any egg, which are suitable for incorporation into a wide range of foods. The process comprises adding to a foodstuff at least 0.2 wt % of a flavor-imparting product produced by a process comprising intimately admixing in the absence of any natural meat or poultry products, a composition consisting essentially of the following:

(a) An enzymatic digest of meat produced by admixing 0.1 to 0.4% enzyme with meat at a pH of 3 to 6, at a temperature of 40° to 75°C for a period of 10 to 150 hr, selected from the group consisting of a Mylase digest of turkey, chicken, beef, lamb or pork;

(b) L-cysteine hydrochloride; and

(c) Thiamine hydrochloride.

The enzymatic digest of meat, L-cysteine hydrochloride and thiamine hydrochloride each are present in an amount sufficient to produce an intense meat flavor, the weight percent of cysteine based on enzymatic meat digest being from 2 to 5%, the ratio of thiamine hydrochloride:cysteine hydrochloride being in the range of 0.25-1:1.

The resulting mixture is heated at a temperature of 100° to 350°C, a pressure of 1 to 10 atm and for a period of time of 0.2 to 10 hr whereby a meat-flavored product is produced. Gum arabic is added to the meat flavor and the resulting mixture is spray-dried. The weight ratio of gum arabic to the meat flavor is 0.2-2:1 based on the solids content of the mixture prior to spray drying. Pepsin digests of meat can replace Mylase digests in the above preparation scheme. The following examples will illustrate ways in which the process can be practiced.

Example 1: Into a 300 ml flask equipped with stirrer, thermometer and heating mantle, the following materials are placed: 50 g of ground raw chicken meat, 100 ml of water, and 0.10 g of Mylase enzyme.

The pH of the contents of the flask is adjusted to exactly 5.0 by addition thereto of phosphoric acid. The contents of the flask is prepasteurized for 30 min at 60°C. The contents of the flask are then incubated at 50°C using agitation (250 rpm) for a period of 120 hr. The resulting enzymatic chicken digest is used in the reaction flavors as set forth in the following examples.

Example 2: The procedure of Example 1 is carried out using the following quantities of chicken meat paste and enzymes except that the pH of each of the mixtures is first adjusted to 4.6 using a buffer which is a one molar solution of sodium citrate and sodium diacid phosphate; 0.1 g enzyme is used in each case.

Quantity of Chicken Meat Paste, g	Enzyme
100	Prolase
100	Bromelain
100	Papain
100	Pilzprotease S
75	Rhozyme P 11
75	Bromelin
50	Pepsin
50	Trypsin
50	J-25
50	Protease 62
40	HT proteolytic 200
25	Fungal protease
25	Prolase MT-7820
25	Molsin
20	Mylase
20	Bakterien

Each of the resulting enzymatic chicken digests are used separately in flavors as set forth in Example 4 following.

Example 3: 68 g of cysteine hydrochloride, 500 g of water, 30 g of enzymatic chicken meat digest according to the process of Example 1 and 34 g of thiamine hydrochloride are intimately admixed. The resulting slurry is then heated to reflux and maintained at reflux with stirring for a period of 4 hr.

The resulting slurry is filtered yielding 0.51 kg of filtrate. An aliquot of the filtrate is withdrawn and dried. Based on the weight of the dry product obtained, sufficient gum arabic is added to the filtrate to provide a composition containing 1 pbw of flavor solids and 1 pbw of gum arabic. The composition is then spray-dried to produce a dry product with an intense chicken flavor having a very close resemblance to natural chicken flavor insofar as flavor nuances, mouthfeel and texture are concerned.

Example 4: Separately, each of the enzymatic chicken meat digests prepared in Example 2 are mixed with 8.8 g cysteine hydrochloride, 29.4 g of thiamine hydrochloride and 400 g of water. Each mixture is refluxed at atmospheric pressure for 4 hr with stirring. Each slurry is then cooled and filtered. 250 g of each flitrate is then mixed with 125 g of a 45% aqueous gum arabic solution and each of the resulting mixtures is spray-dried. Each of the resulting solid spray-dried materials has an excellent natural chicken meat flavor insofar as flavor nuances, mouthfeel and texture are concerned. Several separate chicken noodle soup batches are prepared by mixing:

Ingredient	Grams
Salt	5.00
One of the spray-dried chicken flavors	1.00
Gelatin (180 bloom)	1.00
Monosodium glutamate	0.40
Caramel color	0.40
Garlic powder	0.10
White pepper, ground	0.06
Mixed vegetable pieces	36.00

Three cups of water are added to each mixture and boiled before serving.

Animal Fat plus Brewed Soy Sauce

T. Aishima and A. Nobuhara; U.S. Patent 4,094,997; June 13, 1978; assigned to Kikkoman Shoyu Co., Ltd., Japan have found that a substance having a flavor close to natural beef flavor, particularly to roast beef flavor, can be obtained by admixing an animal fat from domestic animals such as cattle, swine and sheep or other animals such as whale with a brewed soy sauce and then reacting the resulting mixture at elevated temperatures.

Example: 210 g of beef tallow was fused and heated at 160°C for 40 min while air was bubbled to stir the tallow. Then, it was divided into three portions A to C, each 70 g. Portions A, B and C were independently mixed with 40 ml of koikuchi soy sauce (deep colored type of soy sauce), usukuchi soy sauce (thin colored type of soy sauce) and a 10% (w/w) solution of ribose in commercially available amino acid solution (which is a seasoning liquid prepared by hydrolyzing a protein raw material, such as defatted soy beans or a wheat gluten, with hydrochloric acid thereby to isolate amino acids, neutralizing the acids with soda ash and then filtering solid materials), respectively.

Each of the resulting mixtures was reacted under reflux at 160°C for 10 min while introducing air to stir it, after which it was cooled and kneaded thoroughly. Thus, three reaction products A, B and C were obtained.

Products A, B and C were subjected to organoleptic test by a panel of twenty persons, product C being used as a control for the test. The results obtained were shown in the following table.

Item	Number of Persons . . Making Choice of. . . A	C	Statistical Test Results
Agreeableness of meat flavor	18	2	*
Strength of meat flavor	17	3	*

Item	Number of Persons . . Making Choice of. . . B	C	Statistical Test Results
Agreeableness of meat flavor	18	2	*
Strength of meat flavor	18	2	*

*There is a significant difference on a significant level of 1%.

As shown in the above table, products A and B both produced by reacting beef tallow with brewed soy sauce at elevated temperature are far superior to product C obtained by reacting beef tallow with an amino acid solution containing a sugar in agreeableness and strength of meat flavor.

MEAT-FLAVORED SHORTENING FOR DEEP-FAT FRYING

D.S. Kravis; U.S. Patent 4,169,901; October 2, 1979; assigned to The Procter & Gamble Company has provided meat-flavored deep-fat frying compositions containing artificial meaty flavorants which are able to impart meaty flavors to foodstuffs fried therein for extended periods of time. Such compositions comprise from 80 to 98.98 wt % of the composition of a base fat which is an edible triglyceride having acyl groups of from 16 to 22 carbon atoms. The base fat has a smoke point exceeding 350°F and has an I.V. between 30 and 150.

The deep-frying compositions further comprise from 0.02 to 10 wt % of the composition of a volatile artificial meat-like flavorant which is soluble at 70°F in the base fat to an extent of at least 10 wt % of the total weight of the composition. The compositions additionally comprise from 1.0 to 10 wt % of the composition of a stabilizing agent which is an undeodorized oil selected from the group consisting of coconut oil, palm-kernel oil and babassu oil.

Example 1: A fluid deep-frying composition is prepared by combining a triglyceride oil-base fat with additives in the following manner. Approximately 1,330 g of base fat (a bleached and deodorized mixture of 96.5 wt % soybean oil hardened by hydrogenation to an iodine value of 107; 3.5 wt % soybean hardened to an iodine value of 8; the mixture having a solids content index (SCI) value of 3 at 50°F, and a smoke point of 440°F, and approximately 3 to 6 ppm methyl silicone, DC-200, Dow Corning) is placed in a 5-quart, home-type deep-fat fryer. The deep-frying composition is made by adding the following ingredients to the base oil.

Ingredient	Grams	Weight Percent
Base oil (as in above)	1,330.67	97.7
Undeodorized coconut oil	27.24	2.0
Artificial pork flavor*	4.09	0.3
Total	1,362.00	100.0%

*A commercial artificial flavor comprising a mixture of fatty acid/reducing sugar reaction product flavors marketed by Renaud, Ltd. (Stock No. R 2118/A).

The blend is then mixed by hand for approximately 1 min. An aliquot portion is selected for subsequent evaluation. Two control compositions are similarly prepared by adding 27.24 g of undeodorized coconut oil to 1,335 g base oil and by adding 4.1 g artificial pork flavor to 1,358 g base oil.

The fluid deep-frying composition so prepared is effective in imparting a meaty flavor to foodstuffs deep-fried therein. The composition exhibits the flavor-delivery property for extended periods of time including repeated cycles of deep-frying. Thus, the deep-frying composition provides a prolonged flavor-delivery fry-life.

OTHER COMPOUNDS

1,8-Dihydroxy-2,9-Dithia-Tricyclotetradecane

M.J. Greenberg; U.S. Patents 4,094,998; June 13, 1978 and 4,107,342; August 15, 1978; both assigned to The Quaker Oats Company has found that 1,8-dihydroxy-2,9-dithia-tricyclo[8.4.0.0^{3,8}] tetradecane and derivatives thereof having the formula:

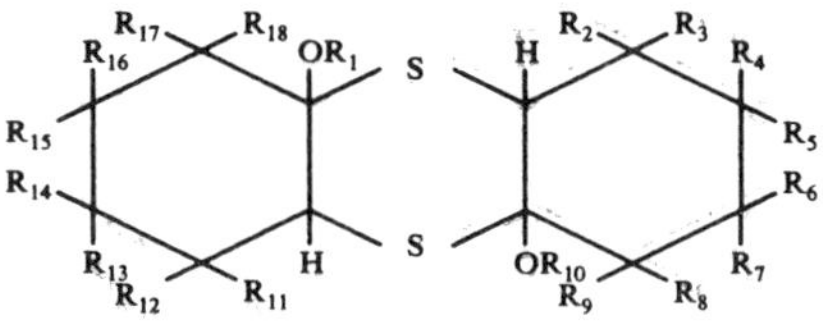

and 1,7-dihydroxy-2,8-dithia-tricyclo[7.3.0.0^{3,7}] dodecane having the formula:

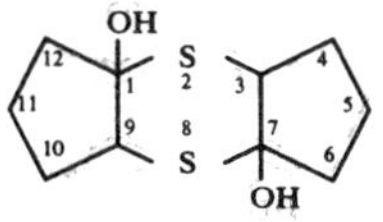

where R_{2-9} and R_{11-18} are the same or different, and are alkyl groups containing from 1 to 4 carbon atoms, a hydroxyl group, or hydrogen; and R_1 and R_{10} are the same or different, and are alkyl groups containing from 1 to 4 carbon atoms, hydrogen, acetyl, phenacyl, or benzyl provide a meat-like taste and aroma for a food when taken singly or in any combination. These compounds either provide meat flavor for food, augment meat flavor for food, or mask otherwise unacceptable flavoring for a food.

An especially suitable compound for this is 1,8-dihydroxy-2,9-dithia-tricyclo-[8.4.0.0^{3,8}] tetradecane. This compound provides a beefy, meaty, and spicy type of aroma. It has the formula as follows.

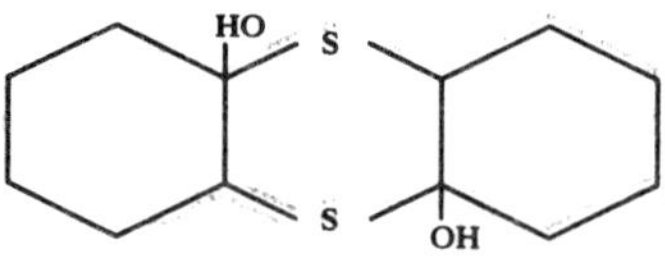

The above compound may be modified by making the acetate ester, the benzoate ester, or the benzyl ether respectively by substituting the appropriate group for H on the oxygen atom which is bonded to carbon atoms 1 and 8.

It is especially preferred, that groups on C_1 be the same as C_8; C_4 be the same as C_{11}; C_5 be the same as C_{12}; C_6 be the same as C_{13}; and C_7 be the same as C_{14}. This requirement simplifies the isolation of a pure compound of a particular type. Otherwise, mixed concoctions of different compounds are achieved and the desired results of this process and the flavorants are not as easily predictable. Especially suitable compounds for the purposes described herein are 2,5-dihydroxy-1,4-dithianes.

The compounds may be used singly or in admixtures comprising two or more thereof. The formulations may combine additional flavoring materials such as furfural, benzaldehyde, and hexanal to simulate a wide variety of organoleptic characteristics. In addition, their derivatives may be admixed with one or more flavorant adjuvants such as stabilizers, thickeners, surface active agents, conditioners, flavorants and flavor intensifiers to provide suitable flavor imitative of roasted meats of various types.

The described compounds can be added to the foods to be flavored by any conventional techniques. Typical conventional techniques include spray drying, blending, stirring, dissolving and the like. The addition of these compounds is carried out in any stage of the preparation of the foodstuff to which the compounds are to be applied.

These flavorants can be added to almost any type of food, but are particularly applicable to pet foods of the moist, semimoist or dry type.

Example 1: The compound 1,8-dihydroxy-2,9-dithia-tricyclo[8.4.0.$0^{3,8}$] tetradecane is prepared as follows. To a 250 ml 3-necked round-bottom flask equipped with a mechanical stirrer, an addition funnel, and a gas inlet tube are added sodium (4.6 g) and ethanol (100 ml). The solution is saturated with hydrogen sulfide at -15°C and with vigorous stirring. To this solution is added 2-chlorocyclohexanone (6.7 g) in a 1:1 ethanol-ether solvent.

The addition is carried out slowly with vigorous stirring and continuing addition of hydrogen sulfide at a temperature of -10°C over a 24 min period. The mixture is stirred for an additional 10 min while maintaining the mixture at 0°C. The resulting precipitate is filtered and washed with warm water. The crude precipitate is then recrystallized from chloroform yielding 5.3 g of product which is a 41% yield. Under a standard, infrared spectroscopy testing process, the pure product in a potassium bromide pellet has infrared spectrum peaks at 3,325 cm^{-1} indicating an O–H stretch, 2,920 and 2,850 cm^{-1} indicating a C–H stretch, 1,360 cm^{-1} indicating an O–H bend and 1,183 cm^{-1} indicating a C–O stretch.

A MP of 145°C combines with infrared spectroscopy data to confirm the structure. This high melting point also shows that the compound would be thermally stable under the process conditions and to produce dry pet foods usually processed at 95° to 140°C, semimoist pet foods, usually processed at 75° to 110°C and canned pet foods usually processed at 95° to 120°C.

Example 2: The flavoring of Example 1 is topically applied to a commercial semimoist pet food. Then two days of tests are conducted using dogs as the test animals to compare preference of the control product containing no flavorant and the same product containing the product of Example 1 in the amount of 15 ppm. The dogs used are of varying size and breed. The results of these tests indicate excellent palatability of the product containing 1,8-dihydroxy-2,9-dithia-tricyclo[8.4.0.$0^{3,8}$] tetradecane at 15 ppm level. Using standard methods of statistical analysis, the experimental product containing the 15 ppm flavorant was judged to be highly significantly preferred over the control product containing no flavorant.

Mixtures of a Terpinenol Propionate and a Methyl Furanthiol

C.J. Mussinan, M.H. Vock and A.L. Liberman; U.S. Patent 4,139,649; February 13, 1979; assigned to International Flavors & Fragrances Inc. have found that the addition of 4-terpinenol propionate having the structure:

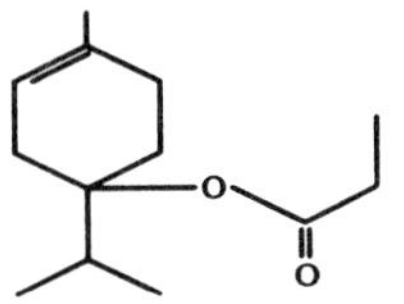

(which is produced by reacting 4-terpinenol with propionic anhydride in the presence of pyridine), and 2-methyl-3-furanthiol having the structure:

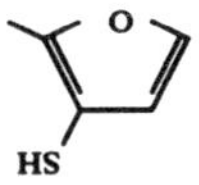

HS

to foodstuffs produces a roast beef aroma and flavor characteristic.

Example 1: *Preparation of 4-Terpinenol Propionate –*

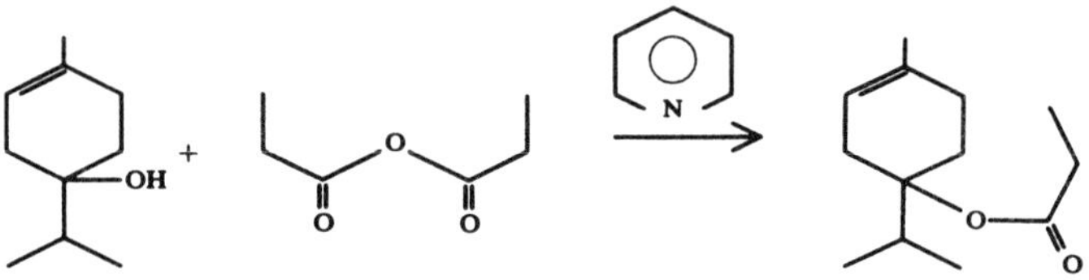

To a 2-liter flask equipped with mechanical stirrer, Friedrich's condenser and heating mantle the following materials are added.

4-terpinenol	154 g (1 mol)
Propionic anhydride	390 g (3 mols)
Pyridine	400 cc

The resulting mixture is heated to reflux for 28 hr, at which time GC indicates complete utilization of the alcohol reactant. The GC-MS profile confirms that the above compound 4-terpinenol propionate along with a small amount of alpha-terpinyl propionate, is present.

4-Terpinenol propionate is purified by means of vacuum distillation from the reaction mixture. The reaction mass is added to a 2-liter reaction flask together with Primol (a hydrocarbon mineral oil from Exxon Incorporated). The fractionation yields 8 fractions.

Fractions 5, 6 and 7 are placed in a 250 ml reaction flask and the lower boiling impurities are further removed by vacuum distillation leaving behind 35.83 g of 90.83% 4-terpinenol propionate and 6.33% alpha-terpinyl propionate. The 4-terpinenol propionate structure is confirmed using NMR and IR analyses.

The resulting product has a black pepper, woody and oriental aroma characteristic and black pepper, woody, oriental and biting flavor characteristics in admixture with the alpha-terpinyl propionate and as pure 4-terpinenol propionate.

Example 2: A 1:1 mixture of 4-terpinenol propionate (prepared according to Example 1) and 2-methyl-3-furanthiol is dissolved in propylene glycol in an amount sufficient to give a propylene glycol solution containing 0.1 wt % of the mixture. 0.9 cc of this solution is added to 7.3 g of a soup base consisting of the following.

Ingredient	Quantity (parts/100 parts)
Fine ground sodium chloride	35.62
Hydrolyzed vegetable protein*	27.40
Monosodium glutamate	17.81
Sucrose	10.96
Beef fat	5.48
Sethness caramel color**	2.73

*4 BE: Nestle's
**Powder B & C

The resulting mixture is added to 12 oz of boiling water to create a soup having an excellent roast beef flavor.

Example 3: 0.5 g of a 1:1 mixture of 4-terpinenol propionate and 2-methyl-3-furanthiol is emulsified in a solution containing the following materials: 100 g of gum arabic, 100 g of water and 0.5 g of 20% solution in ethanol of butylated hydroxyanisole.

The resultant emulsion is spray-dried in a Bowen Lab. Model spray-drier, inlet temperature 500°F, outlet temperature 200°F. 12 g of this spray-dried material is mixed with 29.2 g of the soup base set forth in Example 2. The resulting mixture is then added to 12 oz of boiling water, and an excellent roast beef flavored soup is obtained.

Certain Acyl-Pyrimidines

I. Flament; U.S. Patent 4,166,869; September 4, 1979; assigned to Firmenich, SA, Switzerland have found that certain acyl-pyrimidine derivatives may be used for enhancing, improving or modifying flavor properties of foods, tobacco products, animal feeds, etc., particularly meat-containing and meat-imitating compositions. These compounds have the following formula:

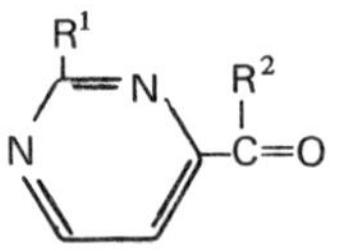

wherein symbol R^1 represents a hydrogen atom or a methyl radical and R^2 stands for a lower alkyl radical containing from 1 to 6 carbon atoms. These pyrimidines may be prepared by conventional processes. One method for making 2-methyl-4-acetyl pyrimidine, for example, is illustrated by the following reaction.

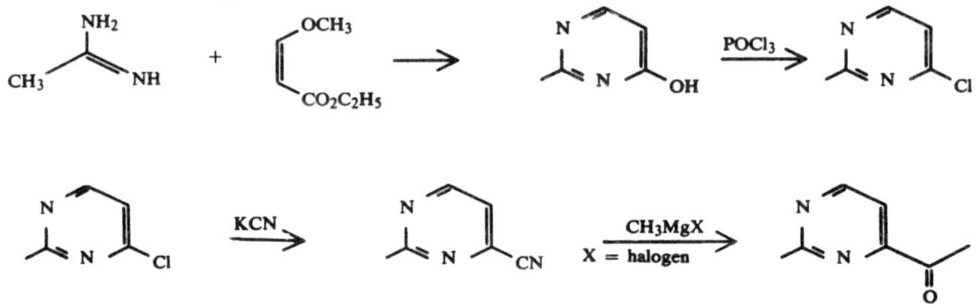

Example 1: A comparative flavor evaluation was carried out by tasting in crystal spring water the following compounds in the concentrations indicated.

Compound	Dosage (ppm)	Evaluation
(1) 2-Methyl-4-acetyl-pyrimidine	0.5	roasted, animal, meat-like, sweetish, caramel, nut
(2) 3-acetyl-pyridine	2.5	less meat-like than (1), hazelnut, roasted cereal-like, less body than (1)
(3) 2-acetyl-pyrazine	1.0	animal, less meaty and roasted than (1), fatty
(4) 5-methyl-quinoxaline	2.5	less meaty and animal than (1), hazelnut, roasted, slightly green

The above given comparison test shows the basic difference inasmuch as the flavor characters are concerned between 2-methyl-4-acetyl-pyrimidine and compounds (2), (3) and (4) known in the art for promoting meat-like flavor notes.

Example 2: A reconstituted beef broth was prepared by mixing together the following ingredients.

Ingredient	Parts by Weight
Commercial beef extract	10
Monosodium glutamate	1
50:50 Mixture of sodium inositate: sodium guanylate	0.0005
Sodium chloride	8

(continued)

Ingredient	Parts by Weight
Lactic acid	0.5
Water	980.495
Total	1,000.000

The broth thus prepared was divided into 3 portions of equal volume. Two of these portions were aromatized with respectively 0.15 and 0.30 ppm of 2-methyl-4-acetyl-pyrimidine and subjected to the evaluation by a panel of flavor experts who expressed their judgement as follows:

(a) nonaromatized broth: bland taste, meaty character, protein, vegetable taste;

(b) aromatized broth (0.15 ppm): the meaty taste was more marked than that of (a); and

(c) aromatized broth (0.30 ppm): strong roasted and animal character.

Example 3: *Tobacco Aromatization* – 0.3 g of a 1 part per thousand solution of 2-methyl-4-acetyl-pyrimidine in 95% ethanol was sprayed onto 100 g of a mixture of tobacco of American blend type. The tobacco thus flavored was used to manufacture test cigarettes, the smoke of which was then subjected to organoleptic evaluation by comparison with nonflavored cigarettes (control). The tobacco used to manufacture the control cigarettes was preliminarily treated with 95% ethanol. The panel of experts declared that the taste of the test cigarettes possessed more body and more defined flavor character reminiscent of the taste of hazelnuts.

ENHANCING MEAT FLAVORS

Alpha-Mercaptoacetophenone

M.J. Greenberg; U.S. Patent 4,096,284; June 20, 1978; assigned to The Quaker Oats Company describes the imparting of meat flavors and aromas to foods, particularly pet foods, by adding to them: (a) α-mercaptoacetophenone or (b) a derivative thereof having the formulas as follows.

(a) $C_6H_5-C(=O)CH_2SH$ or (b) $R-C_6H_4-C(=O)CH_2SH$

Most preferably, these compounds comprise about 1 to 100 ppm by weight of the food. The substituent indicated on the formula permits R to be an alkyl group containing 1 to 4 carbon atoms, hydrogen, a hydroxyl group, alkoxy group containing 1 to 4 carbon atoms, an acetyl group, a phenacyl group, a benzyl group, a halogen, or mixtures thereof. Especially preferred halogens are chlorine and fluorine.

Example 1: To a 250-ml three-necked round bottom flask equipped with mechanical stirrer, addition funnel and gas inlet tube is placed pyridine (7.5 g) saturated with hydrogen sulfide. The solution is vigorously stirred. 2-chloroacetophenone (23.2 g) in ether (100 ml) was slowly added with vigorous stirring and hydrogen sulfide was added at room temperature over a 30 min period. The mixture was stirred an additional 90 min at room temperature, then stirred for 30 min at 50°C, and poured into a mixture of concentrated hydrochloric acid (90 g) and crushed ice (500 g). The mixture is extracted with three 50 ml portions of ether. The ether extract is dried (Na_2SO_4) and distilled yielding a clear colorless oil; 3.0 g (13%); BP 82°C (0.3 mm. Infrared analysis clearly showed the product to be α-mercaptoacetophenone.

Example 2: The flavoring of Example 1 is topically applied to a commercial semimoist pet food. Tests are conducted with dogs of varying breed and size to compare preference of the control product containing no flavorant and the same product containing α-mercaptoacetophenone of varying concentrations. The results of these tests shown below indicate excellent palatability of the product containing α-mercaptoacetophenone, and that the palatability in this case is linear with log concentration.

Flavorant Level in ppm	% Test Material Consumed	% Control Consumed
3	71.3	28.7
5	75.6	24.4
6	81.4	18.6

The average percent consumed of test material for 3 ppm is 71.3% and for 5 ppm it is 75.6%. A statistically significant preference was demonstrated by the pet food flavored with α-mercaptoacetophenone at the 6 ppm level.

Dried Artichoke

V. Nicolosi; U.S. Patent 4,122,207; October 24, 1978 has developed a food flavor enhancing composition comprising cooked, dried and ground artichoke pulp. The composition may further comprise various minor amounts of cooked, dried and ground escarole and/or fava beans.

The composition may be processed by cooking artichokes to reduce the artichokes into pulp and fiber and then to remove the pulp from the fiber. The pulp is then partially dried to form a paste. Similarly prepared fava beans and escarole may be reduced to the same paste-like state in a similar manner and the components mixed to form a composite paste. The composition may also be further processed by completely drying the paste formed from the pulp to form a dry residue. The dry residue is then pulverized to form a powder. At this point, the powdered artichoke may be mixed in appropriate percentages with similarly prepared powdered escarole and fava beans.

This substance is a natural food flavor enhancer which may be effectively applied to a wide variety of foods including meats, fish, foul, vegetables and soups. Its primary ingredient is processed artichoke. Although it is not clearly understood, such processed artichoke when prepared as described above coacts with the flavor-producing compounds of a wide variety of foodstuffs in an unexpectedly beneficial manner. To a large degree, regardless of the type of foodstuff upon which the process is used, the natural flavor of the food upon which it is used is enhanced.

It is speculated that the food flavor enhancer may protect the flavor-bearing oils, extracts, and resins of food-stuffs from oxidation, decomposition or other impairment of flavor and aroma.

The composition can be applied directly to a food-stuff in a dried, powdered state or in a paste or liquid-like state as prepared as described above. Furthermore, it can be used as a component in a sauce or coating which may also be applied to the food-stuff. The amount used will vary depending upon the particular composition chosen, the degree of flavor enhancement desired, the means of application employed, the manner in which the food-stuff is being prepared, and the freshness of the food seasoning.

Substituted Thiazoles

A.O.Pittet and D.E. Hruza; U.S. Patent Reissue 29,843; November 21, 1978; assigned to International Flavors & Fragrances Inc. have found that certain substituted thiazoles are capable of imparting a wide variety of flavors and fragrances to various consumable materials. Briefly, in this process, flavors and/or fragrances of materials are changed by adding thereto a small but effective amount of at least one thiazole having the formula:

R N Y S X

wherein R is hydrogen, alkyl, or acyl; X is alkyl, alkoxy, or hydrogen, and Y is alkyl, acyl, alkoxy or hydrogen, and no more than two hydrogen atoms are substituents on the thiazole ring; with the proviso that X is alkyl only when either R and Y are alkyl or Y is alkoxy. Exemplary of useful thiazoles, with their boiling points, flavor and aroma characteristics are the following:

Name and Formula	Boiling Point	Flavor Characteristics	Aroma Characteristics
$CH_3-C(=O)-$ (N, S) 4-Methyl-5-acetylthiazole	98°C at 9 mm Hg	Sulfury, roasted nut	Earthy peanut
(N, S) Trimethylthiazole	41°C at 2.2 mm Hg	Dark chocolate and light green	Dark chocolate and light green
$CH_3-C(=O)-$ (N, S) 2,4-Dimethyl-5-acetylthiazole	79°C at 1.8 mm Hg	Boiled beef	Meaty, sulfur
CH_3O (N, S) 2-Methyl-5-methoxy-4-isobutylthiazole	69°-70°C at 2.2 mm Hg	Pepper, vegetable	Pepper, onion

(continued)

Name and Formula	Boiling Point	Flavor Characteristics	Aroma Characteristics
 N, S, C_2H_5O 4-Isobutyl-5-ethoxythiazole	80°-81°C at 10 mm Hg	–	Powerful cucumber, green pepper, onion, earthy
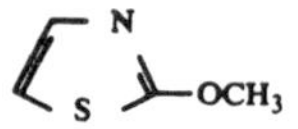 N, S, OCH_3 2-Methoxythiazole	22°C at 1.8 mm Hg	–	Cereal, bread, caramel
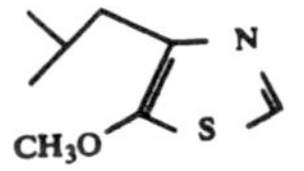 N, S, CH_3O 4-Isobutyl-5-methoxythioazole	92°-93°C at 10 mm Hg	–	Vegetable soup minestrone fragrance and pepper, onion, celery, and green vegetable flavor characteristics
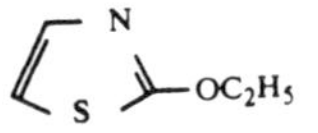 N, S, OC_2H_5 2-Ethoxythiazole	32°C at 2 mm Hg	–	Strong, burnt peanut-roasted meat aroma

Examples are given of the preparation of various of the substituted thiazoles and their use in flavor compositions.

Example 1: *Preparation of Trimethylthiazole* – A 250 ml flask fitted with mechanical stirrer, addition funnel, and dry ice condenser is charged with 25 ml of dry benzene, 14.5 g (0.25 mol) of acetamide, and 13.32 g (0.06 mol) of phosphorus pentasulfide. After heating the flask contents on a steam bath for 5 min, 30 g (0.2 mol) of 3-bromo-2-butanone are added dropwise with continued stirring.

The reaction mixture is then heated on a steam bath to get an exothermic reaction started and then removed from the bath. After the bromobutanone addition is complete, the reaction mixture is refluxed for ½ hr, 100 ml of water is added, the mixture is stirred for an additional hour, after cooling the phases are separated in a separatory funnel, and the aqueous phase is isolated.

This lower aqueous layer is made alkaline with 5 N sodium hydroxide solution and extracted four times with 100 ml volumes of ether. The ether extracts are dried over sodium sulfate, filtered, and evaporated. The residue is distilled at 29°C and 0.8 mm Hg. The distilled product is then redistilled at 41°C and 2.2 mm Hg to provide 8.3 g of colorless liquid trimethylthiazole.

The product is described as characteristic of dark chocolate with additional suitability for coffee, caramel, and related flavors. The threshold level in this context is said to be 0.050 ppm with a use level at 0.2 to 0.5 ppm. It is also described as having a sweet, light green character at 1 ppm in water; sweet, green having a weedy aftertaste at 2 ppm; and strong, fresh stringbean-green with astringent and bitter aftertaste and a light sharp note such as found in Chinese mustard at 5 ppm. In chicken broth at 1 ppm it sharpens the broth and improves it slightly.

Example 2: *Preparation of 4-Isobutyl-5-Methoxythiazole* – A 500 ml flask equipped with heating mantle, condenser, and drying tube is charged with 27.25 g (0.125 mol) of phosphorus pentasulfide, 17.3 g (0.1 mol) of N-formylleucine methyl ester, and 250 ml of chloroform, and the contents are refluxed for 24 hr. The reaction mixture is thereupon cooled to room temperature, washed with 250 ml of 10% potassium hydroxide, and then washed with 250 ml of distilled water.

The chloroform layer is separated from the aqueous layer and dried over sodium sulfate. At this point the chloroform layer has the aroma of green peppers. The chloroform is then evaporated.

The residue is distilled at 55° to 60°C and 1.1 mm Hg, and then redistilled at 92° to 93°C and 10 mm Hg to provide 5.6 g of isobutylmethoxythiazole as a colorless liquid.

This material has an odor in 1% alcoholic solution of fresh sliced onion in vinegar with a green pepper character. At 0.1 ppm in chicken broth it pleasantly accents the green vegetable notes such as parsley. It is suitable for dips, salad dressing, relish dishes, and soups. It is useful for a wide range of green vegetable flavor notes, e.g., peas and beans. Its use level is estimated as 0.01 to 0.05 ppm. Its odor has also been characterized as vegetable compote and minestrone, with a mixture of green pepper, onion, celery, and black pepper.

Example 3: *Preparation of 2-Ethoxythiazole* – A 250 ml single-neck flask supplied with condenser and heating mantle is charged with 11.9 g (0.1 mol) of 2-chlorothiazole, 7 g (0.1 mol) of sodium ethoxide, and 100 ml of absolute ethanol, and the mixture is heated under reflux for 1 hr. After cooling, 100 ml of water is added and the solution extracted three times with 100 ml of diethyl ether. The combined ether extracts are dried over anhydrous sodium sulfate, filtered, and the solvent evaporated.

The residue is distilled under reduced pressure and the 2-ethoxythiazole is obtained as a colorless liquid with a boiling point of 32° to 33°C at 2 mm Hg in a yield of 3.5 g.

Evaluated as a 2.5% ethanol solution, this material has a strong burnt, oven-roasted meat odor which, upon drying, changes to the sweet fruity character of roasted onions. The odor is also described as meaty with a tomato leaf, bay leaf, minty character. In a 0.2 ppm aqueous solution it has a general roasted character, especially suggestive of roasted meat or pot roast. At 0.5 ppm in aqueous solution, the flavor notes of cooked vegetables and cabbage are present. This product has value in meat, milk, cream, vanilla, maple, butterscotch, caramel, roasted nut, and cooked vegetable (onion, cabbage) flavors. The threshold level is about 0.1 ppm, and the use level is 0.2 to 0.5 ppm.

FRUIT AND VEGETABLE FLAVORS

FRUIT FLAVORS

Theaspiran

P. Naegeli; U.S. Patent 4,072,719; February 7, 1978; assigned to Givaudan Corporation has determined that theaspiran (2,6,10,10-tetramethyl-1-oxaspiro[4,5]-dec-6-ene) is an odor- and/or flavor-imparting substance distinguished by special fresh, fruity odor or flavor properties. Of particular interest is a berry-like, green note and a sweetish, woody nuance appearing with increasing concentration. The theaspiran can accordingly be used, for example, for the perfuming or flavoring of products such as cosmetics (soaps, salves, powders etc.), detergents, foods, tobacco and drinks, the theaspiran preferably not being used alone but in the form of compositions which also contain other odor- or flavor-imparting substances.

Because of its very natural notes, theaspiran is especially suited as an odorant for modifying known compositions; for example, those of the Chypre type. Thus, for example, it is very well suited to combinations with flower notes such as, for example, neroli and rose notes.

As a flavor-imparting substance, theaspiran can be used for the production or improvement, intensification, enhancement or modification of fruit or berry flavor in foods, tobacco and drinks. Some effects which can be produced with theaspiran are compiled below.

Aroma	Amount (ppm)	Effect
Tobacco (top flavor)	0.0005*	Better tenacity aroma; intensified fruitier impression
Vanilla	0.03*	Rounding off effect; woody nuance
Raspberry	0.001*	Rounding off effect; pleasant, woody natural nuance

*In the finished product.

Theaspiran can also be oxidized to the flavor-imparting substance theaspirone. The following formula scheme shows the preparation of both theaspiran and theaspirone from β-ionone. Theaspiran is prepared by treating β-ionone with a protonic acid, p-toluene sulfonic acid being preferred. The oxidation of theaspiran produces theaspirone. Formula (4) is theaspiran and formula (5) is theaspirone.

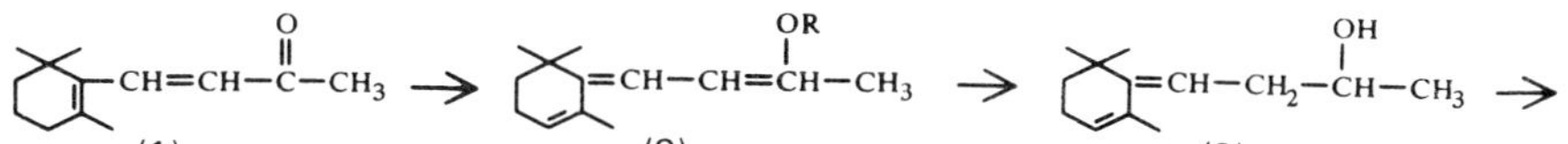

$-CH=CH-C(=O)-CH_3$ → $=CH-CH=C(OR)-CH_3$ → $=CH-CH_2-CH(OH)-CH_3$ →

(1)
β-Ionone

(2)
4-(2,6,6-Trimethyl-2-cyclohexen-1-ylidene)-2-acyloxy-but-2-ene
R = acyl such as lower alkanoyl e.g. acetyl or aroyl, e.g. benzoyl.

(3)
4-(2,6,6-Trimethyl-2-cyclohexene-1-ylidene)-butan-2-ol

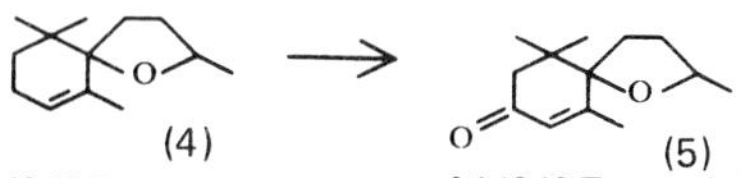

→

(4)
2,6,10,10-Tetramethyl-1-oxaspiro[4,5]-dec-6-ene

(5)
2,6,10,10-Tetramethyl-1-oxaspiro[4.5]-dec-6-en-8-one

Example 1: *Vanilla Flavor –*

Composition	Parts by Weight A	Parts by Weight B
Guaiacol (1% in ethanol)	1.0	1.0
Heliotropin (1% in ethanol)	1.0	1.0
Isoeugenol (1% in ethanol)	2.0	2.0
p-Hydroxybenzaldehyde (1% in ethanol)	3.0	3.0
Vanillin	20.0	20.0
Ethylvanillin	120.0	120.0
Theaspiran (1% in ethanol)	–	3.0
Ethanol	853.0	850.0

Composition B differs organoleptically in a very advantageous manner from composition A which is a conventional vanilla aroma. In particular, the theaspiran imparts a weakly woody and fruity note, by which means the vanilla fragrance is rounded off in a remarkable manner.

100 g of the above vanilla flavor are incorporated (using methods known per se) into 100 kg of caramel (milk/cream) toffees.

Example 2:

Composition	Parts by Weight
Petitgrain oil Paraguay	400
Geraniol extra	200
Phenylethyl alcohol	160
Methyl anthranilate	160
p-Methylquinoline (10% ethanol)	10
Theaspiran (10% in diethyl phthalate)	70

The initially slightly original flowery composition (neroli) has a substantially more rounded-off action and is fuller, softer and sweeter by the addition of theaspiran. The impression of a fresh, natural blossom fragrance is striking.

Preparation of trans-E-Crotonoyl-2,2,6-Trimethylcyclohexane

D.R. de Hann and D.K. Kettenes; U.S. Patent 4,109,022; August 22, 1978; assigned to P.F.W. Beheer BV, Netherlands describe the preparation of the trans-E-1-crotonoyl-2,2,6-trimethylcyclohexane (compound 4) and its use in providing fruity flavors. The preparation of the cis-E stereoisomer of this compound and its utility (as well as that of the trans-E stereoisomer) in perfume compositions is illustrated in U.S. Patent 4,136,066, to be found in the fragrance section of this book. In the synthesis of compound 4, cis-1-acetyl-2,2,6-trimethylcyclohexane (compound 1) is epimerized to the trans isomer (compound 2), which is then condensed with acetaldehyde to the aldol product (compound 3) and dehydrated to produce compound 4, as illustrated by the following reaction scheme, where PPA is polyphosphoric acid and PTS is p-toluene sulfonic acid.

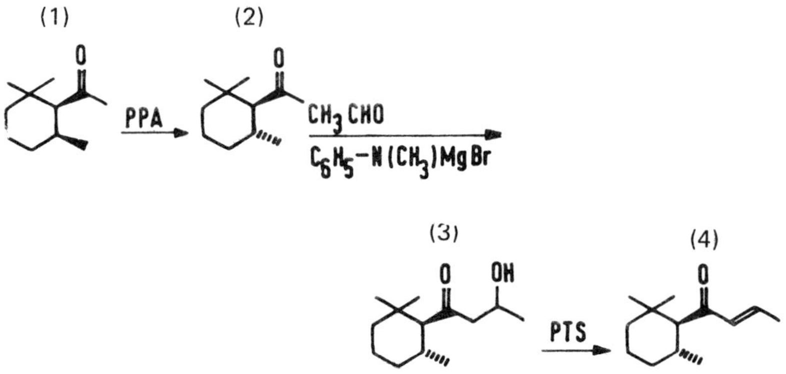

(1) $\xrightarrow{PPA}$ (2) $\xrightarrow[C_6H_5-N(CH_3)MgBr]{CH_3CHO}$ (3) $\xrightarrow{PTS}$ (4)

In flavor compositions, use levels may vary within a wide range; the powerful flavor of trans-E-1-crotonoyl-2,2,6-trimethylcyclohexane is obvious. The percentage of compound can be varied between 0.001% and 1% of the flavoring composition, depending on the choice of the other ingredients. Moreover, it is very effective in improving the flavor of natural extracts such as raspberry concentrates and other berry or stone fruit concentrates. As little as 0.004% of trans-E-1-crotonoyl-2,2,6-trimethylcyclohexane in a natural raspberry concentrate dramatically improves the raspberry flavor. This level corresponds with 0.1 ppm of compound 4 in the finished product. The level in finished products like beverages, ice cream, candy dessert mixes etc. is thus usually below 1 ppm and sometimes as low as 0.1 ppb. This may be higher for chewing gum.

Although many examples of crotonoyl-trimethylcyclohexenes and cyclohexadienes are known as flavoring and perfume materials in the recent patent literature there is no mention in the prior art of analogous compounds with a saturated ring system.

Careful testing of the four stereoisomers of 1-crotonoyl-2,2,6-trimethylcylohexane, has shown them to be not only totally different from the compounds with an unsaturated ring system but also strikingly different from each other. The trans, E isomer is fresh fruity and devoid of any woody and β-ionone character, which is as a rule observed in the aforementioned compounds with an unsaturated ring system. The trans, E isomer has been found to be superior to its stereoisomers in all flavor and perfume applications tested.

Example: A red currant juice concentrate was mixed with the compounds mentioned below, to yield the following products:

(a) Red currant juice concentrate with 1 ppm of trans-E-1-crotonoyl-2,2,6-trimethylcyclohexane.

(b) The same concentrate with 1 ppm of cis-E-1-crotonoyl-2,2,6-trimethylcyclohexane.

(c) The same concentrate with 1 ppm of E-β-damascenone.

(d) The same concentrate with 1 ppm E-β-damascone.

The flavor of the concentrates (a), (b), (c), (d) and a control was compared by tasting drinks prepared by adding 15 ml of concentrated sugar syrup (60%), 0.1 g of citric acid 50% in water, 85 ml of water and 0.3 g of the juice concentrate.

Comparative tests by a test panel gave the following results. Concentrate (a) was found to possess a strongly increased very natural currant taste when compared with the control drink. Concentrate (b) was found to possess a much less natural currant taste. Concentrate (c) was found to have a jammy-raspberry-like character which does not bring the product closer to natural red currant taste. The same is true for concentrate (d) but here the effect of raspberry similarity is even less pronounced.

α-Oxy(oxo)Mercaptans

As has been discussed earlier in this book (see U.S. Patent 4,070,308), certain α-oxy(oxo)mercaptans impart grapefruit-like, green fruity, concord grape and buchu leaf oil-like aromas to perfumes and perfumed articles.

W.J. Evers, H.H. Heinsohn, Jr. and M.H. Vock; U.S. Patent 4,081,480; March 28, 1978; assigned to International Flavors & Fragrances Inc., have found that the compounds described in the abovementioned patent plus several other of the α-oxy(oxo)mercaptans may provide to various foods, chewing gums, medicinal products and toothpastes grapefruit-like, green, citrus, bitter, lemon, buchu leaf oil-like, black currant-like, green vegetable-like, minty and/or cooling flavor characteristics.

Chief among these additional products is 2-mercaptocyclododecanone-1, the preparation and use of which is described in the following examples.

Example 1: *(A) Preparation of 2-Chlorocyclododecanone* – Into a 1,000 ml three-necked, round-bottom flask, equipped with a Y-tube, mechanical stirrer, thermometer, 125 ml addition funnel, vacuum adapter to water aspirator and water bath is placed a solution of 100 g of cyclododecanone in 100 ml anhydrous benzene. The cyclododecanone is dissolved in the benzene with stirring at 25°C and the solution cools by itself as the solution occurs, to 10°C. The solution then is allowed to warm up to 26°C, at which point dropwise addition of SO_2Cl_2 begins with stirring and continues for 1 hour at 26° to 33°C, during which time acidic gases are removed using a water aspirator vacuum. At the end of the addition, stirring is continued for another one hour period. The reaction mass is then transferred to a 1,000 ml, one-neck, round-bottom flask and concentrated on a rotary evaporator during which time the benzene solvent is trapped in a dry ice/acetone trap. GLC analysis on an 8 foot x ¼ inch SE-30 column yields the information that the reaction product contains approximately 24% chlorinated ketone.

The reaction product (117 g) is introduced into a 250 ml, three-necked round-bottom flask, equipped with distillation column, reflux head, magnetic stirrer, heating mantle, high vacuum pump and dry ice trap. Unreacted cyclododecanone (72.5 g) is recovered, distilling at 95° to 113°C vapor temperature and 0.6 to 1.2 mm Hg. The residue (30.3 g) is transferred to a 100 ml, three-necked, round-bottom flask, equipped with an unpacked distillation column and fractionally distilled. Mass spectral, NMR and IR analysis of GLC trapped product confirm its identity.

(B) Preparation of 2-Mercaptocyclododecanone — Into a 250 ml, three-necked, round-bottom flask, equipped with magnetic stirrer, pot thermometer, six inch distillation column with gas outlet at top attached to rubber tubing leading above a stirring 10% NaOH solution, gas inlet tube (for hydrogen sulfide bubbling), gas bubbler (Primol) empty trap between hydrogen sulfide cylinder and bubbler, hydrogen sulfide cylinder, isopropanol/dry ice bath and 50 ml addition funnel is added a solution of 4.65 g of sodium methoxide dissolved in 45 ml anhydrous methanol.

The sodium methoxide solution is cooled to -10°C, at which point hydrogen sulfide bubbling is commenced below the surface of the sodium methoxide solution. The reaction is maintained at a temperature of -5° to -10°C, during which time the hydrogen sulfide bubbling proceeds while stirring the reaction mass for a period of 1½ hours. At this point in time, 90 ml anhydrous methanol is used to dissolve completely the chlorinated ketone starting material at room temperature. While maintaining the reaction mass at 0° to 4°C, the methanolic solution of chlorinated ketone is added dropwise during simultaneous slow flow of hydrogen sulfide. After all the chlorinated ketone solution is added (after another ½ hour) a heavy solid precipitate forms which is stirred under hydrogen sulfide flow.

At the end of an additional three minutes, a sample of the reaction mass is acidified to a pH of between 1 and 2 and extracted with methylene chloride. The methylene chloride extract sample is washed with water, dried and concentrated on a rotary evaporator. The resulting sample contains 2.2% product by GLC. The reaction mass is continued to be treated with hydrogen sulfide for an additional 3 hours while warmed to 17° to 35°C (25° to 35°C for the last hour). The reaction mass is concentrated on a rotary evaporator to 15 ml (a thick yellow slurry). Distilled water (35 ml) and 53 g of 10% NaOH solution is added to the reaction mass and stirred for 20 minutes under a nitrogen blanket at 25° to 30°C.

The reaction mass is then extracted with two 35 ml portions of methylene chloride and the extracts are combined, dried and concentrated yielding 2.6 g oil.

The basic aqueous solution is acidified with 65 ml aqueous 10% hydrochloric acid to pH = 2. The acidified solution is extracted with three 40 ml portions of methylene chloride.

The extracts are combined and washed with 30 ml saturated sodium chloride and dried over anhydrous sodium sulfate. The resulting material is gravity filtered and concentrated on a rotary evaporator to yield 6.2 g of a heavy yellow oil containing 99.3% 2-mercaptocyclododecanone having the structure:

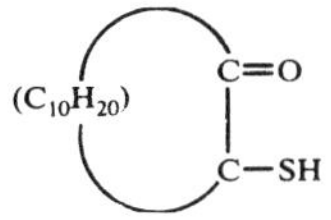

$$(C_{10}H_{20})\left\{\begin{array}{l} C{=}O \\ | \\ C{-}SH \end{array}\right.$$

This structure if confirmed by NMR, IR and mass spectral analyses of GLC trapped material.

Example 2: *Vegetable Flavor Formulation* – 2-Mercaptocyclododecanone-1 is added directly to a food product prior to processing and canning. The following illustrates the beneficial flavor effect when 2-mercaptocyclododecanone-1 prepared according to Example 1 is added directly to several food products just prior to their consumption:

(a) In blended vegetable sauce at approximately 30 ppm, it brings out the green vegetable notes with minty nuances.

(b) In vegetable soup at 40 ppm, it imparts a fresh vegetable flavor. The green notes give the entire vegetable flavor a fuller body.

(c) In bean tomato sauce at approximately 20 ppm, it modifies the flavor by reducing the harsh character of the tomato spice mixture while at the same time adding green fresh notes and developing the cooked tomato note to a fresh tomato note.

The levels of concentration of the 2-mercaptocyclododecanone-1 may be reduced by 25% when 2-isobutyl thiazole is added at the rate of 5 ppm in addition to the 2-mercaptocyclododecanone-1 to the various products set forth above. It should be understood that noticeable differences in the flavor are discernible at other concentrations.

Extracted from Cocoa Shells

In the processing of cocoa beans, the shells are removed after roasting, usually by winnowing techniques. Previously there has been little economic use for the shells, and they have generally been disposed of as refuse or burnt.

I.B. Eggen; U.S. Patent 4,156,030; May 22, 1979; assigned to Societe d'Assistance Technique pour Produits Nestle SA, Switzerland has found that a water-soluble flavoring and coloring material may be prepared by extracting the shells with acidified ethanol. The extracted material has a berry-like taste and a characteristic color and is especially suitable for incorporation in soft drinks, especially carbonated beverages.

The ethanol may be acidified with any nontoxic acid. Preferred acids are hydrochloric acid, phosphoric acid, and organic acids such as tartaric acid and citric acid. Hydrochloric acid leads to an extract having a desirable characteristic red color.

The acidified ethanol solution preferably contains 85 to 90%, advantageously 85 to 87% of ethanol and the balance acid. Extraction is preferably performed under reflux and the cocoa shells are preferably first ground to a particle size of from 50 to 1,000 μ, advantageously 200 to 400 μ. Under the above preferred conditions, a reaction time within the range 30 minutes to 2 hours gives a highly satisfactory extract.

Extraction produces a liquid phase containing acid, ethanol and the extracted components mixed with a residual sludge. After cooling the mixture, the liquid phase is separated from the sludge, preferably in a filter press, and the sludge is discarded. Generally the bulk of the ethanol is removed from the extract to give a concentrated flavoring extract and this is advantageously achieved by distillation, which is preferably performed at atmospheric pressure. The residual extract after distillation typically contains about 5% by weight of ethanol and has a solids content of from 30 to 50%.

The separated sludge contains a considerable portion of ethanol and this is advantageously recovered by distilling the residue. The ethanol so obtained is normally combined with that recovered from the extract and may be used for extracting a fresh batch of cocoa shells.

The liquid extracts obtained by this process may be used in a number of ways. When the concentrated extract is added to water, preferably carbonated, in quantities of the order of 0.3 to 0.9% by weight a highly satisfactory soft drink is obtained. The flavor is further improved by the addition of sugar, citric acid, etc.

The extract may also be used as a flavoring and coloring material for various confectionery and ice cream products. For example, the extract may be incorporated in confectionery coatings which typically contain 30 to 35% fat, at a level of 1.5 to 3.0%. In addition the extract may further be used in gelatin-type desserts, chocolate pudding (tinted red), baked pudding (colored and flavored), ribbon candy (colored and flavored), dietetic coating (colored or flavored), dark chocolate coating (colored or flavored), chocolate cake frosting (tinted red), and cake frosting and filling (colored and flavored).

VEGETABLE FLAVORS

α-Substituted Alkylidene Methionals

Two α-substituted alkylidene methionals have been found by *D.A. Withycombe, A. Hruza, M.H. Vock, C. Giacino, B.D. Mookherjee, A.O. Pittet and W.L. Schreiber; U.S. Patent 4,154,766; May 15, 1979; assigned to International Flavors & Fragrances Inc.* to augment and enhance certain flavors and aromas in foodstuffs.

The structures of the alpha-substituted alkylidene methionals of the process as well as their food flavor properties are set forth in the table on the following page.

Compound	Structure	Flavor Property
2-[(methylthio)-methyl]-3-phenyl-2-propenal	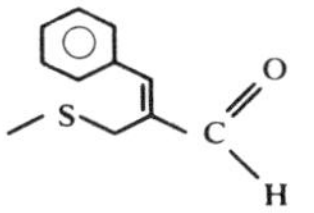	A green weedy, broccoli-like, green bean-like aroma with a green weedy, broccoli-like and green bean-like flavor at 20 ppm.
5-methyl-alpha-[(methylthio)-methyl]-2-furan acrolein	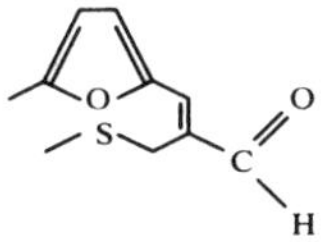	A potato-like, methional-like, green sulfury, hydrolyzed vegetable protein, chicken-like, cabbage, green potato-like aroma character with potato-like, metallic sulfury green, cabbage-like, chicken and mushroom-like flavor nuances at 1 ppm.

Example 1: *Preparation of 5-Methyl-α-[(Methylthio)Methyl]-2-Furan Acrolein* – into a 1 liter reaction flask equipped with mechanical stirrer, 250 ml addition funnel, Y adapter, thermometer and Friedrich condenser is placed 104 g methional and 110 g of 5-methyl furfural. The reaction mass is cooled to 0°C and 100 ml of a 0.5 molar solution of sodium hydroxide is added dropwise with stirring. The reaction mass is permitted to stir at 0°C for 2 hours. The resulting reaction mass is extracted with three 100 ml portions of methylene chloride and the extracts are combined, dried over anhydrous sodium sulfate and evaporated on a rotary evaporator. The resulting product is distilled on a short path column, boiling point 155°C (5 mm Hg).

Example 2: *Potato Flavor* – A potato flavoring material is prepared by admixing the following ingredients:

Ingredients	Parts by Weight
Diacetyl (1% solution)	0.20
Furfural	0.2
2-Acetyl-3-ethyl pyrazine (1% solution)	1.0
2-Ethyl-3-methyl pyrazine	4.0
Methional	2.0
5-Methyl-α-[(methylthio)methyl]-2-furan acrolein*	0.2
Ethanol (95% food grade)	91.6

*Prepared according to Example 1.

A bench panel of five individuals compared the above formulation with one not containing any 5-methyl-α-[(methylthio)methyl]-2-furan acrolein but identical in all other respects. The formulations were compared at the rate of 10 ppm in water solutions. It was concluded that the 5-methyl-α-[(methylthio)methyl]-2-furan acrolein imparted to the flavor a mashed potato-like taste with cabbage nuances.

FRUITY FLAVOR ENHANCERS

α-Oxy(oxo) Sulfides and Ethers

W.J. Evers, H.H. Heinsohn, Jr., and M.H. Vock; U.S. Patent 4,097,615; June 27, 1978; assigned to International Flavors & Fragrances Inc. describe certain α-oxy (oxo) sulfides and ethers used as flavor adjuvants or to augment or enhance flavors and aromas of foodstuffs. These compounds have the structure:

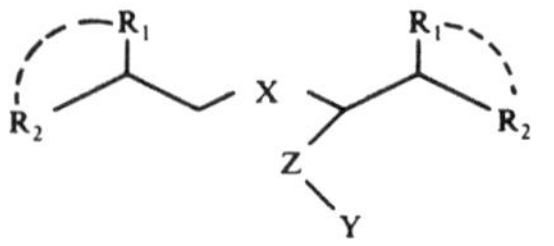

wherein X is

$$-\underset{\underset{O}{\|}}{C}- \quad \text{or} \quad -\overset{\overset{OH}{|}}{C}H-$$

Z is either sulfur or oxygen; when R_1 and R_2 are taken separately, R_1 is hydrogen or methyl, and R_2 is methyl; and when R_1 and R_2 are taken together, R_1 and R_2 form phenyl moieties, and Y is one of C_{1-4} alkyl, C_3 or C_4 alkenyl, acetyl, methoxycarbonylmethyl, or 1,3-diethylacetonyl.

Such α-oxy(oxo) sulfides and ethers are obtained by reacting an alkanone with SO_2Cl_2 to form an α-chloroalkanone; reacting the α-chloroalkanone with either an alkali metal mercaptide or an alkali metal alkoxide (depending on whether Z is sulfur or oxygen) to form either an α-oxo ether which can be used for its food flavor properties; or, if desired, reacting the resulting α-oxo sulfide or α-oxo ether with a reducing agent such as an alkali metal borohydride in order to obtain an α-oxy sulfide or an α-oxy ether.

Specific examples of α-oxy(oxo) sulfides and ethers and their food flavor properties are as follows:

COMPOUND	STRUCTURE	FLAVOR PROPERTY
3-methylthio-4-heptanone	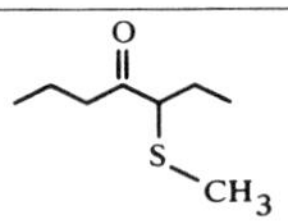 O, S, CH_3	Green, piney, necroli-like, fruity, blackcurrant,bucchu-like and concord grape aroma characteristic and sweet, fruity, blackcurrant, concord grape-like, minty and astringent flavor characteristics at 1.0 ppm.
3-propylthio-4-heptanol	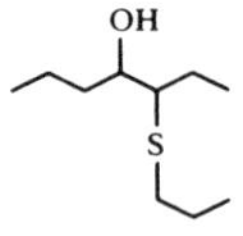 OH, S	"Violet leaves," melon, cucumber, green, fruity, vegetable and floral aroma characteristics with violet leaves melon, cucumber, green, citrus, vegetable, garlic flavor characteristics with a lasting mouthfeel at 5 ppm.

(continued)

COMPOUND	STRUCTURE	FLAVOR PROPERTY
3-isobutylthio-4-heptanone		Sweet/floral, citrus, fruity, necroli, bergamot, jasmin aroma characteristic with a green/floral, minty, petitgrain, fruity, citrus, sulfury flavor characteristic at 2 ppm.
3-propylthio-4-heptanone		Sweet, floral, jasmin and berry-like aroma characteristic with sweet/floral, jasmin, grapefruit and blackcurrant flavor characteristic at 2 ppm.
3-(methallylthio)-2,6-dimethyl-4-heptanone		Citrus, grapefruit, floral, celery stalk-like, rosey aroma characteristic with citrus, grapefruit, floral,spicey,green/fruity, astringent flavor characteristic at 2 ppm.
3-crotylthio-2,6-dimethyl-4-heptanone		Floral, citronellal-like, citrus, grapefruit, woody aroma characteristic and a citrus grapefruit, coriander-like flavor characteristic at 3 ppm.
3-allythio-2,6-dimethyl-4-heptanone		A sweet, grapefruit, floral, citrus, green/spicey,necroli-like aroma characteristic with a sweet, citrus, floral/green,citronellal-like, "decaying fruit-" like flavor characteristic at 4 ppm.
3[(methoxycarbonyl)-methylthio]-4-heptanone		At 10 ppm, a sweet, sulfury, nutty, meaty, cereal aroma character and a sweet,meaty, nutty, cereal flavor characteristic with an outstanding mouthfeel effect.
3-methoxy-4-heptanone		At 3 ppm, a sweet, fruity, gooseberry-like, grape, almond aroma character with a sweet, fruity, gooseberry-like, nutty, grape flavor character.
1-propylthio-1,3-diphenyl-2-propanone		At 0.5 ppm, a green onion aroma with a lachrymating onion and biting effect and a sweet, rubbery, meaty flavor characteristic; at 2 ppm the garlic aroma dominates alongwith the fresh onion notes.
(1,3-diethylacetonyl) (1,3-diisopropylacetonyl) sulfide		At 10 ppm, a grapefruit, floral and woody aroma character with a sweet, sulfury, grapefruit-like, mandarin flavor characteristic and bitter nuances.
3-acetylthio-4-heptanone		At 1 ppm a fresh fruity, blackcurrant-like, buchu leaf oil-like, aroma with meaty and sulfury nuances and a fresh fruit, blackcurrant-like flavor characteristic with an oniony aftertaste.

These compounds must be combined with other flavor adjuvants, ,e.g., geranid, maltol, vanillin, methyl cinnamate, isoamyl acetate, diacetyl, black currant juice, coriander oil, cinnamic alcohol, citral, methional, methyl propyl disulfide, etc.

Example: 3-propylthio-4-heptanol is added to a commercial instant tomato soup mix (Tomatancreme Suppe) at the rate of 2 ppm (based on the weight of the soup as ready to eat, produced by adding 80 g of the dry soup mix to 1,000 ml water and then bringing the resulting mixture to a boil). A second control is prepared which is identical to the initial sample except for the absence of the 3-propylthio-4-heptanol. A four member panel of flavorists compared the control to the soup containing 3-propylthio-4-heptanol. All four members of the panel indicated a strong preference for the tomato soup containing the 3-propylthio-4-heptanol. All four members of the panel indicated that the soup containing the 3-propylthio-4-heptanol includes a fresh tomato note which is not present in the soup prepared without using the 3-propylthio-4-heptanol.

Maltyl-2-Methyl Pentenoates

C.J. Mussinan, B.D. Mookherjee, A.E. Goossens and M.H. Vock; U.S. Patents 4,139,541; February 13, 1979; and 4,169,900; October 2, 1979; both assigned to International Flavors & Fragrances Inc. describe maltyl-2-methyl pentenoates having the structure:

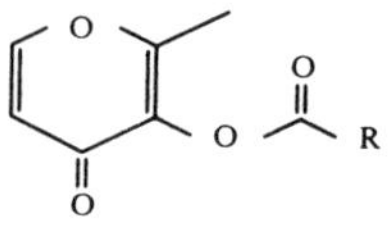

wherein R is a secondary pentenyl moiety having one of the structures:

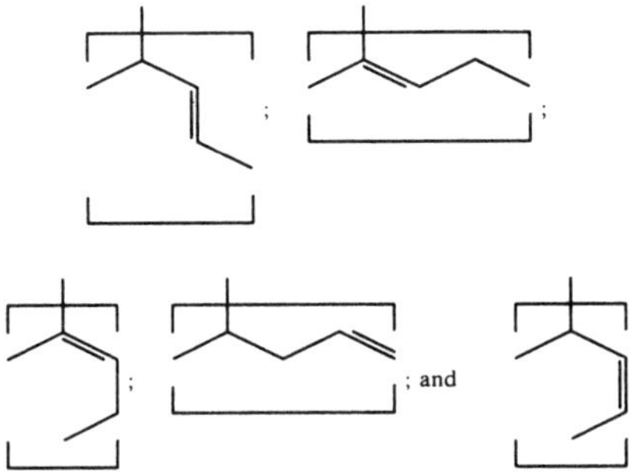

; ; ; ; and

and their use in enhancing flavors and aromas of foodstuffs, chewing gum, medicinal products and chewing tabacco.

Examples of these maltyl-2-methyl pentenoates and their organoleptic properties are shown on the following page.

Structure*	Organoleptic Properties
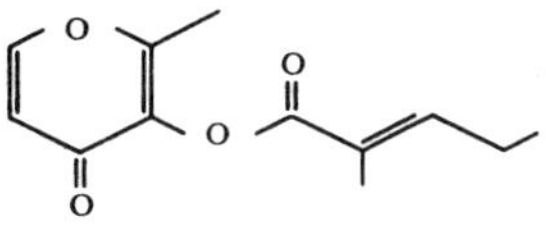 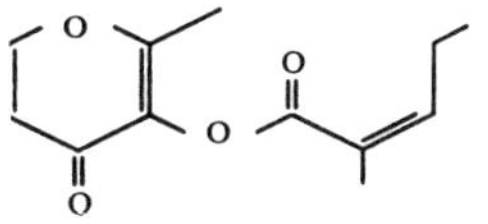 (1) and (2)	A fruity, berry, yeasty, winey, strawberry, maltol-like and cognac aroma with a fruity, berry, yeasty, winey, strawberry-like and allium taste at 10 ppm.
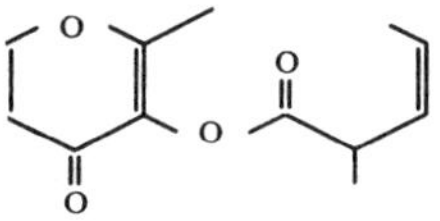 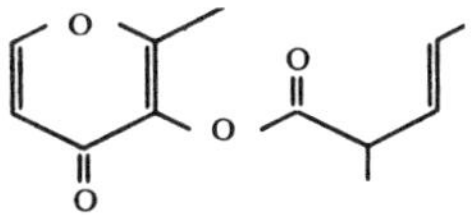 (3) and (4)	A sweet, fruity, strawberry-like and red berry aroma characteristic with sweet, fruity and strawberry flavor characteristics and green nuances at 5 ppm.

*Compounds (1) and (2) are cis- and trans isomers of maltyl-2-methyl-2-pentenoate. Compounds (3) and (4) are cis- and trans isomers of maltyl-2-methyl-3-pentenoate.

Example 1: Into a 250 ml two necked reaction flask equipped with mechanical stirrer, Friedrich's condenser and gas trap, 12 g of a mixture containing a high proportion of 2-methyl-cis-3-pentenoic acid prepared according to Example 7 at column 18 of U.S. Patent 3,984,579 issued on October 5, 1976, in 50 ml benzene is added. Friedrich's condenser side arm is equipped in such a way as to vent the gases released through a trap into a beaker containing a commerical preparation of sodium hypochlorite. With vigorous stirring 11.9 g of freshly distilled thionyl chloride are added to the reaction mixture.

The reaction mixture is then stirred and heated until no more hydrogen chloride or sulfur dioxide gas is released (over a period of 45 minutes). The reaction mass is then allowed to cool and 12.6 g of maltol and 50 ml of benzene are added thereto. The mixture is again heated until no more gas evolves (period of time: 45 minutes). The benzene solvent is then removed on a rotary evaporator. The resulting reaction product, maltyl-2-methyl-3-pentenoate compound (3) containing a high proportion (80%) of maltyl-2-methyl-cis-3-pentenoate and 20% maltyl-2-methyl-trans-3-pentenoate is trapped on a preparative GLC column.

A total of 0.55 g of maltyl-2-methyl-3-pentenoate is collected having a purity of greater than 99%

Example 2: The following basic strawberry flavor formulation is prepared:

Ingredients	Parts by Weight
p-Hydroxybenzyl acetone	2
Vanillin	15
Maltol	20
Ethyl methyl phenyl glycidate	15
Benzyl acetate	20
Ethyl butyrate	10
Methyl cinnamate	5
Methyl anthranilate	5
α-Ionone	1
γ-Undecalactone	2
Diacetyl	2

(continued)

Ingredients	Parts by Weight
Anethole	1
cis-3-Hexenol	17
Ethyl alcohol (95% aqueous food grade)	385
Propylene glycol	500

At the rate of 6%, to half of the formulation given above, compound (3), maltyl-2-methyl-3-pentenoate, prepared according to Example 1 is added. The basic strawberry formulation with compound (3) is compared to the basic strawberry formulation without. Both flavors are compared at the rate of 50 ppm in water and evaluated by a bench panel composed of five individuals. The flavor containing compound (3) is found to have a sweet, fresh, more strawberry-like aroma and taste. Therefore, it is preferred unanimously by the members of the bench panel as being more pleasant, more characteristic and as the better strawberry flavor.

Example 3: 500 g of water are heated to boil and 500 g of dextrin is added with rapid and efficient mixing, using a closed turbine, high shear mixer. Mixing is continued until a homogeneous solution is obtained.

81 g of maltyl-2-methyl-cis-3-pentenoate prepared according to Example 1 is emulsified in 300 g of the shell composition solution (prepared as described above) by means of a homogenizing mixer. At the start the temperature of the matrix composition solution is 20°C and of the maltyl-2-methyl-cis-3-pentenoate 15°C. The mixing vessel is cooled during the operation of the mixer to keep the temperature below 25°C.

100 g of polyethylene glycol (Carbowax 400) at a temperature of 25°C is placed in a vessel equipped with a homogenizing mixer. 100 g of compound (3) produced according to Example 1 is introduced into the polyethylene glycol in a thin stream with steady medium speed operation of the mixer (about 1,500 rpm) shaft speed). By the action of the mixer, the compound is broken up into coarse liquid particles, which in contact with the polyethylene glycol, are rapidly converted into gel particles and finally into virtually anhydrous capsule granules.

The capsule granules are separated from the excess polyethylene glycol by means of a basket centrifuge, mixed with the flavor formulation of Example 2 and added to chewing gum, chewable vitamin tablets and chewing tobacco to provide a strawberry flavor.

FRUIT BITS

Resembling Bits of Natural Fruit Meat

S.F. Ziccarelli; U.S. Patent 4,086,367; April 25, 1978; assigned to Beatrice Foods Company describes a fruit flavored composition, which resembles bits of natural fruit meat. More particularly, the composition has the important properties of being shelf stable and self-preserving.

The fruit flavored bits have the appearance, taste, consistency and mouthfeel of natural fruit meat bits and comprise fruit flavors (natural or imitation) and finely divided fat particles. The fat particles have a melting point not greater than 125°F, especially between 90° and 124°F, and are substantially coated with a fat enrobing agent. The composition must also have a moisture content of no more than 5% by weight.

The fruit flavors may be low moisture natural fruit juice essences or artificially produced fruit flavors or combinations thereof, so long as the flavors have a taste similar to natural fruit meat. Thus, for purposes of this discussion, the term fruit flavors is intended to embrace artifical fruit flavors as well as natural fruit flavors.

The composition may be added to cakes, pies, cereals, pancakes, ice cream, toppings, cookies, sauces, and the like. The composition has a special advantage in that it may be added to a dry composition and will remain shelf stable and self-preserving for extended lengths of time. Accordingly, the dry ingredients of a pancake mix can be packaged in a conventional foil or plastic pouch, including, for example, blueberry bits made as described and the contents of that package will be shelf stable.

Example: 4 parts of imitation blueberry flavor, 18 parts of sugar, 2 parts of dry yeast and whey (FP-37, Beatrice Foods Company, used as the enrobing agent), 2 parts of malic acid, 2 parts of 75° Bloom gelatin, 0.04 part of FD&C Red color No. 40, 0.02 part of FD&C Blue color No. 1, 14 parts of dextrin and 50 parts of 97° to 101°F melt fat were added to a blender and thoroughly mixed at room temperature. The mixture was then placed in a pellet mill and pelletized at a temperature of about 75° to 85°F, which temperature was generated in the composition by the mechanical energy of mixing and extruding. The extruded rods were cut to irregular lengths having a major width of about ⅛ inch and a maximum length of about ¼ inch, although the length was random.

A portion of the product was set aside for shelf stability tests at ambient conditions and after seven months no deterioration of the product was evident. Another portion of the product was stored at a temperature of 110°F for 10 days in an accelerated shelf-life test and no substantial deterioration of the product was observed. The product was then placed in a water-tight container and partially immersed in a water bath maintained at 140°F for a severe accelerated test. The product remained essentially solid for 8 hours. Another portion of the product was placed in the refrigerator and allowed to come to a temperature of approximately 45°F. The product was still most palatable and not unlike refrigerated natural fruit bits. Another portion of the product was placed in commercially available pancake mix and pancakes were prepared therefrom.

The fruit flavored bits remained essentially intact during the cooking of the pancakes and the pancakes had the taste and appearance of pancakes with natural blueberry meat bits therein.

MISCELLANEOUS FLAVORS

SPECIAL FLAVORS AND/OR FLAVOR MODIFIERS

From Glycyrrhizin-Free Fractions of Licorice Root

Heretofore, licorice root has been processed commercially almost exclusively for its water-soluble extracts containing glycyrrhizin. Of course, substantial amounts of solid residue remain after removal of water-soluble material from the root. The solid residue, or spent licorice root, has posed a problem of disposal, and considerable thought has gone into finding acceptable uses for spent root.

Spent root has been treated with alkali to produce an aqueous alkaline solution having a pH of at least 10 to obtain therefrom a complexing agent which has been combined with trace metals to form valuable micronutrients. This is described in U.S. Patent 3,574,592. By and large, however, those in the art have considered extracts containing glycyrrhizin to be the extracts of value from licorice root, and once all appreciable levels of this material have been extracted from licorice root, the spent root is viewed as a by-product.

As environmental considerations heighten, and as the cost of unprocessed licorice root and extraction costs escalate, it has become increasingly imperative to reduce the amount of spent root ultimately discarded, and to isolate other materials of commercial value from spent root.

H.A. Hartung; U.S. Patent 4,163,067; July 31, 1979; assigned to MacAndrews and Forbes Company has developed a process for producing deglycyrrhizinated fractions from spent licorice root, and natural flavorants, flavor potentiators and adjuvants from these fractions.

This process comprises treating water-insoluble residue (spent root) remaining after extraction of fresh licorice root with an alkaline extractant to produce an aqueous alkaline extract having a pH less than 10, preferably a pH of about 7 to 9. Any water-soluble alkaline material may be used for extraction of spent root, so long as sufficient alkali is provided to alkalify the aqueous alkaline extract to a pH less than 10, and include sodium hydroxide, potassium hydroxide, ammonium hydroxide, trisodium phosphate and sodium silicate. Sodium hydroxide is particularly preferred.

The alkaline extraction is most expeditiously carried out at temperatures preferably from about 275° to about 285°F and under pressure of from about 40 to 45 psig. The extract recovered after alkaline extraction is reserved for further treatment.

The aqueous alkaline extractant is further treated with sufficient acid to acidify the extract to a pH of about 2 to 6, preferably below about 5, forming an acid-soluble fraction and an acid-insoluble residue. No special conditions are required for the acidification of the alkaline extract, although it is preferred that acidification be carried out at temperatures from about 100° to about 140°F. Acids useful in acidification of the alkaline extract include mineral acids such as sulfuric, hydrochloric and phosphoric acids, and a large number of organic acids which are sufficiently soluble and possess an ionization constant sufficiently high to provide the required hydrogen ion concentration.

Organic acids, including but not limited to, acetic, butyric, citric, fumaric, glycolic, lactic, malic, oxalic, propionic, succinic, tartaric and vinylacetic acids may be used. It will be appreciated by those in the art, that the selection of acid and the amount of acid useful for acidification at this juncture in the process provide the possibility of considerable variance in characteristics and attributes of the resultant product.

The acid-soluble fraction and the acid-insoluble portion of the alkaline extract may be separated and isolated by conventional methods such as by filtration.

The acid-insoluble residue of the alkaline extract is next treated with sufficient alkali to alkalify the residue to a pH of about 8 to less than 10, preferably about 8.5, forming an alkali-soluble fraction. Again, no special conditions are required for the alkalification of the acid-insoluble residue, although it is preferred that alkalification be carried out at temperatures of from about 90° to about 120°F. Alkalies useful in alkalification of the acid-insoluble residue include ammonium hydroxide, potassium hydroxide and sodium hydroxide, and a large number or organic bases which are sufficiently soluble and possess an ionization constant sufficiently high to provide the required hydroxyl ion concentration.

Organic bases include, but are not limited to, alanine, butylamine, diethylamine, ethanolamine, glycine, methylamine, piperidine, sarcosine and trimethylene diamine. The selection of alkali used for alkalification also provides the possibility of considerable variance in characteristics and attributes of the resultant product.

In the area of potential utility as natural flavorants, flavor potentiators and adjuvants, variation in the characteristics of the deglycyrrhizinated fractions may be achieved depending on the precise agents selected for acidification and alkalification in the process described above. That is, the particular acid or alkali or combination of acids or alkalis used in acidification and alkalification will have a bearing on flavor characteristics and properties of the resultant deglycyrrhizinated fractions.

The complete scope of utility of the acid-soluble and alkali-soluble deglycyrrhizinated fractions as natural flavorants, flavor potentiators and adjuvants has not been determined; however, it is known that the deglycyrrhizinated acid-soluble fraction formed on citric acid acidification has utility in the beverage industry, serving as a flavor enhancer for fermented beverages such as beer and wine; this fraction also has foaming properties and enhances flavors in soft drinks.

The alkali-soluble deglycyrrhizinated fraction formed with citric acid acidification and sodium hydroxide alkalification has flavor enhancing utility in coffee, puddings, pie fillings, syrups and tobacco products. The alkali-soluble deglycyrrhizinated fraction formed with phosphoric acid acidification and sodium hydroxide alkalification has flavor enhancing utility in meats.

Example 1: Water-insoluble licorice root residue remaining after aqueous extraction of fresh licorice root at elevated temperatures and pressure in which the aqueous extractant had a pH of 6 and in which the primary extract of water-soluble materials had a pH of 5.2, was extracted with a dilute aqueous solution of NaOH at a temperature of 285°F and pressure of 45 psig. The weight ratio of water to spent root in the aqueous solution of NaOH is 5, and the amount of NaOH used is 2.7% by weight of the dry root solids. 15% of the spent root mass was solubilized, and a dilute (2.65% by weight) solution of the solute was separated from the insoluble root by filtration. The aqueous alkaline solution recovered had a pH of 8.5, and had a sweet flavor with resinous character and only a very mild hint of licorice flavor by taste test.

Example 2: The aqueous alkaline solution of Example 1 was concentrated to 10% solids by weight and treated at a temperature of 120°F with 92 pounds of food grade citric acid (100%) per 100 pounds of solids in the solution to reduce the pH from 8.4 to 3.2. After brief stirring to insure homogeneity, the resultant slurry was separated by filtration producing an acid-soluble fraction and an acid-insoluble residue. The acid-soluble fraction was recovered, and the acid-insoluble residue was mixed with an equal weight of water and food grade NaOH was added at a temperature of 110°F with stirring to form an alkali-soluble fraction having a pH of 8.5.

The acid-soluble fraction recovered had a pH of 3.2 and a slight sweet-sour taste, but no flavor characteristic of licorice was detected by taste test. The alkali-soluble fraction recovered had a pH of 8.5, but no flavor characteristic of licorice was detected by taste test. Combinations of the acid-soluble fraction and the alkali-soluble fraction in various proportions produced mixtures having a spectrum of flavors including mixtures in which the distinct flavor characteristic of licorice was found by taste test.

Example 3: An imitation vanilla was formed by mixing 1 part by weight, ethyl vanillin, food grade; 2.5 parts by weight, vanillin, food grade; 40 parts by volume, propylene glycol USP; and 56 parts by volume, of a 25% by weight aqueous solution of the alkali-soluble fraction of Example 2. The mixture was formed by heating the propylene glycol at 140°F, adding the vanillin and ethyl vanillin with stirring until completely dissolved and adding, with stirring, the solution of the alkali-soluble fraction of Example 2. The resultant imitation vanilla was five-fold stronger in vanilla flavoring than household strength baking vanilla extract. For comparison, a vanillin composition was prepared without the alkali-soluble fraction of Example 2. Each formulation was added to simple syrup at levels of 0.5 to 1% by weight and in side by side taste tests the formulation containing the alkali-soluble fraction of Example 2 was found to possess improved flavor and aroma of vanilla.

Pyridine Sulfur Compounds for Coffee Flavor

A comprehensive review of chemical compounds which may be used for adding or enhancing flavors of various foodstuffs, primarily coffee, is given by *M. Winter, F. Gautschi, I. Flament, and M. Stoll; U.S. Patent 4,085,109; Apr. 18, 1978; assigned to Firmenich & Cie, Switzerland*. Thirty-seven categories of compounds are given, and tables showing the organoleptic evaluations of representative compounds of each group are included. Most of these compounds have been described in detail in an earlier book, but the specific group of compounds covered by this particular patent will be covered here. These compounds are pyridine sulfur compounds of the general formula:

$$\text{(pyridyl-2)}-(CH_2)_nSR$$

wherein R stands for hydrogen, alkyl, acyl or pyridyl, and n is 0 or 1. As examples there can be mentioned:

(a)	(pyridyl-2)methanethiol	c.a.*
(b)	2-mercaptopyridine	c.a.
(c)	2-methylthiopyridine	n.c.**
(d)	2-ethylthiopyridine	n.c.
(e)	(pyridyl-2)thiol acetate	n.c.
(f)	di(pyridyl-2)sulfide	*J. Chem. Soc.*, (1942), 239
(g)	2-(pyridyl-2)ethanethiol	*J. Org. Chem.*, 26, 82, (1961)
(h)	2-(pyridyl-2)ethyl methyl sulfide	see below
(i)	2-(pyridyl-2)ethyl ethyl sulfide	n.c.
(j)	2-(pyridyl-2)ethanethiol acetate	see below
(k)	2-(pyridyl-2)ethyl furfuryl sulfide	n.c.
(l)	(pyridyl-2)methyl methyl sulfide	*Helv.* 47, 1754 (1964)
(m)	(pyridyl-2)methyl ethyl sulfide	n.c.
(n)	(pyridyl-2)methanethiol acetate	n.c.

*c.a. = commercially available
**n.c. = new compound

The method used for preparing the compound (h), 2-(pyridyl-2)ethyl methyl sulfide was as follows: 2-vinylpyridine was reacted with methylmercaptan by the action of UV light in the presence of trace amounts of benzoyl peroxide and diphenyl sulfide. The product has a BP of 48°C/0.03 mm Hg.

The same method was used for preparing the known compound (j), except that thioacetic acid was used instead of methylmercaptan. The product has a BP of 80°C/0.02 mm Hg.

The new compounds included in this group can be obtained as follows: Compound (c), 2-methylthiopyridine was prepared according to the method described in Houben-Weyl, 4th ed., vol. 9, 7 (1955) by alkylating 2-mercaptopyridine with methyl halide. The resulting pyridinium salt was neutralized with NaOH and the base thus obtained extracted and distilled. The product had a BP of 67° to 68°C/10 mm Hg.

Compound (d), 2-ethylthiopyridine was prepared by the same method as used for compound (c), except that ethyl halide was used instead of methyl halide. The product had a BP of 77° to 77.5°C/8 mm Hg.

Compound (e), (pyridyl-2)thiol acetate was prepared by reacting acetic anhydride with 2-mercaptopyridine in alkaline medium according to the method described in Houben-Weyl, 4th ed., vol. 9, 753 (1955) and in *J.A.C.S.* 59, 1089 (1937). The product has a BP of 117° to 118°C/9 mm Hg.

Compound (i), 2(pyridyl-2)ethyl ethyl sulfide was prepared by the same method as used for compound (h), except that ethylmercaptan was used instead of methylmercaptan. The product has a BP of 62°C/0.005 mm Hg.

Compound (m), (pyridyl-2)-methyl ethyl sulfide was prepared by the same method as used for compound (l). The product has a BP of 107° to 110°C/10 mm Hg.

Compound (n), (pyridyl-2)-methanethiol acetate was prepared by reacting acetyl chloride with 2-mercaptomethylpyridine in alkaline medium. The product has a BP of 102° to 103°C/ 9 mm Hg.

Evaluation test data are reported in the table that follows.

The compounds were subjected to organoleptic evaluation tests either in a syrup base (A), or one of the two soluble coffee bases (B) and (C). The following table gives the results of these organoleptic evaluations. The column headed "Test" gives the method of the test, and the column headed "Quantity" sets out the amount of the test compound used in grams per 100 liters of the base material.

Compound	Test	Quantity	Organoleptic Characterization
a	A	5.0	Fortifies the bitter taste
a	C	0.093	Popcorn, nutty, caramel, cereal taste
b	A	0.25	Enhances the burnt note
c	A	0.25	Enhances the phenolic note
d	A	5.0	Enhances the burnt note
d	C	0.025	Green, acid, cereal, bitter sour note
e	A	0.2	Enhances the roast note
f	A	6.0	Weak note
g	B	0.30	Roasted, astringent, earthy note
h	B	0.12	Mushroom, bitter, green taste
i	B	0.25	Astringent, fatty, green taste
j	B	0.40	Astringent, roasted taste
k	B	0.40	Astringent, green note
l	B	0.30	Bitter, green, earthy note
m	B	0.12	Metallic note
n	B	0.40	Bitter fatty note

Benzofuran Hydrocarbon Flavor Modifiers

M. Winter, F. Gautschi, I. Flament, M. Stoll and I.M. Goldman; U.S. Patent

4,113,891; September 12, 1978; assigned to Firmenich & Cie, Switzerland describe the enhancement of foodstuffs effected by the addition of a small but effective flavor-modifying amount of one of the compounds having the general formula

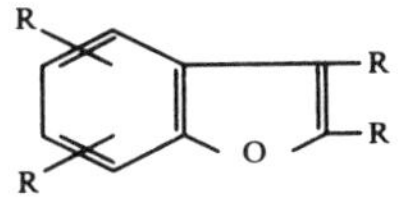

wherein R is hydrogen, an alkyl group having 1 or 2 carbon atoms, or an alkenyl group having 2 or 3 carbon atoms, provided the sum of the carbon atoms of the substituent groups does not exceed 3.

Specific compounds included in this group of compounds are:

(a)	Benzofuran	Commercially available
(b)	2-Methylbenzofuran	*Soc.,* 3689, (1955)
(c)	2-Ethylbenzofuran	*J.A.C.S.,* 73, 754, (1951)
(d)	2,3-Dimethylbenzofuran	*Soc.,* 3689, (1955)
(e)	2-Vinylbenzofuran	*J.A.C.S.,* 73, 754, (1951)
(f)	2-Isopropenylbenzofuran	New compound
(g)	7-Methylbenzofuran	*J. Chem. Soc.,* (1920), 1534
(h)	7-Ethylbenzofuran	New compound
(i)	2,7-Dimethylbenzofuran	New compound

Compound (f), 2-isopropenylbenzofuran is prepared according to the method described in *J.A.C.S.* 73, 754 (1951). 2-Acetylbenzofuran is reacted with methylmagnesium bromide to form 2-(2-hydroxyisopropyl)benzofuran which is converted to its acetate. Pyrolysis of the acetate yields 2-isopropenylbenzofuran of BP 81° to 83°C/0.001 mm Hg.

Compound (h), 7-ethylbenzofuran is prepared by the method described in *J. Chem. Soc.,* (1920), 1534, but using o-ethylphenol instead of o-cresol. The MS of the product thus obtained shows the following ion peaks with the relative intensities given within brackets: 131 (100%), 146 (38%) and 77 (10%).

Compound (i), 2,7-dimethylbenzofuran is prepared by subjecting 7-methylbenzofuran to a Wilsmeyer reaction to form 7-methylbenzofuran-2-aldehyde which is converted into 2,7-dimethylbenzofuran by a Wolf-Kishner reaction by the method described in *Bull. Soc. Chim. France* 29, 1875, (1952). The product thus obtained has the following peaks in its MS: 146 (100%), 145 (92%) and 131 (32%).

Organoleptic evaluations of this group of compounds are set out in the table below: These compounds were subjected to organoleptic evaluation tests either in a syrup base (A), or a soluble coffee base (B). The following table gives the results of these organoleptic evaluations. The column headed "Test" gives the method of the test and the columnn headed "Quantity" sets out the amount of the test compound used in grams per 100 liters of the base material.

These compounds may be added to substances in varying amounts to alter or to modify the flavor of the substance by masking or blanking out undesirable flavors, by enhancing or fortifying desirable flavor or flavor notes, or by adding to the original substance an entirely new and different flavor.

Compound	Test	Quantity	Organoleptic Characterization
a	A	0.05	Styrene-like, aromatic
b	A	1-3	Slightly phenolic, burnt taste
b	B	0.4	Enhancement of the bitter note
c	A	0.3	Salicylate-like taste
d	A	0.25	Earthy flavor note
e	A	1.0	Burnt, caramel taste
f	A	0.25	Phenolic, saffron-like
g	B	0.12	Earthy, mushroom-like, hazelnut
h	B	0.50	Burnt, green taste
i	B	0.095	Earthy flavor note

As will also be apparent to those skilled in the art, various mixtures or blends of the flavor agents described may be used to achieve a desired flavor or flavor note. If, for example, one wishes to enhance a certain flavor note, or group of flavor notes present in a substance such as coffee, one needs only mix together certain of the described flavor agents to obtain the desired result.

It should be kept in mind, as will be appreciated by those skilled in the art, that with many flavors it is possible to imitate the natural flavor by selecting a limited number of the flavor-enhancing substances exemplified above. Coffee aroma, on the other hand, is much more complex than the ordinary flavoring materials and may necessitate the combination of many more of the exemplified ingredients for reproduction.

It will be understood that whereas the main purpose is directed toward the enhancement or modification of coffee flavors, the concept has much wider application. While some of the compounds may be characterized by terms which are not directly related to coffee flavors when used in more complex formulas, they may contribute desirable flavor notes to the overall flavor and aroma.

COFFEE

Thiophene Sulfur Compound Flavor Modifiers

M. Winter, F. Gautschi, I. Flament, and M. Stoll; U.S. Patent 4,138,410; Feb. 6, 1979; assigned to Firmenich & Cie, Switzerland describe two members of the group of thiophene sulfur compounds of the formula:

S —X—S—R

wherein X is ethylidene and R is hydrogen or a methyl group; and wherein X is a carbonyl group and R is an ethyl or furfuryl group.

2-(Thienyl-2)ethanethiol was prepared as follows: 2-Vinylthiophene [obtained by the method described in *Org. Synth.* 38, 86 (1958)] was reacted with thioacetic acid according to the method described in *J. Org. Chem.* 27, 2853 (1962), and the resulting addition product was subjected to hydrolysis with an acid. The product has a BP of 55°C/0.1 mm.

2-(Thienyl-2)ethanethiol acetate was obtained as the intermediate product obtained by reacting 2-vinylthiophene with thioacetic acid in the preparation of

2-(thienyl-2)ethanethiol above. The product had a BP of 90°C/0.7 mm Hg.

When 2-(thienyl-2)ethanethiol was tested in a dilute sugar syrup, it was characterized by a burnt, coffee ground, onion taste. 2-(Thienyl-2)ethanethiol acetate gave a burnt, onion taste.

LIQUID SMOKE

Addition of Acidulating and Solubilizing Agents

The use of natural liquid smoke solutions in lieu of direct smoking to impart smoky flavor and appearance to food products such as meat, cheese, fish and the like, has been known and commercially practiced for some time. Liquid smokes are generally aqueous solutions capable of imparting a smoky hue or coloration and/or a smoky flavor to comestibles exposed to the liquid smoke or its vapor phase. Conventional techniques for preparing liquid smoke solutions involve the burning of wood (e.g., hardwoods such as hickory or maple) and the condensation or extraction of the constituents of the smoke formed from the burning of such woods. Typically, the smoke constituents are extracted by condensation or absorption in an aqueous medium followed by concentration of the resulting solution.

Regardless of the method by which the natural liquid solutions are made or the procedures by which they are applied to the food products, a universal problem associated with the use of liquid smoke is the tendency of the solutions to form polymeric solids which settle to the bottom of the solution containers during storage or deposit on equipment surfaces or clog liquid smoke spray nozzles.

G.V. Rao, F.K. Shoup and G.R. Popenhagen; U.S. Patent 4,112,133; Sept. 5, 1978; assigned to Far-Mar-Co, Inc. have provided a liquid smoke composition which includes a solubilizing additive therein for preventing solids formation in the composition. The additive consists essentially of a food grade acidulant and a solubilizing agent selected from the group consisting of sorbitan fatty compounds, polyoxyethylene sorbitan fatty compounds, polyoxyethylene fatty compounds, polyoxyethylene glycols, and mixtures thereof.

The acidulant and solubilizing agent are present in the composition in amounts keyed to the quantity of phenols and carbonyls in the liquid smoke composition, which amounts range from about 1.5 to 3 wt % acidulant and from about 3 to 8 wt % solubilizing agent for known commercially available liquid smoke solution, the percentage based on the total liquid smoke composition, i.e., the liquid smoke solution, acidulant, solubilizing agent and any trace quantities (generally less than about 0.1% by wt) of noninterfering additives.

Example 1: To 20 parts of a natural liquid smoke solution having a phenol concentration of 1,700 mg/l and a carbonyl concentration of 3,000 mg/l was added 80 parts of water. Upon addition of the water, the solution immediately became very turbid. The solution was heated to 110°F, a temperature commonly employed when treating food products with liquid smoke. The liquid smoke remained turbid and, after 48 hours, an undesirable heavy precipitate had formed in the container.

Example 2: A liquid smoke composition was formed consisting of 94 parts by weight of the undiluted liquid smoke solution of Example 1, 4 parts by weight Tween 80 and 2 parts by weight glacial acetic acid. The composition was admixed thoroughly. The resulting solution was clear when formed and remained clear after 48 hours heating at 110°F. The solution was placed in a closed container in prolonged storage (3 months) following which the solution was examined. No turbidity or undesirable precipitate was present in the storage container.

The same composition, consisting of liquid smoke solution, acidulant and Tween 80 was diluted with water to 80:20 by volume water to composition and checked for turbidity and/or solids formation after 48 hours heating at 110°F and after 3 months storage; no turbidity or precipitate was noted.

Example 3: The procedure of Example 2 was repeated three times using the undiluted liquid smoke composition of Example 2 except that in lieu of glacial acetic acid, the acidulants employed were: formic acid, phosphoric acid and vinegar (In this case sufficient vinegar was added until the equivalent of 2% by weight glacial acetic acid was present. Since vinegar is predominantly water, the resulting liquid smoke composition was quite diluted).

In each case, no turbidity or precipitate was noted upon formation of the composition, after 48 hours heating at 110°F, or after 3 months storage.

For Food Casings

Tubular food casings are used extensively for processing a great variety of meat products and other food items.

It has been suggested, as for example disclosed in U.S. Patent 3,330,669, that application of a viscous liquid smoke solution to the inside surface of a tubular food casing by the food processor immediately prior to the stuffing thereof with a sausage emulsion results in preparation of processed food products that after cooking and removal of the casing exhibit good color and smoky flavor.

However, liquid smoke solutions are highly acid and thus have a detrimental effect on cellulosic food casings, causing them to lose their strength.

H.S.-G. Chiu; U.S. Patent 4,104,408; August 1, 1978; assigned to Union Carbide Corporation has developed an aqueous liquid smoke composition that is in a generally neutralized condition, having a pH greater than 5, and is suitable to impart smoke color and flavor characteristics to a cellulosic food casing and to food products processed therein comprising a uniform mixture, and preferably an aqueous solution of smoke coloring and flavoring constituents, an alkaline neutralizing agent in an amount sufficient to maintain the solution at a pH greater than 5, and an amount of a short chain alcohol solubilizing agent sufficient to maintain the smoke constituents in solution.

It has been discovered that this substantially neutralized aqueous liquid smoke solution may be applied to the surface of a cellulosic food casing in an amount that will be suitable for imparting smoke flavor and color characteristics to the food casing and to a wide variety of meat products processed therein.

The liquid smoke that is preferred for use with this process is a solution of natural wood smoke constituents. This liquid smoke is produced by the limited burning of hardwoods and the absorption of the smoke so generated into an aqueous solution under controlled conditions.

Certain liquid smokes have been approved for use in foods by the U.S. Food and Drug Administration and the Meat Inspection Division of the U.S. Department of Agriculture. Exemplary of suitable commercially available liquid smokes are Charsol from Red Arrow Products Co.; Liquid Hickory Smoke from Hickory Specialties, Inc.; Griffith's Natural Smoke Flavor from Griffith Laboratories, Inc.; and Smokaroma Liquid Smoke Code 10 from Meat Industry Suppliers, Inc.

Alkaline neutralizing agents suitable for use are any of the well known water-soluble alkaline materials such as, for example, potassium hydroxide, sodium hydroxide, ammonium hydroxide, sodium carbonate, sodium bicarbonate, sodium phosphate, disodium hydrogen phosphate, trisodium phosphate, and the like. The alkaline materials may be in solid form or as a concentrated solution thereof. The most highly alkaline materials are most advantageously employed in order that the concentration of smoke constituents will be maintained as high as possible in a substantially neutralized uniform mixture thereof.

Suitable solubilizing agents are short chain monohydric and polyhydric alcohols that are water-soluble, nontoxic, and preferably are approved for use in food related applications. Exemplary of suitable materials are ethyl alcohol, glycerine, propylene glycol, triethylene glycol and the like.

The amount of alkaline neutralizing agent present in the uniform aqueous mixture of smoke constituents of the process is largely determined by acidity and composition of the particular liquid smoke solution that is to be neutralized and the particular neutralizing agent that is employed. In general, however, it has been found that at least about 6% by weight of a neutralizing agent, such as sodium hydroxide, based on the weight of liquid smoke solution and preferably at least about 7% by weight will be required.

The amount of alcohol solubilizing agent that is present is also largely determined by the composition of the particular liquid smoke solution that is used and the desired pH of the neutralized liquid solution, but, in general, the solubilizing agent will be present in an amount of at least about 10% by weight of solution and at least about 11% by weight of the liquid smoke solution, and preferably in an amount of at least about 15% by weight of solution and at least about 20% by weight of liquid smoke solution.

Example: A series of liquid smoke compositions were prepared from a variety of commercially available liquid smokes using the following proportion of ingredients:

	Composition (g)										
Ingredients	A	B	C	D	E	F	G	H	I	J	K
Liquid smoke A	25	25	25	25	–	–	–	–	–	–	–
Liquid smoke B	–	–	–	–	25	25	25	25	–	–	–
Liquid smoke C	–	–	–	–	–	–	–	–	25	25	25
Solid NaOH	–	1.3	1.3	1.9	–	0.8	0.8	2.0	–	2.5	3.0
Propylene glycol	–	–	2.8	8	–	–	2.5	5	–	5	7

Liquid Smoke A was Smokaroma Pure Liquid Hickory Smoke, Code 10 from Meat Industry Suppliers, Inc. Liquid Smoke B was Charsol C-10 from Red Arrow Products Co. Liquid Smoke C was Liquid Smoke Flavoring 18%, from Mallincrodt Chemical Works.

The pH and appearance of each of the liquid smoke compositions of this example are reported in the table.

Composition	pH	Composition Appearance
A	2.2	clear
B	4.9	tarry precipitate
C	5.0	clear
D	6.5	clear
F	2.2	clear
F	4.4	tar separation
G	4.4	clear
H	6.4	clear
I	2.2	clear
J	5.8	clear
K	6.9	clear

The results reported in the table show the importance of using an alcohol solubilizing agent in preparing liquid smoke compositions having a pH greater than 5.

EXTRACTING NATURAL FLAVOR AND AROMA COMPONENTS

Coffee

G.S. Hurlow, J.R. Blain, M. Coombes, J.-C. Richard and P.W. Hitchinson; U.S. Patent 4,101,681; July 18, 1978; assigned to General Foods Limited, Canada have developed a process for the separation, recovery and concentration of roasted coffee aroma constituents and the subsequent incorporation of the aroma constituents into food products.

Some problems that hampered extensive commercial use of the Feldman et al. process (Canadian Patent No. 603,954) were that (1) the condensate could not be removed from the condenser without undesirable prolonged contact with air resulting in the loss and degradation of valuable aroma constituents; (2) the condenser had to be emptied of the cryogenic coolant and warmed every time the condensate had to be removed thus putting a condenser out of action for a prolonged period of time, wasting valuable coolant material, and due to the condenser's large mass it remained cold for extended periods of time thus hampering attempts to remove the condensate; and (3) the physical state of the condensate in the condenser could not be effectively controlled thus making handling difficult as efficient operation makes desirable the formation of solid condensate on the collection surface.

This process entails subjecting coffee oil in a distillation chamber to distillation at subatmospheric pressures and mild temperatures to separate aroma constituents therefrom and condensing the aromatic constituents in a condensing chamber on the outer surface of a heat-conductive sleeve removably and snugly surrounding a heat-conductive container filled with a coolant, and then removing the sleeve with the condensed constituents thereon.

The condensed aromatic constituents in the condensing chamber are preferably collected as a solid by maintaining a sufficient pressure in the condensing chamber through the introduction of an inert gas.

The sleeve with the condensed aromatic constituents thereon is then preferably immersed into an edible substance contained in a folding chamber in order to combine the condensed aromatic constituents with the edible substance.

Figure 10.1 is a cross-sectional view of the condenser used in the process.

Figure 10.1: Cross-Section of the Condenser

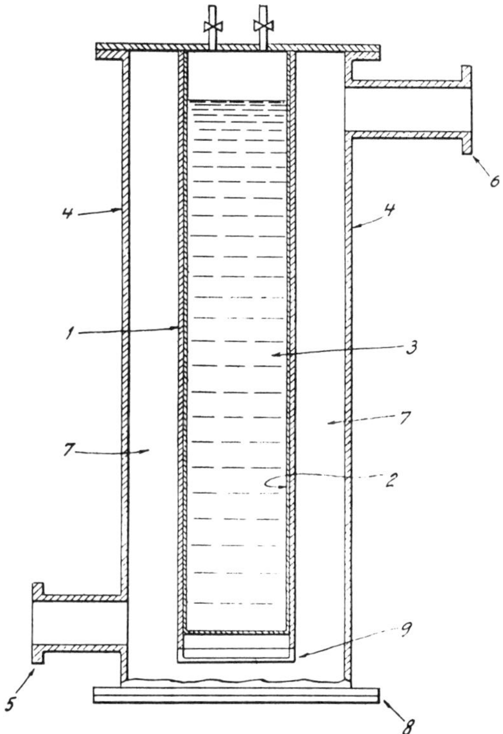

Source: U.S. Patent 4,101,681

Coffee aroma constituents are separated by distilling coffee oil at subatmospheric pressures and under mild temperatures and then condensing the separated aroma constituents. The coffee aroma constituents are separated from the oil in such a manner that various desirable aroma constituent fractions can be isolated and subsequently incorporated in a final product in the proportions desired. Low-boiling aroma constituents having a fragrance like that of roasted coffee grinder gas and lacking body form the bulk of the initial distillate. Medium boiling aroma constituents and some higher boiling aroma constituents having a burnt or smoky roasted character are also separated from the oil. As a result, the various aroma fractions can be proportioned by controlling the temperature, pressure and period of distillation to provide the desired levels of low, medium, and high boiling aroma constituents to obtain a desired coffee aroma.

Generally, the coffee oil is obtained by expressing roasted coffee at a temperature below 150°C and preferably at about 130°C. This is a sufficiently elevated temperature compatible with satisfactory yields of coffee oil and aroma therein. The pressures exerted on the coffee to provide a high yield of coffee oil and quality aromas cannot be stated precisely or directly. However, coffee oil is expelled in an acceptable condition using apparatus estimated at exerting 5,000 to 20,000 psi (approximately 350 to 1,400 kg/cm^2) on the coffee.

The pressures for adequate oil yield and aroma quality can be determined in terms of the temperatures of the expeller cake or meal and the expressed oil. The expeller cake (expressed roasted coffee or coffee meal) should have a temperature below 150°C and preferably, a temperature of at least about 75°C. The expressed oil should have a temperature ranging from about 25° to 120°C when measured immediately after expression. Adherence to such expeller cake and oil temperatures assures that the necessary pressures have been applied to the coffee to recover oil therefrom without undesirable modification of the aroma constituents therein.

The distillation of the aroma constituents from the expressed coffee oil is preferably carried out within a temperature range of 15° to 100°C with the preferred aroma constituents being distilled at a temperature below about 60°C. The oil is maintained under a reduced pressure in the distillation chamber of generally less than 50 mm and preferably less than 10 mm. The most preferred distillation condition involves subjecting coffee oil at ambient temperatures (i.e., 20° to 35°C), to a pressure ranging from 10 mm to about 60 μ.

The oil is preferably introduced into the distillation chamber by a method that maximizes the liquid/gas interfacial area. This allows a maximum of aroma constituents to be distilled in a minimum of time. This method is preferably carried out by atomization of the oil into fine droplets. This can be accomplished by using an impingement-type atomizing nozzle that operates by having the oil stream impinge, at high velocity, upon a pin held directly at the exit of the nozzle orifice. Alternatively the oil can be provided in a thin film for distillation; preferably in a rapidly moving film having a thickness ranging from 10 to 100 μ.

The process of desorbtion can also be employed to increase the efficiency and degree of separation of the aroma constituents from the oil. Desorbtion involves sweeping the coffee oil with an inert gas such as, for example, nitrogen, carbon dioxide, helium, etc. in the distillation chamber and thereby collecting the more fugitive aromatic constituents.

Referring to Figure 10.1, the aroma constituents are then carried off as a gas stream and are condensed onto the outer surface of a heat-conductive sleeve **1** removably and snugly surrounding a heat-conductive container **2** which has a coolant **3** therein. The heat-conductive sleeve is enclosed by a condensing chamber **4** which has ports **5** and **6** to allow the passage of the aromatic constituents within the space **7** between the condensing chamber and the sleeve. The condensing chamber also has a removable cover **8** which seals an opening through which the sleeve is removed from the container.

The sleeve is preferably a thin gauge metal and can be kept in place snugly surrounding the container by a number of means. This can include the sleeve being screwed, latched, hooked, etc. in place, or being held in place by friction between the container and the sleeve or by the force of gravity. The container is preferably of metal and is in contact with the sleeve so as to maximize heat conduction between the container and the sleeve.

The coolant within the heat-conductive container can have a temperature ranging from about 0° to about -269°C depending upon what aroma constituents one wants to collect. It is possible though to condense all the aroma constituents with a coolant on the order of -196°C such as liquid nitrogen.

As the gas stream of volatile aroma constituents enters the condensing chamber through an inlet port, it comes into contact with the cooled surface of the sleeve and the aroma constituents which condense form a frost thereon. The constituents of the gas stream which are not condensed exit out of the condensing chamber through an outlet port **6**. After sufficient frost has been collected on the sleeve, the condensing chamber preferably is isolated, vented (preferably with an inert gas), the cover opened and the sleeve removed, preferably by use of handle **9** on the sleeve.

Through use of this sleeve no significant amount of frost condenses on the container and therefore it is not necessary to remove the coolant from the container to recover the frost. Previously, due to the large mass of the container which collected the frost on its surface, after the coolant was removed the container retained the cold longer and when it was contacted with oil to incorporate the frost into the oil, the oil would freeze onto the container's outer surface and would have to be scraped off. As well as being time consuming, this led to the use of a complicated mechanical scraper and unwanted exposure of the coffee aroma constituents to air.

Due to the sleeve's low mass and high heat transfer coefficient, the frost can be easily and quickly recovered when contacted with oil since there is little if any freezing of the oil on the sleeve's surface, and the exposure of the condensed aroma constituents to air can be minimized due to the ease with which the sleeve may be handled. The condenser can also be quickly reactivated after inserting a clean sleeve onto the container.

The condensed aroma constituents on the sleeve can be collected by venting the condensing chamber preferably with an inert gas, then removing the cover of the condensing chamber and taking out the sleeve with the aroma constituents condensed thereon in the form of frost and contacting the constituents with an edible nonvolatile fluid substance, e.g., an oil. The edible substance serves as an aroma carrier enabling the aroma constituents to be conveniently added to a food product such as instant coffee. The coffee aroma is preferably folded into the edible substance at a sufficiently high level to enable the concentrated aroma and aroma carrier to be incorporated into the food product without imparting a wet appearance or causing poor flowability.

The preferred aroma carrier is coffee oil (expressed, solvent extracted or otherwise obtained) which, due to the presence of natural emulsifiers therein, enables large quantities of aroma frost to be combined with or incorporated into the oil. The aroma is preferably added to the coffee oil to obtain a 2- to 10-fold coffee aroma level (aroma from 1 to 9 parts coffee oil combined with 1 part coffee oil), the level also depending upon how much aroma is to be added later to the food product such as instant coffee. Preferably at least a 5-fold level of aromatized oil is added to obtain a desired aroma level in a product such as instant coffee without adversely affecting its appearance and flowability.

Both Hydrophilic and Lipophilic

N. Yano, I. Fukinbara and M. Takano; U.S. Patents 4,069,351, January 17, 1978; and 4,136,065; January 23, 1979; both assigned to Asahi Kasei Kogyo KK, Japan have devised a method for obtaining mixtures containing the hydrophilic and lipophilic flavor and odor producing components of natural product substrates such as foods and flowers.

In accordance with the process, the substrate is extracted with defined liquid mixtures of water and dimethyl ether at a temperature of about -25° to 40°C and the extraction residue separated. At least a portion of the dimethyl ether is then removed, most conveniently by evaporation. The balance of the dimethyl ether and the water can be similarly removed. The weight ratio of dimethyl ether to substrate is typically from 1:1 to 5:1. The weight ratio of water to dimethyl ether in the extraction mixture is such that in the phase diagram of the system water-dimethyl ether, the amounts of each component are described by the solubility curve defining dimethyl ether saturated with water at from about -25° to 0°C, and by the curves defining dimethyl ether saturated with water and water saturated with dimethyl ether at a temperature of from about 0° to 40°C.

Dimethyl ether is gaseous under standard atmospheric conditions. It can be readily liquified by cooling to below -25°C at atmospheric pressure or by compression to above about 5 atmospheres at room temperature. Liquified dimethyl ether readily dissolves most oils and also dissolves about 6.3% by weight of water at 20°C. It has a specific gravity of 0.661 and a latent heat of vaporization of 96.6 cal/g. Its physical properties are especially important to this process. Since it dissolves both water and oil, it dissolves substantially all of the oil and many of the water-soluble ingredients in the complex mixture so that the extract is closer to the natural mixture. Since the amount of water it dissolves is not too great, it freezes during evaporation of the dimethyl ether at room temperature or well below room temperature. As a result, the volatile components in the complex mixture do not evaporate during vaporization of the solvent phase.

This process is especially valuable in the production of instant coffee or instant tea. For this particular utility, the original substrate is first extracted with a dimethyl ether-water extraction mixture. The residue is then extracted in the usual way with water (usually hot), and the instant coffee or tea product is produced by evaporation procedures such as spray drying or other suitable techniques. This material is deficient in odor and flavor, since these have been extracted with the dimethyl ether. The flavor and odor of the instant coffee can be reimparted by the addition of the mixture obtained by evaporation of the extraction liquid. A special advantage is that the degree of odor and flavor can be controlled by the amount of mixture added. Another is that the odor and flavor can be blended by the addition of a variety of odor and flavor constituents from a number of different coffee or tea species to one basic instant product obtained by water extraction.

Still another advantage of the process is that the aroma and flavor components can be readily deposited on a preservative or carrier directly from the dimethyl ether-water solution. For example, a solid adsorbent or a liquid solvent can be added to the extraction solution prior to the evaporation of the extraction liquid. Upon evaporation of the ether, the solute will be adsorbed on, or will dissolve in, the additive. Typically useful carriers include talc, kaolin, celite, white carbon, lactose, sucrose, starch, microcrystalline cellulose (e.g., Avicel), and powdered cereals. The most useful liquid carrier is ethanol, and this is used principally with flower extracts in the preparation of perfume. The constituents may be preserved on the adsorbent or in the solvent until ready for use, or they may be used in the adsorbed or dissolved form.

Example 1: A 100 cc microcylinder was charged with 10 g of pulverized lemon skin containing about 6 g of water and 90 cc of liquefied dimethyl ether and hermetically sealed. The mixture in the microcylinder was shake extracted (60 cycles/min) at 18°C under 4 atmospheres for 15 minutes. The microcylinder was turned upside down and the dimethyl ether containing the extract was drawn out. When the solution was heated to 18°C under atmospheric pressure to vaporize the dimethyl ether, there was obtained 1.0 g of extract in the form of an oleo-resin. After three extractions, there were obtained 2.5 g of extract. When this extract was extracted with 20 cc of 95% ethanol and 16 cc of water to remove terpene, there were obtained 38 mg of terpine-free lemon essence. Orange and citron essence are obtained by the same procedure.

A dimethyl ether solution, 90 cc in volume, containing the extract obtained by treating 10 g of powdered lemon skin in accordance with the above procedure was sprayed onto 10 g of sucrose and left standing at 18°C under a reduced pressure to permit vaporization of the dimethyl ether. Consequently, there was obtained 12 g of lemon sugar.

Example 2: An agitator-type autoclave having a volume of 2 liters was charged with 300 g of Ceylon black tea containing 6 g of water, 200 g of water (5°C) and 600 g of liquid dimethyl ether. The mixture was held at a pressure of 4.5 atmospheres for 20 minutes to effect extraction, and thereafter filtered to separate the extracted solution and the residue. A hot water extract was obtained by passing separate 500 g portions of hot water three times through the residue, and it was then vacuum concentrated to produce 150 g of concentrate. This concentrate was blended with the formerly collected liquid dimethyl ether-water extracted solution.

The dimethyl ether content of the resultant mixture was evaporated at -5°C. The remaining mixture in its frozen state was transferred to a freeze drying apparatus and the water removed. Consequently, there was obtained 74.5 g of instant black tea powder. A solution of 1 g of this powder in 150 ml of hot water gave a black tea which was tasty and had the natural odor and flavor of tea.

Example 3: 1 kg of roasted coffee beans (Peru) containing 20 g of water was placed in an extraction tower, 500 g of water at 5°C were added and the tower was tightly sealed. Thereafter, 2.5 kg of liquid dimethyl ether were added, and the resultant mixture in the tower was left standing for 20 minutes under a pressure of 5 atmospheres to effect extraction. After the extraction, the mixture was filtered. The dimethyl ether content of the resultant extracted solution was recovered by vaporization at −10°C, with the result that a frozen extract having the desired odor and flavor was obtained. The residue of extraction was extracted with 3 liters of hot water, and the resultant hot water-soluble extract was concentrated to 500 g.

This extract was frozen and then combined with the abovedescribed frozen extract. The mixture was dried by freeze-drying to produce 280 g of instant coffee powder. When 1.5 g of this powder were added to 150 ml of hot water, the solution had an odor and flavor resembling those of coffees prepared in a percolator. It was free from the odor peculiar to the usual marketed instant coffee.

Example 4: A 1 liter pressure proof container was charged with 100 g of jasmine petals containing about 8 g of water and 700 cc of liquid dimethyl ether and hermetically sealed. Thereafter, the contents of the container were subjected to shake extraction (60 cycles/min) at 0°C under 2.5 atmospheres for 20 minutes. The extract was transferred in a liquid form into another 1 liter pressure proof container to separate the solution from jasmine petals. When this solution was left standing at 5°C under atmospheric pressure, there was obtained 10 g of extract. When this extract was combined with 50 cc of 95% ethanol to effect alcohol extraction and then held at a lowered temperature under a reduced pressure to effect removal of alcohol, there was obtained 0.2 g of jasmine essence oil.

Example 5: A 100 cc microcylinder was charged with 10 g of powdered nutmeg containing 0.2 g of water, 50 g of liquefied dimethyl ether (−30°C) and 2 g of water. It was then hermetically sealed. The mixture in the microcylinder was subjected to shake extraction (60 cycles/min) at 18°C under 3.5 atmospheres for 15 minutes.

The microcylinder was turned upside down to allow the dimethyl ether solution containing the extract to spurt out through a filter cloth.

This solution was heated with water at 10°C under atmospheric pressure to separate the extract from dimethyl ether through vaporization of dimethyl ether. Consequently, there was obtained 20 g of extract.

This was left to stand at 10°C under reduced pressure (50 mm Hg) for 30 minutes, with the result that the residual dimethyl ether content of the extract was lowered to below 10 ppm.

The extraction procedure was repeated at −30°C. With both extraction procedures, an extraction mixture was obtained with a strong nutmeg odor and flavor.

Spice

J. Chiovini, J.-P. Marion and S. Adamer; U.S. Patent 4,158,708; June 19, 1979; assigned to Societe d'Assistance Technique pour Produits Nestle SA, Switzerland describe a process for the production of an aromatic spice extract, which comprises:

(a) Grinding a spice and collecting an aromatic fraction A consisting of the gases given off during grinding;

(b) Treating the spice with an apolar organic solvent so as to obtain an aromatic fraction B contained in this apolar solvent; and

(c) Treating the spice with at least one polar solvent so as to obtain an aromatic fraction C contained in this polar solvent.

The aromatic spice extracts obtained by this three-step process are found to have superior aromatic quality as compared with extracts prepared by conventional processes such as pressing, steam distillation or extraction with a single solvent.

In a first embodiment of the process, a spice or a mixture of spices selected, for example, from cardamom, caraway, coriander, cumin, curcuma, cloves, laurel, nutmeg, paprika, pimentoes, chillies and pepper, is finely ground at a low temperature of the order of -40°C (a dry grinding). The grinding gases are liberated by heating the powder to around 20°C and are entrained by passing a gentle stream of inert gas, preferably under reduced pressure, over the powder for a period ranging, for example, from 2 to 12 hours. They are condensed in cold traps kept at a temperature of -80°C. The condensate obtained, which represents from 1 to 6% by weight of the ground spice, constitutes the aromatic fraction A. The powder is then treated by conventional solid-liquid extraction at 15° to 35°C with an apolar solvent, for example, petroleum ether or hexane, in a quantity of from 2 to 6 liters of liquid per kg of powder either once or several times.

After separation of the undissolved fractions, the liquid phase containing the aromatic fraction B, which represents from 2 to 20% by weight of the ground spice, is collected. The extraction treatment is then repeated on the drained powder with a polar solvent, preferably ethanol for reasons of food legislation.

After separation of the undissolved fractions, the liquid phase containing fraction C, which represents from 1 to 15% of the ground spice, is collected. The required aromatic extract is obtained by combining the three aromatic fractions A, B and C and may be concentrated under reduced pressure. Fractions B and C are preferably concentrated before mixing. In this way, it is possible to obtain concentrated extracts of which the residual solvent contents are below the detection limit.

In a second embodiment, the spice which has been ground and freed from its grinding gases in the same way as described above (dry grinding), is treated with a mixture of an apolar solvent and at least one polar solvent, preferably with an azeotropic mixture, for example hexane/ethanol or hexane/ethanol/water. The liquid phase obtained after separation of the undissolved fractions and, optionally, concentration is enriched with fraction A to form the aromatic extract.

In a third embodiment, the crude spice is ground in the presence of an organic solvent (wet grinding). For example, the spice is immersed in hexane in a quantity of from 15 to 30% by weight and then ground therein. After separation of the undissolved fractions, a liquid phase containing the grinding gases is collected. The ground powder is then treated with an apolar solvent, preferably the same as that used for grinding, and then with a polar solvent in the same way as described above. The combination of the three liquid phases obtained, optionally after concentration, constitutes the required aromatic extract.

In a fourth embodiment, the crude spice is again ground in the presence of an organic solvent (wet grinding) so as to obtain a liquid phase containing the grinding gases. The ground powder is then treated with a mixture of solvents of which one is preferably the solvent which was used for grinding, this mixture advantageously being an azeotropic mixture. The combination of the two liquid phases obtained, optionally after concentration, constitutes the required aromatic extract.

Example 1: 350 kg of nutmeg are finely ground (average grain size approximately 0.8 mm) in an Urschel mill kept at -40°C. The powder obtained is transferred to a tank where it is heated to 20°C. During heating, the tank is evacuated, after which a gentle stream of nitrogen under a pressure of 12 mm Hg is passed through it for 12 hours. The grinding gases are condensed in three cold traps which are kept at -80°C and which are situated downstream of the tank. 6.2 kg of fraction A are thus obtained.

The powder is then treated with 500 liters of rectified petroleum ether 40 to 60 (i.e., a fraction having a BP of 40° to 60°C) which is circulated in a closed loop by pumping the solvent to the bottom of the tank and reintroducing it at the top. The solvent is renewed after 2 hours and is then renewed once again after another 2 hours. The three volumes of solvents are then combined and concentrated in vacuo at approximately 30°C so as to obtain a fraction B of 63.7 kg which contains approximately 22% of solids. The concentration time is of the order of 2 hours for 10 liters to be concentrated.

The same operation is then repeated for the treatment with ethanol, giving a fraction C of 28.0 kg with a solids content of 50%.

The required aromatic extract is obtained by combining fractions A, B and C (97.9 kg). 280 kg of grounds are also obtained and approximately 1,500 liters of petroleum ether and 1,500 liters of ethanol are recovered, being recycled after rectification.

The extract of Example 1 (nutmeg) on the neutral base was found to be powerful and full-bodied. Four tasters preferred it to a comparable extract, but freed from the grinding gases. It was recommended as a flavoring agent for a mashed potato in a quantity of 0.07 to 1.5 g/kg of puree.

Example 2: A trigonal mill (Siefer-type SM 180) is filled with freshly rectified hexane to remove the air from the grinding chamber, after which the mill is switched on and is fed with black pepper so as to obtain a suspension containing around 200 g of pepper per liter of hexane.

After 10 minutes' operation, the mill is stopped, after which the grounds are separated and the liquid fraction recovered for concentration under reduced pressure. The grounds are then introduced into an extraction tank and are then treated with fresh hexane under the same conditions as in Example 1. The same operations are then repeated for the treatment with ethanol of the grounds obtained.

The combination of the various liquid phases obtained constitutes the required aromatic extract. From 90 to 110 g of this extract are obtained for 1 kg of black pepper. This extract is judged by the tasters to be green, fresh and intense.

Hops

M. Moll, R. Flayeux, P. Dicesare, and B. Gross; U.S. Patent 4,160,787; July 10, 1979; assigned to G.I.E. Tepral, France have developed a process for converting the α- and β-acids of hops into the iso-α acids which are preferred for use as a bitter flavoring ingredient for beers. The α-acids content of the hops is directly isomerized to the iso-α-acid, the bitter principle ingredient. The β-acids in the hops extract are transformed to the α-acid either after a direct separation step before the α-acid isomerization or the residual β-acids, after the isomerization, are then transformed and the resultant α-acid is isomerized. The β-acid transformation results from a radiation activation followed by an oxidation step with a peracid.

The isomerization of the α-acid to the iso-α-acid is accomplished by reaction with an alkaline earth metal ethoxide preferably the magnesium ethoxide.

Figure 10.2 shows the different stages and solutions for obtaining the iso-α-acids which are the desired bitter principle.

Starting from hops or a crude hop extract **1**, the α-acids and β-acids are extracted therefrom by processes **2** using an organic solvent such as hexane, pentane, petroleum ether or the like. This mixture of acids is primary extract **3**.

Starting with the primary extract, two different methods may be used to obtain the same resultant product, primarily iso-α-acids. The first process involves the drect isomerization of the α-acids and the separation and transformation of the β-acids to α-acids followed by further isomerization. The other process consists of the direct separation of the α-acids from the β-acids, transformation of the β-acids to α-acids and then isomerization of the combined α-acids.

In accordance with a first method, the primary extract is subjected to a direct isomerization step **4** by the use of an alkaline earth metal ethoxide, preferably magnesium ethoxide.

The magnesium ethoxide is obtained by reacting anhydrous ethyl alcohol with pure magnesium. The ethoxide transforms only the α-acids into iso-α-acids, without modifying the β-acids.

One can thus proceed in accordance with the following scheme.

The primary extract is dissolved in anhydrous ethanol. To this solution of α-acids there is added slightly more than twice the stoichiometric amount of α-acids of the magnesium ethoxide suspended in ethyl alcohol to isomerize the α-acids. The isomerization operation is carried out under an inert, preferably nitrogen atmosphere and the mixture composed of the solution of α-acids and magnesium ethoxide, suspended in ethyl alcohol, is refluxed, while avoiding oxidation. At atmospheric pressure, the temperature in the solution is about 80°C, for example 78°C.

Figure 10.2: Converting Hops Extracts into Flavor Components

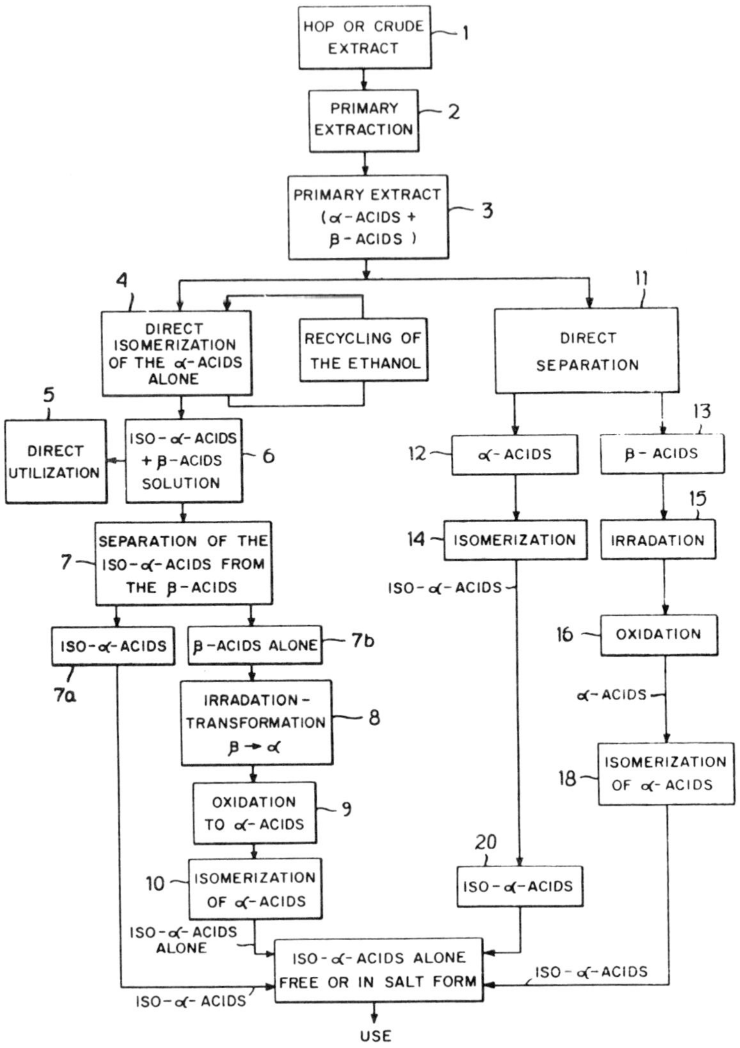

Source: U.S. Patent 4,160,787

The mixture of isomerized α-acids and the unreacted β-acids **6** from the first procedure can be directly utilized as a bitter flavoring **5** or the mixture after isomerization can be separated by a separation procedure **7** into iso-α-acid **7a** and β-acid **7b**. The β-acids are activated for transformation by an irradiation step **8** and then the activated β-acids are oxidized by peracids in step **9** to α-acids. The resultant α-acids are then isomerized by the magnesium ethoxide procedure utilized in step **4** to the iso-α-acids.

According to the second procedure the primary extract of α- and β-acids is subjected to a direct separation step **11** to yield a separate α-acid fraction **12** and a separate β-acid fraction **13**.

The α-acid fraction is then subjected to the alkaline earth metal ethoxide isomerization **14** to yield the iso-α-acids or salts **20** which are useable products of this process. The β-acids are activated by an irradiation step **15** and then oxidized by peracids **16** to yield α-acids **17** which are then isomerized **18** via alkaline earth ethoxides to iso-α-acids **19** which are combined with the iso-α-acids **20** from the isomerization of the original α-acids **12**.

The isomerization of α-acids from hops to the iso-α-acids bitter principles is performed in anhydrous ethanol by an alkaline earth ethoxide, preferably the magnesium ethoxide.

The magnesium ethoxide is prepared by reacting pure magnesium metal with anhydrous ethanol. The presence of water is to be avoided as magnesium hydroxide, which would form, impedes the reaction of the alcohol with the magnesium. The reaction of the ethanol with magnesium is initiated by warming the mixture, then additional alcohol and magnesium may be added to the refluxing mixture once the reaction has started.

The isomerization reaction to convert the α-acids of hops to the iso-α-acid form is carried out at reflux temperatures at about normal pressure. At least twice the stoichiometric amount of magnesium ethoxide to α-acid is preferred. Under such conditions at least 85 and generally 90 to 95% of the α-acids are isomerized.

It has been noted that the isomerization reaction with the alkaline earth ethoxide has no effect at all on any β-acids which may be present in the starting extract. However, the isomerization reaction, as mentioned above, may be carried out on either a mixture of the unseparated α- and β-acids or upon the α-acids obtained after a separation step.

Upon completion of the isomerization, the solution is cooled to room temperature. It contains the iso-α-acids in the form of their magnesium salts and β-acids. The ethanol is distilled and recycled.

The magnesium salts of the iso-α-acids may be used directly after the evaporation of the ethanol in countries which permit the addition of metal ions to beers; or the magnesium or other alkaline earth salts may be the object of additional treatments to break up the salts so that the sole resultant bitter flavor product is essentially pure iso-α-acids.

As mentioned above, the isomerization of the α-acids by the magnesium ethoxide suspended in ethanol can be carried out either on pure α-acids such as humulone or on the α-acids mixed with β-acids. The resultant solution of the iso-α-acids after isomerization may be used directly as a bittering principle by addition, before or after boiling of the wort and either before, during or after either the primary or secondary fermentations as desired by the brewmaster.

However, as certain countries permit only the addition of natural principles or extracts to beers, the direct addition of the magnesiums salts of the iso-α-acids is forbidden. To meet such requirements, the magnesium salts are converted to the free acids and the magnesium ions are removed. This is readily performed by passing the ethanolic solution of the magnesium salt of the iso-α-acid through a suitable strong, ion-exchanger in the acidic phase. The effluent from such an ion exchanger bed or column is the free iso-α-acid.

The separation of the α- or the iso-α-acids from the β-acids by the solvent partition method is carried out in an apparatus known as a pulsed column, consisting of a vertical tubular central portion which is provided with uniformly spaced, stationary, perforated, horizontal trays and with two flared, cylindrical ends which are without trays and of small height.

The column has a heavy-phase inlet at the upper part and a heavy-phase outlet at the lower part as well as a light-phase inlet at the lower part and a light-phase outlet at the upper part.

The operation of this apparatus consists in causing the two fluids in liquid phase to flow countercurrent under the action of their differences in densities. A pulsing mechanism assures a reciprocating contact of the total liquids contained in the column. This movement provides and maintains a homogeneous mixture of the two phases throughout the entire central portion or contact zone. Regulating the feed rates and the pulsing permits the regulating of the interphase and makes it possible to maintain the latter in such a position that each of the liquid phases of different density emerges entirely clear.

The use of this pulsed column as separator for the separating of the α-acids or iso-α-acids from the β-acids makes it possible to obtain substantially complete separation thus leading to satisfactory results.

The separation treatment by this interphase contact comprises the following operations: evaporation of the ethanol from the mixtures of iso-α-acids and β-acids to obtain a dry residue of the acids, treatment of the residue with an organic solvent of the hexane-type at room temperature after acidification of the residue with a strong acid such as hydrochloric acid in order to free the magnesium.

The separation proper, is carried out countercurrent in the pulsed column between the resultant organic phase and an aqueous phase consisting of a solution of salts, for example, sodium and potassium salts. Control of the separation is achieved by the variable relative proportions of liquids, their rates of flow, and the concentrations of the hops acids and salts in the liquids. The operation is preferably carried out in the dark, under an inert gas and at a temperature slightly above room temperature. The components during separation are light-sensitive.

The β-acids are transformed into α-acids by means of two successive stages of treatment. The first step consists in ultraviolet irradiation at a wavelength of 3500 to 3600 A. The β-acids recovered either before or after isomerization of the α-acids are dissolved in ethyl alcohol or methyl alcohol and are irradiated for a length of time which is variable with the concentration of β-acids and as an inverse function of the power of the lamps employed. During the radiation, the temperature is kept between 10° and 50°C. The radiation-activation products obtained are deoxyhumulones, the term humulone being a general term employed herein for designating α-acids. The yield of deoxyhumulones in the transformation is from 60 to 75% of the β-acids used. This is in essence an activation of the β-acids used for the next step.

The second step comprises the oxidation of the deoxyhumulones or activated β-acids. The oxidation which is carried out in a basic, aqueous, medium at a pH of 10 to 12 buffered by dissolving the trisodium, disodium phosphates or other buffers in water. The oxidation is effected by means of a peracid such as monopersulfuric acid, p-nitro perbenzoic acid or peracetic acid at a temperature below 25°C. The yield of α-acids is at least 60% based on deoxyhumulones or at least 35 to 50% of the total β-acids. The α-acids obtained are isomerized under the conditions described above, utilizing magnesium ethoxide.

COMPOUNDS GIVING COOLING EFFECTS

Sulfones, Sulfoxides, Phosphine Oxides and Amides

D.G. Rowsell and R. Hems; U.S. Patent 4,070,449; Jan. 24, 1978; and D.G. Rowsell and D.J. Spring; U.S. Patent 4,070,496; Jan. 24, 1978; and D.G. Rowsell, D.J. Spring and R. Hems; U.S. Patent 4,153,679; May 8, 1979; all assigned to Wilkinson Sword Limited, England describe three different groups of compounds which have a physiological cooling effect similar to that observed with menthol. Since these compounds do not have the high volatility and strong minty odor of menthol they are useful in applications where an odorless cooling agent is desired.

The first such group of compounds consists of the aliphatic, acyclic sulfones and sulfoxides of the formula:

$$R_1-\underset{R_3}{\overset{R_2}{\underset{|}{\overset{|}{C}}}}-SO_xR^1$$

where R_1 is H or C_{1-5} alkyl; R_2 is C_{1-6} alkyl; R_3 is C_{1-8} alkyl; R_1, R_2 and R_3 together provide a total of from 3 to 15 carbon atoms, preferably 5 to 10 carbon atoms; R^1 is alkyl, optionally containing hydroxy, carboxy, or alkylcarboxy substituents, R_1, R_2, R_3 and R^1 together providing a total of from 6 to 18 carbon atoms; and x is 1 or 2.

Preferred compounds are those in which x is 1 or 2, R^1 is alkyl and contains up to 8 carbon atoms, R_1 is hydrogen or C_{1-4} alkyl, and R_2 and R_3 are each C_{1-6} alkyl, R^1, R_1, R_2 and R_3 together providing a total of from 8 to 14 carbon atoms.

Generally speaking, the sulfoxides, i.e., compounds where x is 1, are to be preferred over the corresponding sulfones, i.e., the compounds where x is 2.

These sulfoxides and sulfones may be readily prepared by conventional methods, such as by the oxidation of the corresponding sulfide, e.g., with hydrogen peroxide in glacial acetic acid.

The second such group of compounds consists of the phosphine oxides of formula:

$$(R_1)(R_2)(R_3)P=O$$

In the formula above R_1 is an alkyl radical containing at least 3 carbon atoms, R_2 is an alkyl radical containing at least 3 carbon atoms or a cycloalkyl radical and R_3 is an alkyl or cycloalkyl radical, R_1, R_2 and R_3 together presenting a total of from 3 to 17 carbon atoms, and at least one of R_1, R_2 and R_3 having branching in a α, β or γ position relative to the phosphorus atom.

Branching in this context is to be taken to include cyclic structures, as well as branched chain acyclic groups, i.e., this condition will be met by compounds in which either of R_2 and R_3 is cycloalkyl, the carbon atom of R_2 or R_3, as the case may be, in the α-position relative to the phosphorus atom being part of the ring and, therefore, being considered branched.

Preferably R_1, R_2 and R_3 are such that any two when taken together present a total of at least 6 carbon atoms. Compounds of highest activity and especially preferred, are those where R_1 is a straight chain alkyl group of from 4 to 8 carbon atoms, more especially 5 to 8 carbon atoms, R_2 is a branched chain alkyl group of from 3 to 5 carbon atoms, and especially isopropyl, sec-butyl, isobutyl or isopentyl, and R_3 is an alkyl group, preferably a branched chain alkyl group, of from 3 to 6 carbon atoms, preferably 4 or 5 carbon atoms, or a cyclopentyl group, R_1, R_2 and R_3 providing a total of from 13 to 16 carbon atoms.

These phosphine oxides may be readily prepared by conventional routes. If two alkyl groups are the same, then dialkylphosphinyl chlorides, R_2POCl, may be prepared by the action of thionyl chloride on tetraalkyldiphosphine disulfides, or preferably by the action of chloride on secondary phosphine oxides (prepared from Grignard reagents and diethyl phosphite). These dialkylphosphinyl chlorides will react with Grignard reagents to form the desired tertiary phosphine oxides.

If all three alkyl groups of a tertiary phosphine oxide are different then the following route is satisfactory. Reaction between a Grignard reagent, RMgX, and diethyl chlorphosphite gives ethyl alkylphosphinite. This latter compound, when allowed to react with a Grignard reagent, R'MgX, will give the unsymmetrical secondary phosphine oxide, RR'P(O)H. Tertiary phosphine oxides can be prepared from these secondary phosphine oxides as described previously.

The third such group of compounds consists of the amides of the formula:

$$R_2-\overset{\displaystyle R_1}{\underset{\displaystyle R_3}{\overset{|}{\underset{|}{C^*}}}}-CONR'R''$$

where R' and R", when taken separately, are each hydrogen, C_{1-5} alkyl or C_{1-8} hydroxyalkyl and provide a total of no more than 8 carbon atoms, with the proviso that when R' is hydrogen, R" may also be alkylcarboxyalkyl of up to 6 carbon atoms; R' and R", when taken together, represent an alkylene group of up to 6 carbon atoms, the opposite ends of which group are attached to the amide nitrogen atom thereby to form a nitrogen heterocycle, the carbon atom chain of which may optionally be interrupted by oxygen; R_1 is hydrogen or C_{1-5} alkyl; and R_2 and R_3 are each C_{1-5} alkyl. R_1, R_2 and R_3 together should provide a total of at least 5 carbon atoms, preferably from 5 to 10 carbon atoms, and when R_1 is hydrogen, R_2 is C_{2-5} alkyl and R_3 is C_{3-5} alkyl and at least one of R_2 and R_3 is branched, preferably in an alpha or beta position relative to the carbon atom marked (*) in the formula.

Where the compounds have an asymmetric carbon atom, either optical isomer may be used in pure form but generally a mixture of optical isomers will be used. In some cases the degree of cooling produced by the compounds on the skin will differ between the optical isomers, in which case one or the other isomer may be preferred.

The preferred amides for use are the tertiary compounds, i.e., those where each of R_1, R_2 and R_3 is C_{1-5} alkyl, especially those where R_1 is methyl, ethyl or n-propyl and at least one of R_2 and R_3 is a branched chain group having branching in an alpha or beta position relative to the C atom marked (*) in the formula. Also preferred are monosubstituted amides, i.e., where R' is H, and disubstituted amides where R' and R'' are methyl or ethyl. A further preferred group consists of amides of the formula given where R_1 is hydrogen and at least one of R_2 and R_3 is branched in an alpha position relative to the carbon atom marked (*) in the formula.

The amides may readily be prepared by conventional techniques. for example, by reaction of an acid chloride of the formula $R_1R_2R_3COCl$ with an amine of the formula HNR'R'' in the presence of a hydrogen chloride acceptor. Such reactions are entirely conventional and the procedures involved will readily be understood by persons skilled in the art.

Compounds from all three groups may be used to provide a physiological cooling effect in a wide variety of edible and topical preparations.

Example 1: *Eye Lotion* – An eye lotion was prepared containing the following ingredients:

	Percent
Witch hazel	12.95
Boric acid	2.00
Sodium borate	0.50
Allantoin	0.05
Salicylic acid	0.025
Chlorobutol	0.02
Zinc sulfate	0.004
Water	to 100

To the formulation was added 0.005%, based on the total composition of 3-methylpent-3-yl-2-ethylcarboxyethyl sulfoxide. When used to bathe the eyes, a cool fresh sensation is apparent on the eyeball and eyelids.

Example 2: *Antiseptic Ointment* – An ointment was prepared according to the following formulation:

	Percent
Cetyltrimethyl ammonium bromide	4.0
Cetyl alcohol	6.0
Stearyl alcohol	6.0
White paraffin	14.0
Mineral oil	21.0
Water	to 100

The ingredients were mixed, warmed to 40°C and emulsified in a high speed blender. Added to the mixture during blending was 3.0% of 3-methylhex-3-yl-n-octyl sulfoxide. The final ointment when applied to the skin gave rise to a cooling effect.

Example 3: *Antipruritic Ointment* – The following ingredients were warmed together to form a homogeneous melt:

	Percent
Methyl salicylate	50.0
White beeswax	25.0
Anhydrous lanolin	25.0

To the melt was added 3.0% of 3-methylpent-3-yl-pent-2-yl sulfoxide and the mixture then allowed to solidify. A soft ointment resulted having a soothing effect on the skin accompaned by a cooling effect.

Example 4: *Cleansing Tissue* – A cleansing liquid was prepared having the formulation:

		Percent
Triethanolamine lauryl sulfate		1.0
Glycerol		2.0
Perfume		0.95
Water	to 100	

To this liquid was added 1.0% of diisopentyl-n-pentyl phosphine oxide. A paper tissue was then soaked in the liquid.

When the impregnated tissue was used to wipe the skin, a fresh cool sensation developed on the skin after a short interval.

Example 5: *Cigarette Tobacco* – A proprietary brand of cigarette tobacco was sprayed with an ethanolic solution of sec-butyl-n-hexyl-isopropyl phosphine oxide and was rolled into cigarettes each containing approximately 50 micrograms of active compound. Smoking the impregnated cigarettes produced a cool effect in the mouth characteristic of mentholated cigarettes but without any attendant odor other than that normally associated with tobacco.

Impregnation of the filter tip of a proprietary brand of tipped cigarette with 0.05 mg of sec-butyl-n-hexyl-isopropyl phosphine oxide produced a similar effect.

Example 6: *Toothpaste* – The following ingredients were mixed in a blender:

		Percent
Dicalcium phosphate		48.0
Sodium lauryl sulfate		2.5
Glycerol		24.8
Sodium carboxymethylcellulose		2.0
Citrus flavorant		1.0
Sodium saccharin		0.5
Water	to 100	

Shortly before completion of the blending operation, 0.1% by weight of n-hexyl-isopentyl-isopropyl phosphine oxide was added to the blender.

When applied as a toothpaste, a pleasant cooling effect was noticed in the mouth.

Example 7: *Deodorant Composition* – A deodorant composition suitable for formulation and dispensing as an aerosol under pressure of a suitable propellent was formulated according to the following recipe:

	Percent
Denatured ethanol	96.9
Hexachlorophene	2.0
Isopropyl myristate	1.0
Perfume	0.1

To the composition was added 2% by weight of N,N,2,2,-tetraethyl-3-methyl-butanamide. Application of the final composition gave rise to a definite cooling sensation on the skin.

Example 8: *Hair Shampoo* – Sodium lauryl ether sulfate, 10 g, was dispersed in 90 g of water in a high speed mill. To the dispersion was added 2% by weight of N-(1,1-dimethyl-2-hydroxyethyl)-2-isopropyl-2,3-dimethylbutanamide. When the hair was washed using the shampoo, a fresh, cool sensation was noticed on the scalp.

Example 9: *Solid Cologne* – A solid cologne was formulated according to the following recipe:

		Percent
Denatured ethanol		74.5
Propylene glycol		3.0
Sodium stearate		5.0
Perfume		5.0
Water	to 100	

The sodium stearate was dissolved by stirring in a warm mixture of the ethanol, propylene glycol and water. To the solution was added the perfume and 2% of N,N,2-dimethyl-2-isopropylhexanamide and the mixture then allowed to solidify into a waxy cake.

When applied to the forehead a distinct cooling effect was noticeable.

Example 10: *Mouthwash* – A concentrated mouthwash composition was prepared according to the following recipe:

		Percent
Ethanol		3.0
Borax		2.0
Sodium bicarbonate		1.0
Glycerol		10.0
Flavorant		0.4
Thymol		0.03
Water	to 100	

To the composition was added 0.1% of N-(1,1-dimethyl-2-hydroxyethyl)-2,2-diethylbutanamide.

When diluted with approximately 10 times its own volume of water and used to rinse the mouth, a cooling effect was obtained in the mouth.

Amides, Substituted Ureas and Sulfonamides

D.G. Rowsell and R. Hems; U.S. Patents 4,137,305; January 30, 1979; 4,136,164; January 23, 1979; and 4,137,304; January 30, 1979; all assigned to Wilkinson Sword Limited, England describe a group of cyclic and acyclic amides, substituted ureas and sulfonamides which stimulate the cold receptors of the body's nervous system to produce a cold sensation and which may be used for this purpose in various edible and topical preparations. These compounds do not have the strong minty odor of menthol; they are inexpensive and easily synthesized from readily available starting materials.

These compounds may be represented by the formula:

$$(1) \qquad \begin{matrix} R_1 \diagdown \\ \quad N{-}X \\ R_2 \diagup \end{matrix}$$

R_1, when taken separately, is H, C_{1-7} alkyl or C_{3-6} cycloalkyl;

R_2, when taken separately, is C_{3-8} alkyl or C_{3-8} alkylcycloalkyl, alkylcycloalkylalkyl, cycloalkyl, or cycloalkylalkyl, with the proviso that R_2 is branched at an alpha carbon atom relative to to the N atom when R_1 is H or at an alpha or beta carbon atom where R_1 is alkyl or cycloalkyl, this condition to be satisfied, in the case of cyclic groups, when the carbon atom alpha or beta to the N atom is part of the cycle;

R_1 and R_2, when taken together, represent a straight or branched chain alkylene group forming with the N atom to which they are attached a 5 to 10 membered heterocycle, and preferably having branching at an alpha or beta carbon atom relative to the N atom;

R_1 and R_2 when separate groups and when taken together provide a total of at least 5 carbon atoms; and

X represents R_3CO-, R_4SO_2- or R_5R_6NCO-; where

R_3 is H, C_{1-6} alkyl, C_{3-6} cycloalkyl, C_{2-8} hydroxyalkyl, C_{2-8} carboxyalkyl or C_{3-8} alkylcarboxyalkyl, with the proviso that when R_3 is C_6 alkyl it is primary in structure;

R_4 is C_{1-6} alkyl or C_{3-6} cycloalkyl, with the proviso that when R_4 is C_5 or C_6 alkyl it is primary in structure;

R_5 and R_6, when taken separately, are each H, C_{1-6} alkyl, C_{3-6} alkylcycloalkyl, cycloalkyl, or cycloalkylalkyl, C_{2-8} hydroxyalkyl, C_{2-8} carboxyalkyl or C_{3-8} alkylcarboxyalkyl; or together represent a straight or branched chain C_{3-10} alkylene group optionally containing an ether oxygen atom; and where

R_1, R_2 and X provide a total of from 7 to 16 carbon atoms.

Broadly speaking, the compounds of highest activity are the substituted ureas which may be represented by the formula (2)

$$(2) \qquad \begin{matrix} R_1 \\ R_2 \end{matrix} \!>\! N - \overset{O}{\overset{\|}{C}} - N \!<\! \begin{matrix} R_5 \\ R_6 \end{matrix}$$

where R_1, R_2, R_5 and R_6 are as defined above. Preferred substituted ureas are those compounds where R_1 is H or C_{1-7} alkyl, R_2 is C_{3-8} alkyl or C_{3-8} cycloalkyl, cycloalkylalkyl or alkylcycloalkyl, R_2 having branching in an alpha position relative to the N atom when R_1 is H, or at an alpha or beta position when R_1 is alkyl, or where R_1 and R_2 are joined to form an alkylene group having up to 10 carbon atoms and having branching at an alpha or beta position relative to the N atom, and forming together with the nitrogen atom a 5- to 7-membered ring; and where R_5 is H or C_{1-6} alkyl and R_6 is C_{1-6} alkyl, or C_{3-6} cycloalkyl, or where R_5 and R_6 jointly represent an alkylene group, optionally containing an ether oxygen atom and forming with the N atom a 5- or 6-membered ring.

The substituted ureas of formula (2) may be easily prepared by reaction of an appropriate amine with an isocyanate or carbamoyl chloride.

The useful cyclic and acyclic amides may be represented by formula (3):

$$(3) \qquad \begin{matrix} R_1 \\ R_2 \end{matrix} \!>\! N\overset{O}{\overset{\|}{C}}R_3$$

where R_1, R_2 and R_3 are as defined above in connection with formula (1). Preferred values for R_1 and R_2 are as set out above for formula (2) and the preferred values for R_3 are H, C_{1-6} alkyl, C_{2-6} hydroxyalkyl, carboxyalkyl or alkylcarboxyalkyl and C_{3-6} cycloalkyl.

The amides may readily be prepared by reaction of the appropriate acid chloride and a substituted amine in accordance with procedures well known for the preparation of amides.

Although less preferred than the substituted ureas and amides hereinbefore described, but nevertheless still possessing utility in these compositions are sulfonamides of formula (4):

$$(4) \qquad \begin{matrix} R_1 \\ R_2 \end{matrix} \!>\! NSO_2R_4$$

where R_1, R_2 and R_4 are as above defined in connection with formula (1). Preferred values of R_1 and R_2 are as defined above in connection with formula (2), while the preferred values for R_4 are C_{1-4} alkyl.

The sulfonamides of formula (4) may readily be prepared from the corresponding sulfonyl chloride and substituted amine by procedures well known in the art.

The patent lists dozens of specific compounds (defining the R_{1-6} substitution groups) and gives their boiling or melting points and a subjective measurement of their cooling effect arrived at by testing on the tongue using menthol as a control.

These cold receptor stimulants may be incorporated into a wide range of edible and potable compositions comprising an edible or potable base and usually one or more flavoring or coloring agents. The particular effect of the cold receptor stimulant is to create a cool or fresh sensation in the mouth, and in some cases, even in the stomach, and therefore, the compounds find particular utility in sugar-based confectionery such as chocolate, boiled sweets, mints and candy, in ice cream, jellies and in chewing gum.

The formulation of such confections will be by traditional techniques and according to conventional recipes and as such, forms no part of this process. The cold receptor stimulant will be added to the recipe at a convenient point and in amount sufficient to produce the desired cooling effect in the final product. The amount will vary depending upon the particular compound, the degree of cooling effect desired and the strength of other flavorants in the recipe. For general guidance, however, amounts in the range of 0.01 to 1.0% by weight based on the total composition will be found suitable.

Similar considerations apply to the formulation of beverages. Generally speaking the compounds will find most utility in soft drinks, e.g., fruit squashes, lemonade, cola etc., but may also be used in alcoholic beverages. The amount of compound used will generally be in the range of 0.01 to 1.0% by weight based on the total composition.

Because of their cooling effect on the skin and on the mucous membranes of the mouth, throat and nose and of the gastrointestinal tract, the cold receptor stimulants may also be used in a variety of oral medicines.

A particular utility for the compounds is in the formulation of antacid and indigestion remedies, and especially those based on sodium bicarbonate, magnesium oxide, calcium or magnesium carbonate, aluminum or magnesium hydroxide or magnesium trisilicate. In such compositions, the compound will usually be added in an amount of from 0.1 to 2.0%.

Example 1: *Toothpaste* – The following ingredients were mixed in a blender:

	Percent
Dicalcium phosphate	48.0
Sodium lauryl sulfate	2.5
Glycerol	24.8
Sodium carboxymethylcellulose	2.0
Citrus flavorant	1.0
Sodium saccharin	0.5
Water	to 100

Shortly before completion of the blending operation, 1.0% by weight of N-(2,2-dimethylpropionyl)-2,6-dimethylpiperidine was added to the blender.

When applied as a toothpaste a pleasant cooling effect was noticed in the mouth.

Example 2: *Soft Drink* – A soft drink concentrate was prepared from the following recipe as shown below:

	Percent
Pure orange juice	60
Sucrose	10
Saccharin	0.2
Orange flavoring	0.1
Citric acid	0.2
Sulfur dioxide	trace amount
Water	to 100

To the concentrate was added 0.10% of N-(2,4,4-trimethylpent-2-yl)-2-methylpropionamide.

The concentrate was diluted with water and tasted. An orange flavor having a pleasantly cool after-effect was obtained.

Example 3: The tip of a wooden toothpick was impregnated with an alcoholic solution containing N,N-di-sec-butyl methanesulfonamide in an amount sufficient to deposit on the toothpick 0.05 mg of the compound. The toothpick was then dried.

When placed against the tongue a cool sensation is noticed after a short period of time.

Example 4: *Soft Sweet* – Water was added to icing sugar at 40°C to form a stiff paste. 0.2% of N-n-propyl-N-(1-ethyl-n-propyl)-N'-cyclopropylurea was then stirred into the paste and the mixture allowed to set. A soft sweet mass resulted having the characteristic cooling effect in the mouth of peppermint but without the minty flavor or odor.

3-Substituted-p-Menthanes

H.R. Watson, D.G. Rowsell and J.H.D. Browning; U.S. Patent 4,157,384; June 5, 1979; and H.R. Watson, D.G. Rowsell and D.J. Spring; U.S. Patent 4,136,163; January 23, 1979; both assigned to Wilkinson Sword Limited, England report that a number of 3-substituted-p-menthanes have a pronounced physiological cooling activity and are substantially nontoxic. They also have low volatility and little or no odor and can, therefore, be used as substitutes for menthol in applications where the physiological cooling action of menthol is desired but its strong odor is not.

These compounds are of the formula:

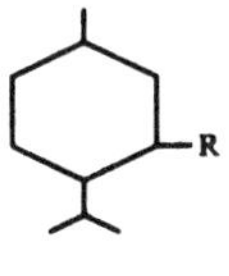

R

where R is (1) $-CONH_2$ or (2) $-COOR'$ or (3) $-CONR''R'''$.

When R is $-COOR'$, R' may be hydrogen; an alkali (eg., Na, K etc.) or alkaline earth Ca, Mg, etc.) metal atom, or an ammonium or substituted ammonium radical (e.g., trimethylammonium, β-hydroxyethylammonium); or a radical containing from 2 to 10 carbon atoms and selected from hydroxyaliphatic radicals having a hydroxyl substituent in a 2- or 3-position and a hydrogen atom in the 1-position; a lower alkylene oxide (e.g., ethylene oxide, propylene oxide, etc.) adduct of such a hydroxyaliphatic radical; a ketal derivative of such a hydroxyaliphatic radical with a lower ketone (e.g., acetone); a lower acyl (e.g., acetyl) derivative of such a hydroxyaliphatic radical; a hydroxyaryl radical having a hydroxyl substituent in a 2- or 3-position relative to the ester grouping; a carboxyaliphatic radical having a carboxyl group in a 1-, 2- or 3-position; an alkali metal (e.g., Na, K), alkaline earth metal (e.g., Ca, Mg), ammonium or substituted ammonium (e.g., trimethylammonium, β-hydroxyethylammonium) salt of such a carboxyaliphatic radical; or a lower alkyl (e.g., methyl, ethyl, etc.) ester of such a carboxyaliphatic radical.

When R is –CONR''R''', R'', when taken separately, is hydrogen or an aliphatic radical containing up to 25 carbon atoms; R''', when taken separately is hydroxy, or an aliphatic radical containing up to 25 carbon atoms, with the proviso that when R'' is hydrogen, R''' may also be an aryl radical of up to 10 carbon atoms and selected from the group consisting of substituted phenyl, phenalkyl or substituted phenalkyl, naphthyl and substituted naphthyl, pyridyl; and R'' and R''', when taken together with the nitrogen atom to which they are attached, represent a cyclic or heterocyclic group of up to 25 carbon atoms, e.g., piperidino, morpholino, etc.

In the above definitions, aliphatic is intended to include any straight-chained, branched-chained or cyclic radical free or aromatic unsaturation, and thus embraces alkyl, cycloalkyl, alkenyl, cycloalkenyl, alkynyl, hydroxyalkyl, acyloxyalkyl, alkoxy, alkoxyalkyl, aminoalkyl, acylaminoalkyl, carboxyalkyl and similar combinations.

Typical values for R'' and R''' when aliphatic are methyl, ethyl, propyl, butyl, isobutyl, n-decyl, cyclopropyl, cyclohexyl, cyclopentyl, cycloheptylmethyl, 2-hydroxyethyl, 3-hydroxy-n-propyl, 6-hydroxy-n-hexyl, 2-aminoethyl, 2-acetoxyethyl, 2-ethylcarboxyethyl, 4-hydroxybut-2-ynyl, carboxymethyl, etc.

When R''' is aryl, typical values are benzyl, naphthyl, 4-methoxyphenyl, 4-hydroxyphenyl, 4-methylphenyl, 3-hydroxy-4-methylphenyl, 4-fluorophenyl, 4-nitrophenyl, 2-hydroxynaphthyl, pyridyl, etc.

These 3-substituted-p-menthanes may be readily prepared by conventional methods. Thus, the p-menthane-3-carboxylic acid and its salts may readily be prepared by carbonation of a Grignard reagent derived from menthol. The carboxylic acid may then readily be converted into its acid chloride, for example, by reaction with thionyl chloride, and the acid chloride converted into the amide or an ester derivative by reaction with ammonia or an appropriate alcohol. Other methods for the preparation of the acid, the amide or the esters used in this process will be apparent to those skilled in the art.

These compounds exhibit both geometric and optical isomerism and, depending on the starting materials and the methods used in their preparation, the compounds may be isomerically pure, i.e., consisting of one geometric or optical isomer, or they may be isomeric mixtures, both in the geometric and optical sense.

As is well known, the basic p-menthane structure is a chair-shaped molecule which can exist in cis or trans forms. Substitution of the carboxyl or amide group into the 3-position gives rise to four configurational or geometric isomers depending upon whether the substitution is axially or equatorially into the cis or trans isomer, the four isomers being related as menthol is to neomenthol, isomenthol, and neoisomenthol. In general it is found that the equatorially substituted derivatives have the greater cooling effect than the axial compounds and are to be preferred.

Substitution of the carboxyl or amide group in the 3-position of the p-menthane structure also gives rise to optical isomerism, each of the abovementioned four geometric isomers, existing in d, l and dl forms. The physiological cooling effect is found, in most cases, to be greater in the l-form than in d-form, and in some cases substantially greater. The l-acid and derivatives of the l-acid are, therefore, preferred.

Example 1: *After Shave Lotion* – An after shave lotion was prepared according to the following recipe by dissolution of the solid ingredients in the liquid and cooling and filtering.

		Percent
Denatured ethanol		75
Diethylphthalate		1.0
Propylene glycol		1.0
Lactic acid		1.0
Perfume		3.0
Water	to 100	

Into the base lotion was added 2.0% by weight based on the total composition of p-menthane-3-carboxylic acid.

When applied to the face a clearly noticeable cooling effect became apparent after a short interval of time.

Example 2: *Aerosol Shaving Soap* – An aerosol shaving soap composition was formulated according to the following recipe:

		Percent
Stearic acid		6.3
Lauric acid		2.7
Triethanolamine		4.6
Sodium carboxymethyl-cellulose		0.1
Sorbitol		5.0
Perfume		0.4
Water	to 100	

The composition was prepared by fusing the acids in water, adding the triethanolamine, cooling and adding the other constituents. To the mixture was then added 1.0%, based on the total composition of N,N-dimethyl-p-menthane-3-carboxamide. The composition was then packaged in an aerosol dispenser under pressure of a butane propellant.

When used in shaving, a fresh cool sensation was distinctly noticeable on the face.

COMPANY INDEX

The company names listed below are given exactly as they appear in the patents, despite name changes, mergers and acquisitions which have, at times, resulted in the revision of a company name.

INVENTOR INDEX

U.S. PATENT NUMBER INDEX

NOTICE

Nothing contained in this Review shall be construed to constitute a permission or recommendation to practice any invention covered by any patent without a license from the patent owners. Further, neither the author nor the publisher assumes any liability with respect to the use of, or for damages resulting from the use of, any information, apparatus, method or process described in this Review.

HAZARDOUS AND TOXIC EFFECTS OF INDUSTRIAL CHEMICALS 1979

by Marshall Sittig

There is a continuing need to assess the status of potentially dangerous substances including those now available and those that may reach commercial availability in the future. This should be done with a predictive view to avoid or at least to ameliorate catastrophic episodes similar to those that have occurred with methyl mercury, polychlorinated biphenyls, vinyl chloride monomer, dioxin and a number of pesticides.

This handbook is intended to be a working guide for the industrial hygienist and other interested persons. Most of the individual chemical listings contain a number of literature references.

This book is intended to be a guide to these hazardous chemicals, to give first warning signals, and to serve as an introduction to the published literature and to government agencies offering detailed guidance.

Information (where available) under each chemical listing includes:

Description. Derivation, chemical structure and systematic chemical name.
Synonyms. Other chemical names, generic names and trademarks.
Potential Occupational Exposures.
Permissible Exposure Limits.
Routes of Entry.
Harmful Effects. Local, systemic.
Medical Surveillance.
Special Tests.
Personal Protective Methods.
Bibliography. References to articles in journals and government publications.

The book consists of about 250 individual monographs arranged alphabetically by the common name of each substance described. It is impossible to indicate here all of the chemical entries in this book. In order to give an indication of the vast scope of information dealt with here, the names of most entries in the beginning of the book are reproduced here.

Acetaldehyde
Acetates
Acetic Acid
Acetic Anhydride
Acetone (See Ketones)
Acetonitrile
Acetylaminofluorene
Acetylene
Acridine
Acrolein
Acrylamide
Acrylonitrile
Alkanes
Allyl Alcohol
Allyl Chloride
Aluminum & Compounds
Aminodiphenyl
Ammonia
Amyl Alcohol
Aniline
Antimony & Compounds
Arsenic
Arsine
Asbestos
Asphalt Fumes
Barium & Compounds
Benzene
Benzidine & Salts
Benzoyl Peroxide
Benzyl Chloride
Beryllium & Compounds
Bis(Chloromethyl) Ether
Bismuth & Compounds
Boron & Compounds
Boron Hydrides
Boron Trifluoride
Brass
Bromine
1,3-Butadiene
Butyl Alcohol
Butylamine
Cadmium & Compounds
Calcium Cyanamide
Calcium Oxide
Carbaryl
Carbon Dioxide
Carbon Disulfide
Carbon Monoxide
Carbon Tetrachloride
Carbonyls
Cement Dust
Cerium & Compounds
Chlorinated Benzenes
Chlorinated Lime
Chlorinated Naphthalenes
Chlorine
o-Chlorobenzylidene Malonitrile
Chlorodiphenyls & Derivs.
Chloroform
Chloromethyl Methyl Ether
Chloroprene
Chromium & Compounds
Coal Tar Products
Cobalt & Compounds
Copper & Compounds
Cotton Dust
Creosote
Cresol
Cyclohexanone
Diacetone Alcohol
2,4-Diaminoanisole
4,4'-Diaminodiphenylmethane
Dibromochloropropane
1,2-Dibromoethane
3,3'-Dichlorobenzidine
1,2-Dichloroethane
1,2-Dichloroethylene
Dichloroethyl Ether
Diisobutyl Ketone
N,N-Dimethylacetamide
4-Dimethylaminoazobenzene
N,N-Dimethylformamide
Dimethyl Sulfate
Dinitrobenzene
Dinitro-o-Cresol
Dinitrophenol
Dinitrotoluene
Dioxane
Diphenyl
Epichlorohydrin
Ethanolamines
Ethyl Alcohol
Ethylbenzene
plus 160 other chemicals

ISBN 0-8155-0731-3

460 pages

DETERGENT MANUFACTURE INCLUDING ZEOLITE BUILDERS AND OTHER NEW MATERIALS 1979

by Marshall Sittig

Chemical Technology Review No. 128

This new text demonstrates that great advances continue to occur in the field of synthetic detergents. After World War II, when polypropylene benzene sulfonates held the major share of the detergent market, their surfactant action was enhanced by tetrasodium pyrophosphate or trisodium polyphosphate builders to buffer pH, disperse soil, and prevent soil redeposition. Once biodegradability became a prime concern, however, it became imperative that both surfactant and builder meet environmentally acceptable standards. Linear chain materials replaced branched chains as key ingredients. But one of the most promising innovations came with the use of zeolitic ion-exchange builders to replace the ecologically unsatisfactory phosphates. Zeolites receive extensive coverage in this new volume.

Sittig gives universal treatment to the many classes of detergents, detailing about 650 processes. Chapter headings with **examples of some** subtitles are given in the partial table of contents below. The number of processes per topic are in parentheses.

ISBN 0-8155-0749-6

565 pages

SYNTHETIC OILS AND ADDITIVES FOR LUBRICANTS 1980
Advances Since 1977

Edited by M. William Ranney

Chemical Technology Review No. 145

Fully synthetic lubricants are highly resistant to oxidation and sludge formation having anti-wear and water tolerance not equalled by straight petroleum oils.

Synthesis and production of pure synthetics are still costly and complex however. Moreover it is claimed that fossil oils have certain lubricant qualities that could be superior to those of the synthetics.

A solution to the dilemma is to take the much less expensive natural hydrocarbon oil stock and make it resistant to deterioration by addition of synthetics without sacrificing any inherent lubricant power. This is the essence of this book.

Our prolific chemical industry has made available so many diverse fluids and additives intended to upgrade lubricating oil stock that almost any desired property can be emphasized. One need only to examine the following partial table of contents which gives **examples of some chapter headings and subtitles.** Numbers in ( ) indicate numbers of processes per topic.

ISBN 0-8155-0781-X

409 pages

FOOD PROCESSING ENZYMES 1979
RECENT DEVELOPMENTS

by Nicholas D. Pintauro

Food Technology Review No. 52

This book presents developments in enzyme technology according to pertinent food categories, as listed below. Subject matter is organized to demonstrate means to solve a manufacturing problem, to improve storage, or to provide greater convenience in recipe preparation.

Enzyme technology is still in its infancy, but chemical elucidation of protein structure and binding mechanisms of enzymes with substrates are beginning to unfold as researchers probe for answers to problems of enzyme reactions. The 185 processes described in this book attest to the progress already made in various sectors of the food industry. Chapter headings and **examples of some** important subtitles are given in the condensed table of contents below. Numerals in parentheses specify the number of processes per topic.

ISBN 0-8155-0748-8

420 pages